Third Edition

AIR QUALITY

Third Edition

Thad Godish

LEWIS PUBLISHERS

Boca Raton   New York

Library of Congress Cataloging-in-Publication Data

Godish, Thad.
    Air quality / Thad Godish. — 3rd ed.
      p. cm.
    Includes bibliographical references and index.
    ISBN 1-56670-231-3
    1. Air—Pollution. 2. Air quality management. I. Title.
TD883.G57   1997
363.739'2—dc21   97-10860
                                                          CIP

© 1997 by CRC Press LLC
Lewis Publishers is an imprint of CRC Press LLC

No claim to original U.S. Government works
International Standard Book Number 1-56670-231-3
Library of Congress Card Number 97-10860
Printed in the United States of America 1 2 3 4 5 6 7 8 9 0
Printed on acid-free paper

The Author

Thad Godish is Professor of Natural Resources and Environmental Management at Ball State University, Muncie, IN. He received his doctorate from Pennsylvania State University, where he was affiliated with the Center for Air Environment Studies.

Dr. Godish has conducted research in a variety of air pollution-related areas. He is best known for his work on indoor air quality. He is internationally known for his research and public service activities on the subject of formaldehyde contamination of buildings. He is also a leader in the area of indoor air-quality research and training associated with school buildings. He has published two well-received books: *Indoor Air Pollution Control* (1989) and *Sick Buildings: Definition, Diagnosis and Mitigation* (1995).

Dr. Godish has taught and continues to teach a variety of environmental science courses including air quality, indoor air-quality management, occupational/industrial hygiene, building asbestos and lead management, and hazardous materials health and safety.

He is a member of Sigma Xi, national honorary research society, and is a Fellow, Indiana Academy of Science. He has been Visiting Scientist, Monash University, Gippsland, Australia, and Harvard University, School of Public Health.

He is a member of the Air & Waste Management Association, American Industrial Hygiene Association, American Conference of Governmental Industrial Hygienists, International Society of Indoor Air Quality and Climate, and American Association for the Advancement of Science. He has served as chairman of the East Central section and Indiana chapter of the Air Pollution Control Association.

And the Ether's Breath Filled the Valley

Acknowledgments

This and previous editions of *Air Quality* reflect the encouragement of my colleague Clyde Hibbs, Professor Emeritus, Natural Resources & Environmental Management at Ball State University. He started me on the writer's path, and I shall forever be grateful to him.

This edition (as has been the case for previous ones) owes its existence to my colleagues in the various colleges and universities in the U.S., as well as in other countries, who have elected to use it in their classes and those who have used it to acquire a better understanding of ambient and indoor air quality. My wish is that it provides instructors of air quality classes with a better tool for their teaching than previous editions or competing titles. As a university professor who uses *Air Quality* in my own course, I am very much aware of the rare opportunity that I have in reaching beyond the walls of my classroom to the many classrooms where our future environmental scientists, technicians, engineers, and other environmental and health practitioners are being trained. I am grateful for the opportunity that you have afforded me.

Preface

This edition of *Air Quality*, as well as those which preceded it, attempts to provide readers (instructors, students, others) a comprehensive overview of air quality, the science and management practices. As always it attempts to be both thorough and readable.

Both the science and practice of air-quality management continue to evolve. As a consequence, it is desirable (nay, necessary) to periodically update that which has been previously published. Additionally, when looking back on one's work, it is evident that one can always improve on it, present material in a new way, and add material which enhances its usability for readers.

Regulation of air quality will soon be in its fourth decade, soon in a new century. We will continue to struggle with air-quality concerns that were identified long ago and the more recent issues associated with stratospheric O_3 depletion and global warming. What the new century will bring is unknown. Of course, the challenges of the present day will continue, and new ones will surely arise.

This edition is characterized by significant revisions. This has been particularly the case in Chapters 2, 4, 7, 8, 10, and 11. In both Chapters 2 and 4, chemistry of atmospheric pollutants is more extensively treated. Expanded treatment of atmospheric chemistry began in the second edition. This continued emphasis on atmospheric chemistry reflects our increased understanding of the science and the author's sense that readers need a fuller understanding of the complexities involved in major air-quality problems such as photochemical air pollution, stratospheric O_3 depletion, global warming, and atmospheric scavenging processes.

As the second edition was being written, the U.S. Congress was in the final stages of passing the Clean Air Act Amendments of 1990. These amendments have had significant impact on present day air pollution control efforts and practices. Revisions to Chapter 8 reflect the very important regulatory and public policy changes that the 1990 CAA Amendments have wrought.

Indoor air quality (Chapter 11) is a rapidly evolving field. Our understanding of indoor air-quality problems is increasingly becoming mature. The significant revisions in Chapter 11 reflect the increasingly good science that indoor air quality has become.

Chapter 13 represents a significant evolution of this book's writing. In the first edition, the author deliberately avoided the use of equations and quantitative materials so characteristic of texts written at that time by individuals with engineering backgrounds. The very limited focus on the quantitative aspects of air quality and its management without any doubt enhanced the book's readability and thus its use by students with a broad range of backgrounds.

As the field of environmental science/management continues to evolve, it is evident that graduates of such programs (non-engineering) increasingly need to be familiar with a variety of quantitative concepts and techniques. Chapter 13,

Quantitative Aspects, is an initial effort to focus on the various quantitative aspects of air quality and its management. By organizing this material in a single chapter, course instructors have the option of using these materials as they deem desirable in presenting their courses.

The third edition of *Air Quality* is designed as a text for advanced level undergraduate and beginning level graduate students in environmental science, health, engineering and industrial hygiene programs. It may also be used as a supplement in engineering courses where the primary focus is the design and operation of control equipment. It is also designed to provide a broad overview of the science of air quality and its management to a variety of nonstudent readers who may have a professional or personal interest in the field.

Thad Godish
Muncie, Indiana
May 1997

Table of Contents

1   THE ATMOSPHERE

The earth, the planet that we share with a large variety of living things, is immersed in an invisible sea of gases and condensed water vapor. The atmosphere is maintained in its place by the earth's gravitational pull. The earth's atmosphere is unique to the solar system because it has made life, as the earth has known it, possible for the past 3 billion years.

CHEMICAL COMPOSITION

The atmosphere is a mixture of gases which vary from trace levels to nitrogen (N_2), which comprises approximately 78% of its mass and volume. This gaseous mixture may be characterized by those substances whose concentrations remain stable or constant over many millennia (Table 1.1) and those that vary daily, seasonally, or over the time course of human experience (Table 1.2).

The atmospheric gases whose concentrations remain relatively constant include N_2, oxygen (O_2), argon (Ar), neon (Ne), helium (He), krypton (Kr), hydrogen (H_2), and xenon (Xe). Though N_2 is the single most abundant gas in the atmosphere, it has a relatively limited direct role in atmospheric or life processes. It serves as a precursor molecule for the formation of nitrate nitrogen (NO_3^-) which is required by plants to make amino acids, proteins, chlorophyll, and nucleic acids — molecules that are essential, directly or indirectly, to all living things. The conversion of N_2 to NO_3^- occurs as a result of symbiotic biological processes and chemical processes in the atmosphere.

Nitrogen can, as a result of biological and atmospheric processes, combine with O_2 to produce nitrogen oxides (NO_x), which include nitric oxide (NO), nitrogen dioxide (NO_2), and nitrous oxide (N_2O). Nitrogen oxides formed by the oxidation of N_2 occur in the atmosphere at trace levels (Table 1.2), and, unlike their precursors (N_2 and O_2), they appear to vary with time. This is particularly true for NO and NO_2. Until recently, N_2O was thought to be a nonvarying gas.

Table 1.1   Atmospheric Gases —
             Constant Concentrations

Gas	Concentration (ppmv)
Nitrogen	780,840.00
Oxygen	209,460.00
Argon	9,340.00
Neon	18.18
Helium	5.24
Krypton	1.14
Hydrogen	0.50
Xenon	0.09

From National Oceanic and Atmospheric
Administration, 1976. NOAA S/T 76-1562.

Nitrous oxide concentrations have been increasing as a result of increasing disturbance of soils and use of nitrogen fertilizers which can be denitrified by soil microorganisms to produce N_2 and N_2O.

Molecular O_2 comprises approximately 21% of the atmosphere's mass. Its presence in elevated concentrations has resulted in the evolution of oxidative metabolism, the series of energy-transferring chemical reactions that sustain most life forms. As a result, O_2 is vital to almost all living things. It is also important because its evolving abundance over billions of years has provided the precursor molecules necessary for the formation of the ozone (O_3) layer, which protects organic molecules and all living things from high-energy ultraviolet (UV) radiation incident on the earth's atmosphere. Background surface air O_3 levels are relatively low [circa 0.02 parts per million volume (ppmv)]. Concentrations, however, change significantly with height with peak concentrations at 30 to 40 km based on ppmv mixing ratios and at about 25 km based on partial pressure (Figure 1.1). These two measures of concentration differ because the molecular mass thins out exponentially with altitude. Mass-based mixing ratios indicate that

Table 1.2   Atmospheric Gases —
             Variable Concentrations

Gas	Concentration (ppmv)
Water vapor	0.1–30,000.00
Carbon dioxide	360.00
Methane	1.72
Nitrous oxide	0.33
Carbon monoxide	0.11
Ozone	0.02
Ammonia	0.004
Nitrogen dioxide	0.001
Sulfur dioxide	0.001
Nitric oxide	0.0005
Hydrogen sulfide	0.00005

From National Oceanic and Atmospheric Administration, 1976. NOAA S/T 76-1562.

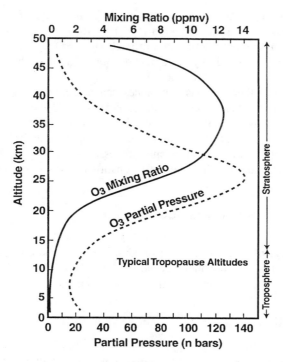

Figure 1.1 Average O_3 concentrations in the atmosphere expressed as mixing ratios (ppmv) and partial pressure (n bars). (From NRC, 1991. National Academy Press, Washington, DC. With permission.)

average peak O_3 levels occur at about 35 km. About 10 to 15% of the total atmospheric O_3 is found in the troposphere.

In contrast to N_2 and O_2, the concentration of carbon dioxide (CO_2) in the atmosphere is relatively low — about 0.036% or 360 ppmv. Carbon dioxide is one of the two principal raw materials from which green plants (in the process of photosynthesis) make the food that most living things need. Life is carbon based and CO_2 is the source of that carbon. Carbon dioxide is also a major greenhouse gas, and because of its thermal absorptivity it is responsible in good measure for maintaining a favorable global heat balance.

Water (H_2O) vapor is the atmospheric constituent with the highest degree of variability (from 0.1 to 30,000 ppmv). Like CO_2, it is a major greenhouse gas, absorbing thermal energy radiated from the earth's surface. Water vapor is significant in the atmosphere because it readily changes phase. On cooling it condenses to form large masses of air which contain tiny droplets of liquid H_2O.

Hydrogen, He, Ne, Kr, and Xe are called noble gases. They are inert and, as a result, do not appear to have any major effect on, or role in, the atmosphere. Hydrogen and He are the lightest of all gases and are the most likely to escape the earth's gravitational pull and be lost to space.

The atmosphere contains trace gases produced by biological and/or geologic processes. Among these are ammonia (NH_3), methane (CH_4), hydrogen sulfide (H_2S), carbon monoxide (CO), and sulfur dioxide (SO_2). Ammonia, CH_4, and H_2S are primarily produced by biological decomposition. Methane absorbs thermal energy and serves as a greenhouse gas.

A variety of other gaseous substances can be found in the atmosphere in addition to those listed in Tables 1.1 and 1.2 and described above. Such substances are often released by living things. During periods of active growth, plants release large quantities of volatile compounds such as isoprenes and pinenes. Because of their relatively rapid removal by sink processes, such biogenic substances are not generally considered to be "normal" constituents of the atmosphere.

PHYSICAL CHARACTERISTICS

The atmosphere is characterized by the gases which comprise it and the various physical phenomena that act on and within it. These include solar radiation, thermal energy, atmospheric density and associated pressure characteristics, gravitational forces, water, and the movement of air molecules.

Solar Radiation

The sun radiating energy approximately as a blackbody (perfect absorber and emitter of radiation at all wavelengths) at an effective temperature of 6000°F "showers" the atmosphere and the earth's surface with an enormous quantity of electromagnetic energy. The total amount of energy emitted by the sun and received by the earth and its atmosphere is constant. Approximately 1.92 cal/cm²/sec (1400 w/m²/sec), the solar constant, is received on a perpendicular surface at the extremity of the atmosphere.

Spectral characteristics of solar radiation both external to the earth's atmosphere and at the ground can be seen in Figure 1.2. More than 99% of the energy flux from the sun is in the spectral region ranging from 0.15 to 4 μm, with approximately 50% in the visible light region of 0.4 to 0.7 μm. The solar spectrum peaks at 0.49 μm — the green portion of visible light. Because of the prominence of visible light in the solar spectrum, it is not surprising that life processes such as photosynthesis and photoperiodism are dependent on specific visible wavelength bands.

Incoming solar radiation is absorbed by atmospheric gases such as O_2, O_3, CO_2, and H_2O vapor. Ultraviolet light at wavelengths <0.18 μm (180 nm) are strongly absorbed by O_2 at altitudes above 100 km. Ozone below 60 km absorbs most UV between 0.2 and 0.3 μm (200 to 300 nm). Atmospheric absorption above 40 km results in the attenuation of approximately 5% of incoming solar radiation. Under clear sky conditions, another 10 to 15% is absorbed by the lower atmosphere or is scattered back to space, with 80 to 85% reaching the ground.

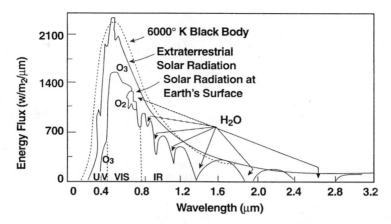

Figure 1.2 Solar spectra. (From Kondratyev, K.Y. 1969. Academic Press, New York. With permission.)

The earth, however, has considerable cloud cover so that under average conditions of cloudiness only 50% of incident solar radiation reaches the earth's surface. Cloud droplets and other atmospheric aerosols reflect and scatter incoming solar radiation. The reflection/backscattering of solar radiation by atmospheric aerosols and ground surfaces is described as the earth's albedo. The albedo of clouds and ground surfaces varies considerably. Clouds, snow, and ice have albedos that range from 0.5 to 0.8; fields and forests from 0.03 to 0.3; and from water 0.02 to 0.05 depending on the angle of incidence of the sun's energy.

The albedo of the earth and its atmosphere varies from 30 to 39%, with an average of 35%. Eighty-five percent of this value is due to backscattering associated with clouds. It is reflected light that allows us to see the moon and the nearby planets in our solar system.

The amount of solar energy actually received at the earth's surface is considerably less than the solar constant. Differences between the extraterrestrial solar curve and the one near the earth's surface (Figure 1.2) are due to absorption, and backscattering of the sun's energy by the atmosphere.

Molecular constituents such as H_2O vapor, CO_2, and O_3 contain numerous bands which absorb incoming solar radiation. Water vapor and CO_2 absorb in the infrared region; O_3 absorbs in both the UV and infrared regions.

Thermal Radiation

Solar radiation incident on the atmosphere and on the earth's surface is absorbed and reradiated at longer wavelengths. The earth and its atmosphere reradiate nearly as a blackbody with an effective temperature of 290°K. This reradiation results in an emission spectrum in the infrared range of 3 to 8 μm with a peak of about 11 μm. A nighttime thermal emission spectrum is illustrated in Figure 1.3. (Note the significant absorption characteristics of both CO_2 and

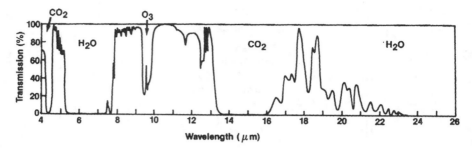

Figure 1.3 Nighttime thermal emission spectrum of the earth/atmosphere. (From Gates, D.M. 1962, Harper & Row Monographs, New York. With permission.)

H_2O vapor.) The atmosphere is mostly opaque to wavelengths <7 μm and >14 μm. Such absorption produces what is known as the "greenhouse effect."

The atmosphere radiates thermal energy to space directly through the atmospheric window (particularly in the spectral region between 8.5 and 11 μm) and successive reradiations from greenhouse gases.

Temperature Variation

Because the amount of solar radiation received at the earth's surface is a direct function of the angle of inclination of the sun, more energy is received in the equatorial regions than anywhere else in the world. Conversely, the polar regions receive the least solar energy. Though the equator receives the most solar energy, heat or thermal energy does not accumulate there, nor is there increased emission of infrared energy back to space. Rather, thermal energy is transferred poleward by both air and ocean currents. Warm currents move poleward and cold currents move toward the equator. Considerable heat transfer also occurs as a result of evaporation and subsequent condensation. These heat transfer mechanisms result in no net heat gain in equatorial regions.

The unequal distribution of solar energy on the earth's surface and the various heat transfer processes that exist cause considerable variation in surface air temperatures, and consequently differences in regional climate. Changes in the earth's inclination to the sun that occur over a year's time result in temporal and climatic differences that we associate with the seasons.

Significant variations in temperature also occur in the vertical dimension. A vertical temperature profile can be seen in Figure 1.4. Differences in temperature are used to describe various zones or atmospheric strata.

In the troposphere, temperature decreases with height at an average rate of −6.5°C/km. The troposphere has an average depth of 10 km varying from 12 km over the equator to 8 km over the poles. Because of the intense movement of thermal energy and significant temperature differences, the troposphere is a relatively unstable part of the atmosphere. This is the atmospheric layer in which day-to-day temperature, cloud cover, and precipitation changes occur — changes which we call weather.

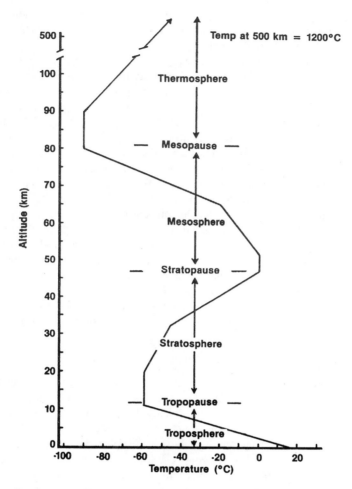

Figure 1.4 Vertical temperature profile of the earth's atmosphere.

At the top of the troposphere and below the stratosphere, temperature is isothermal (i.e., does not change with height). This isothermal region is called the tropopause. Similar isothermal regions are located between other layers of the upper atmosphere, including the stratosphere and the mesosphere, and the mesosphere and the thermosphere.

The stratosphere is characterized by a profile in which the temperature increases with height to about 0°C or 273°K. It has a depth of approximately 30 to 35 km. In addition to temperature, it is characterized by relatively high O_3 concentrations, with peak levels occurring between 25 and 40 km (Figure 1.1). Although the concentration of O_3 begins to decrease above 35 km, at these altitudes O_3 efficiently absorbs UV light accounting for the higher temperatures observed at the top of the stratosphere.

The stratosphere, in contrast to the troposphere, is a relatively stable region. It has very little manifestation of what we call weather. There are very few clouds, and winds are primarily horizontal.

The mesosphere is the atmospheric region above the stratopause. Its temperatures decrease up to 80 km altitude. In some classifications of atmospheric zones or strata, the mesosphere is described as the exosphere, as it is from this part of the atmosphere that light molecules such as H_2 and He are lost to space.

The thermosphere extends from an altitude of approximately 90 km to more than 1000 km. The thermosphere has no defined upper limit. It contains only a minute fraction of the atmosphere's mass. The relatively few molecules of N_2 and O_2 which are present absorb very-short-wave solar radiation. The high temperatures (which may exceed 1000°C) result from this absorption. Although the temperature may be high, an object in the thermosphere is unlikely to become hot, because there are too few molecules present to transfer their energy to it.

The atmosphere is electrically charged in the altitude range of 90 to 400 km. It is here that molecules of O_2 and N_2 become photoionized, producing positive ions, as electrons are set free to produce electric currents. The ionosphere, as it is called, reflects AM radio signals back to earth during nighttime. Following intense solar activity, the aurora borealis (northern lights) and aurora australis (southern lights) may occur. These phenomena are produced by solar storms which emit protons and electrons which are captured by the earth's magnetic field.

Atmospheric Density and Pressure

Concentrations of molecules that comprise the atmosphere decrease with height. As molecules become fewer, their mass per unit volume (density) decreases. This decrease in atmospheric density (Figure 1.5) is exponential. Ninety percent of the mass of the atmosphere is below an altitude of 12 km; 99% below 33 km. The total mass of the atmosphere is estimated to be about 5×10^{18} kg.

As atmospheric density decreases with height, atmospheric pressure decreases as well. Pressure (defined as force per unit area) is produced when air molecules in constant motion strike an object, rebound, and transfer momentum to it. The amount of momentum transferred is a function of the average kinetic energy of air molecules and is proportional to the absolute temperature. At elevated temperatures, molecules have more kinetic energy and thus move more rapidly. As they strike an object, more momentum is transferred and as a result the force experienced is greater; i.e., the pressure is greater.

An object on the earth's surface supports a vertical column of air which overlays it. This column exerts a force on the object equal to its weight. This force or atmospheric pressure is on the average at its greatest value near sea level where the weight of the atmosphere above it is also at a maximum. Atmospheric pressure at sea level is equal to 1.013×10^5 N/m^2 or 14.7 lb/in^2 (PSI). Atmospheric pressure is more commonly expressed in millibars, and in millimeters or inches of mercury. At sea level atmospheric pressure is equal to 1010 mbar, 760 mm Hg, and 29.92 in Hg. Changes in atmospheric pressure with height parallel those of density changes (Figure 1.5).

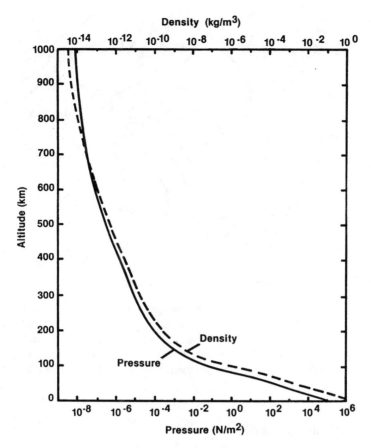

Figure 1.5 Density and pressure changes with altitude. (From National Oceanographic and Atmospheric Administration. 1976. NOAA S/T 76-1562.)

Horizontal and vertical pressure differences occur in the troposphere as a result of changes in thermal conditions. As air warms, it expands, and its density as well as pressure decreases. Conversely, as air cools it becomes more dense, and pressure increases. Significant pressure differences in the horizontal dimension occur in both cyclonic and anticyclonic air mass systems (Figure 1.6).

Gravity

The atmosphere is held closely to the earth's surface by gravitational forces. According to Newton's law of universal gravitation, every body in the universe attracts another body with a force equal to:

$$F = G \frac{m_1 m_2}{r^2} \tag{1.1}$$

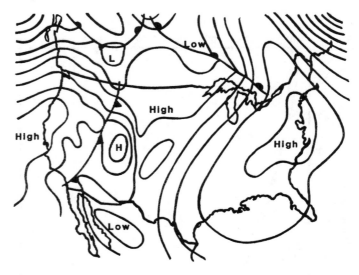

Figure 1.6   Regional differences in atmospheric pressure.

where m_1 and m_2 are the masses of the two bodies, G is a universal constant of 6.67×10^{-11} N · m^2/kg^2, and r is the distance between the two bodies. The force of gravity decreases as an inverse square of the distance between them.

Because of density changes associated with thermal conditions, gravity affects vertical air motion and planetary air circulation. It also plays a significant role in removing atmospheric aerosols.

ATMOSPHERIC MOTION

One defining feature of the atmosphere is its constant motion which occurs in both horizontal and vertical dimensions. Thermal energy (produced from the heating of the earth's surface and the air molecules above) is responsible for the atmosphere's motion. The differential heating of the earth's surface described previously results in energy flows from the equator poleward.

Pressure Gradient Force

Horizontal air movements, or advective winds, result from temperature gradients which give rise to density gradients and subsequently pressure gradients. The force associated with these pressure variations is described as the pressure gradient force. The direction of this force is perpendicular to lines of equal pressure (isobars) and is directed from high to low pressure. Regional differences in atmosphere pressure are indicated by isobars in Figure 1.6. If the isobars are close together, the pressure gradient force is relatively large, and such areas are characterized by high wind speeds. If isobars are widely spaced

Figure 1.7 Effect of earth's rotation on the trajectory of a propelled object.

as is the case in high pressure systems, the pressure gradient force is small, and winds are light.

Coriolis Effect

In large-scale north or south movements, air appears to be deflected from its expected path. Air moving poleward in the northern hemisphere appears to be deflected toward the east; air moving southward appears to be deflected toward the west. This apparent deflection is due to the earth's rotation.

The rotation of the earth makes it appear that a force is acting on air molecules as they travel north or south over the earth's surface. This phenomenon is illustrated in Figure 1.7. If an object is projected from Point A to Point B, it will actually reach Point B' because as it is moving in a straight line, the earth rotates beneath it from east to west.

Air flowing from Point A to Point B will also appear to be deflected. This apparent deflection is called the Coriolis effect or Coriolis force. Because of the Coriolis effect, air moving toward the equator in the northern tropical Hadley cell (Figure 1.8) appears to be deflected to the southwest. These are the northeast trade winds. In the southern hemisphere these winds are deflected toward the northwest. These are the southeast trade winds. Where the tradewinds converge at the equator a region of atmospheric calm called the doldrums is produced.

As previously indicated, the earth's rotation has a very significant effect on global circulation patterns. It, along with friction, is responsible for the zonal patterns evident in Figure 1.8.

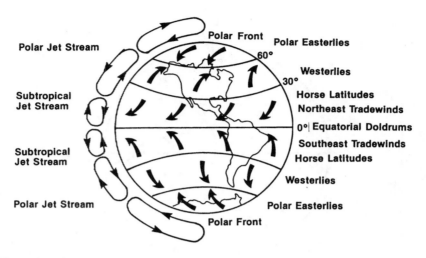

Figure 1.8 Three-zone global circulation model.

Geostropic Wind

At high altitudes (>700 m) where friction is not a significant factor, large-scale air movements reflect a balance between the pressure gradient force and the Coriolis effect. When the Coriolis effect is opposite the direction of the pressure gradient force and equal in magnitude, winds develop which are perpendicular to these forces with their direction along lines of constant pressure. When isobars are straight, the balance between the pressure gradient and Coriolis forces results in the formation of the geostropic wind, which is parallel to adjacent isobars. In the northern hemisphere low pressure is to the left of the geostropic wind; in the southern hemisphere it is to the right.

Gradient Wind and Friction

A balance occurs in low pressure systems when the pressure gradient force equals the combined effects of the centrifugal and Coriolis forces and the wind continues parallel to the isobars. A balance occurs in high pressure systems when the combined effects of pressure gradient and centrifugal forces equal the Coriolis force. A larger pressure gradient force is needed in air flows around low pressure systems (as compared to high pressure systems) to maintain a given gradient wind speed. The wind associated with curved isobars is called the gradient wind.

Air flow is also affected by frictional forces in the lowest 1 km of the atmosphere. These frictional forces increase toward the earth's surface. Friction reduces wind velocity which in turn decreases the Coriolis effect. The wind will turn toward low pressure until frictional and Coriolis forces balance the pressure gradient force. Friction turns the wind in the direction of low pressure, decreasing upward. The extent to which the gradient wind is reduced and its direction

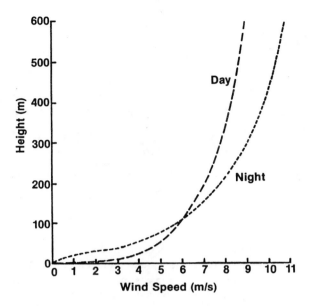

Figure 1.9 Day/night wind speeds as a function of height. (From Turner, D.B. 1969. EPA Publication No. AP-26.)

changed depends on the roughness of the terrain (surface roughness) and temperature gradients in the frictional layer. Day and nighttime gradient wind speeds as a function of height are illustrated in Figure 1.9. Frictional effects are less at night, resulting in the day–night differences in wind speeds indicated.

Cyclones and Anticyclones

The effects of the pressure gradient, Coriolis, and centrifugal and frictional forces in the proximity of curved isobars produce wind flow patterns associated with low and high pressure systems in the northern hemisphere (Figures 1.10 and 1.11). As previously indicated, friction diverts winds near the surface toward regions of low pressure. Air flows into the low pressure system and rises. In the northern hemisphere this cyclonic motion is counterclockwise; in the southern hemisphere it is clockwise. Migratory cyclones are common in tropical and temperate zones, and are accompanied by cloudy skies, precipitation, and considerable turbulence.

Anticyclones are produced in regions of high pressure where cool air descends from aloft and diverges outward in a clockwise motion in the northern hemisphere and a counterclockwise motion in the southern hemisphere.

Large-scale cyclonic and anticyclonic systems are major global air movement phenomena. They play significant roles in dispersion of atmospheric pollutants, the buildup of elevated pollutant levels, and in pollutant sink processes.

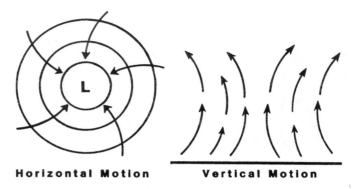

Figure 1.10   Air-flow patterns associated with a low pressure system.

Jet Streams

Above 3 km, west to east air flows occur in a wavelike fashion (in contrast to the closed pressure systems, cyclones and anticyclones, near the earth's surface). The highest wind speeds occur in distinct zones (300 to 500 km wide) at 11 to 14 km above the ground. Wind speeds of 150 km/hour are common.

Two well-defined jet streams occur in the northern hemisphere. The polar jet stream forms at the convergence of polar and temperate air masses. Because the polar front advances and retreats in different regions, the polar jet has a meandering pattern following polar fronts. The flow of the polar jet stream over the central U.S. is illustrated in Figure 1.12. A second jet stream, the subtropical, occurs at about 30°N latitude where tropical and temperate air masses converge.

Jet streams have significant effects on surface air flows. When jet streams accelerate, divergence of air occurs at the altitude of the jet stream which promotes convergence near the surface and the formation of cyclonic motion. On the other hand, deceleration causes convergence aloft and subsidence near the surface, causing an intensification of high pressure systems.

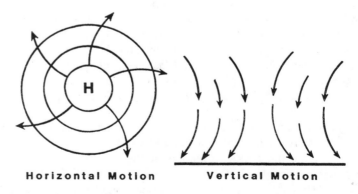

Figure 1.11   Air-flow patterns associated with a high pressure system.

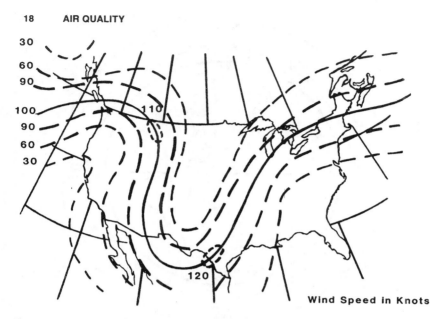

18   AIR QUALITY

Wind Speed in Knots

Figure 1.12   Flow of polar jet stream across North America. (From Anthes, R.A. et al. 1984. Charles E. Merrill Publishing Co., Columbus, OH. With permission.)

Jet streams are caused by strong temperature gradients. As a result, it is often above the polar front which occurs between the mid-latitude and polar circulation cells. The position of this jet stream varies, shifting poleward with the intrusion of warm tropical air and equatorially with the intrusion of cold polar air into mid latitudes.

Global Air Flow Models

In 1735 Sir George Hadley attempted to explain global air circulation using a relatively simple model. In Hadley's model sensible heat would be transported from the equator to each pole by large, simple, convective cells. At the equator, intense solar heating of land and water surfaces would produce warm air that would rise to the top of the troposphere where it would move poleward (Figure 1.13). At the poles, warm air would lose its heat by radiative and convective cooling, subside and flow to the equator where it would complete a cell. Because other factors also influence global air movement, Hadley's model did not adequately describe real-world global air circulation patterns.

In Hadley's model, global air circulation would occur along the lines of longitude. Because of the earth's rotation and the apparent deflection due to the Coriolis force, longitudinal motions are deflected and become zonal (along latitudinal lines). As a result, global atmospheric circulation is best described by the three-zone model illustrated in Figure 1.8. Here, Hadley-type cells are formed

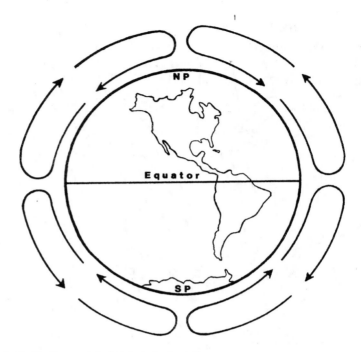

Figure 1.13 Hadley model of global air circulation.

on both sides of the equator. In these tropical Hadley cells, warm air rises at the equator and then proceeds poleward at high elevations in the troposphere. Thermal energy or heat is lost by radiation and convection along the way with a subsidence of cool air near latitudes of 30° N and S. As air subsides, it flows toward the equator to complete the cell. Hadley-type circulation cells also appear to form in the polar regions (>60° N and S latitude). These cells are characterized by rising air that moves poleward, subsidence at the poles and near surface flows south-ward/northward to complete the cell. In contrast to these more or less persistent circulation patterns, atmospheric flows in middle latitudes are more variable. These regions are influenced greatly by sharp temperature contrasts associated with polar and tropical fronts and the many migrating high and low pressure systems which cause frequent changes in weather.

In the latitudinal areas where air sinks to complete a cell, high pressure systems and subsidence inversions develop. The areas around 30° N and S latitudes represent the earth's great desert belts because of persistent high pressure systems and low rainfall. Populated areas along these belts such as the southern coast of California may be subject to persistent inversions and significant air-quality problems.

Water

Water is a unique chemical and physical factor in the earth's atmosphere that can exist simultaneously as a solid, liquid, or gas. In certain clouds all three phases may be found in close association. This transformation of water from one phase to another is significant in the phenomenon that we call weather.

We discussed earlier in this chapter how air is a mixture of gases. These gases exert a common pressure which we call atmospheric pressure. According to Dalton's law, each gas exerts a partial pressure, the sum of which equals the total atmospheric pressure. The partial pressure of water is called its vapor pressure. Vapor pressure increases as the concentration of H_2O vapor increases. The actual concentration of H_2O vapor is called absolute humidity and is expressed as mass of H_2O vapor per unit volume of air. At the saturation vapor pressure, air contains as much H_2O vapor as it can hold. Condensation occurs when the saturation value is exceeded. Evaporation, on the other hand, occurs at vapor pressures less than the saturation value.

Vapor pressure and saturation vapor pressure depend on temperature. At higher temperatures the atmosphere holds more H_2O vapor. The H_2O vapor present compared to the saturation value at a given temperature is called relative humidity. This value indicates the likelihood of condensation if unsaturated air is cooled. If the air is cooled below the dew point temperature, condensation will occur. The dew point temperature is the temperature to which a parcel of air has to be cooled to reach the saturation value.

As warm, moist air rises, it cools below the dew point. The subsequent condensation produces clouds. These clouds are air masses of condensed water droplets or ice particles whose size is too small to let them fall at an appreciable rate. As these droplets or ice particles grow to sufficient size, precipitation may result. As indicated earlier, clouds have a significant physical purpose in that they reflect sunlight back to space, making the earth a brighter planet.

EVOLUTION OF THE ATMOSPHERE

Although it was once widely believed that the earth's primordial atmosphere was composed of gases originating from the sun which gave it birth, there is increasing evidence that the early atmosphere was produced by thermal outgassing from early planetary matter. It appears that the earth and its sister planets were produced by the accretion of small particles and larger aggregates from the condensation of solar plasma at relatively moderate temperatures. Various gases such as the noble gases, H_2, CO_2, H_2O, N_2, HCl, CO, SO_2, and H_2S were believed to have been trapped within these accretions which formed the earth, and subsequently were emitted to form the early atmosphere.

Given the present mass of the earth's oceans, H_2O vapor must have been a significant degassing product of the early earth. Given the amount of water in the oceans, it is probable that average surface temperatures on the primitive earth were less than 300°K, since higher temperatures would have produced H_2O vapor concentrations so excessively high that an irreversible greenhouse effect would have occurred. The earth, with its favorable distance from the sun and associated temperatures, had an appropriate environment for the condensation of H_2O and formation of a huge ocean.

The formation of this large ocean provided a large reservoir for the absorption of relatively soluble gases such as CO_2, HCl, and SO_2. This is why the atmosphere of the earth contains considerably lower concentrations of CO_2 than the atmospheres of Mars and Venus. The formation of carbonate sediments removed significant quantities of atmospheric CO_2.

The earth's oceans formed very early during the earth's history. As a consequence, CO_2 levels in the atmosphere have been in a relatively small range in the past 500 million years, with variations associated with the formation and use of the earth's vast deposits of coal, oil, and natural gas.

As previously indicated, N_2 is the predominant gaseous constituent of the atmosphere. Its abundance is primarily due to its low solubility, relatively low (compared to O_2) chemical reactivity, and limited direct use by life forms.

Free O_2 is not a thermal emission product of planetary materials. It was essentially absent from the atmosphere during the first 2.5 billion years of the earth's history. Its source appears to have been the photodissociation of H_2O and CO_2, producing O_2 and H_2, with H_2 escaping to space and O_2 accumulating. If one assumes a constant generation rate and the absence of any major sinks, sufficient O_2 could have been produced over the past 4.5 billion years to account for the present mass concentration of O_2 in the atmosphere. However, O_2 is rapidly removed by mineral oxidation processes.

Our present understanding of the evolution of O_2 in the atmosphere recognizes the significant role of plant photosynthesis in O_2 production. Because the same amount of O_2 is used to oxidize organic matter to CO_2 as is released in photosynthesis, the biogenic production and accumulation of O_2 in the atmosphere is a relatively complex process. When a small fraction of organic matter is deposited in sediments, it creates an imbalance between biogenic O_2 production and consumption resulting in increased atmospheric concentrations.

Primitive life began to evolve in the earth's oceans approximately 3.5 billion years ago. Two billion years later, O_2 levels apparently reached 1%. By the time life evolved on land approximately 400 million years ago, it was approximately 10%, abundant enough to produce the O_3 layer in sufficient depth to protect terrestrial life from the lethal effects of exposure to UV light.

Noble gases such as Ar and He, which are present in the early and today's atmosphere, originated from radioactive decay products in the earth's mantle and crust. Both gases have accumulated over time with Ar being retained because of the earth's gravitational pull. Helium, on the other hand, because of its low mass, escapes slowly to space. The latter is also true for H_2.

The atmosphere has undergone enormous changes over the 4.5 billion years of the earth's history. Such changes have occurred over enormously long time scales compared to the relatively brief period of human history. However, even within the modern context, the composition of the atmosphere continues to undergo change. What is significant about such change is that it is primarily due to human activities. Changes include increases in: (1) CO_2 from the burning of fossil fuels and the clearing of tropical forests; (2) CH_4 from livestock and rice production, use of gas deposits, and termite decomposition of forest-clearing residue; (3) N_2O from soil disturbance and the agricultural use of nitrogen-based fertilizers; and (4) a variety of long-lived halogenated hydrocarbons from refrigerants, foaming and degreasing agents, etc. Humans have gained the power to affect the evolution of the atmosphere in the course of only a century or so of our existence.

READINGS

1. Anthes, R.A., J. J. Cahir, A. B. Frasier, and H. A. Panofsky. 1984. *The Atmosphere.* 3rd ed. Charles E. Merrill Publishing Co., Columbus, OH.
2. Boubel, R.W., D. L. Fox, D. B. Turner, and A. C. Stern. 1994. *Fundamentals of Air Pollution,* 3rd ed. Academic Press, New York.
3. Gates, D.M. 1962. *Energy Exchange in the Biosphere.* Harper & Row Monographs, New York.
4. Kondratyev, K.Y. 1969. *Radiation in the Atmosphere.* Academic Press, New York.
5. Lutgens, F.K. and E.J. Tarbuck. 1982. *The Atmosphere — An Introduction to Meteorology.* 2nd ed. Prentice Hall, Englewood Cliffs, NJ.
6. Lyons, T.J. and W.D. Scott. 1990. *Principles of Air Pollution Meteorology.* CRC Press, Boca Raton, FL.
7. Munn, R.E. 1966. *Descriptive Micrometeorology.* Academic Press, New York.
8. National Oceanic and Atmospheric Administration. 1976. *U.S. Standard Atmosphere.* NOAA S/T 76-1562.
9. National Research Council. 1991. *Rethinking the Ozone Problem in Urban and Regional Air Pollution.* National Academy Press, Washington, DC.
10. Turner, D.B. 1969. *Workbook of Atmospheric Dispersion Estimates.* EPA Publication No. AP-26.
11. Warneck, P. 1988. *Chemistry of the Natural Atmosphere.* Academic Press, New York.
12. Williamson, S.J. 1973. *Fundamentals of Air Pollution.* Addison-Wesley Publishing Co., Reading, MA.

QUESTIONS

1. There is a considerable difference between the amount of solar energy that strikes the extremity of the atmosphere and that which actually reaches the ground. What factors are responsible for this attenuation?

2. When you drive up a high mountain your vehicle may perform poorly. Explain why.

3. In the stratosphere, temperature increases with height. What is the physical/chemical basis for this increase?

4. Infrared emissions from the earth in the spectral region from 7 to 13 μm are very strong. What is the reason for this?

5. What phenomenon causes the ionization of the atmosphere?

6. Distinguish among absolute humidity, relative humidity, and saturation vapor pressure. How are they related?

7. Air flowing from the equator northward has a southwesterly flow. Explain.

8. What are jet streams? How are they formed? Where are they located?

9. Global air flow appears to behave as a three-zone model. How does this differ from global air circulation described by Hadley?

10. What causes the atmosphere to move?

11. How does gravity affect the flow of air in the atmosphere?

12. What is the origin of hydrogen and helium in the earth's atmosphere?

13. Though biogenic emissions from plants introduce a variety of substances into the atmosphere, they are not considered to be normal constituents. Why?

14. What forces affect horizontal air movement?

15. Thermal emissions from the earth and the atmosphere are characterized by electromagnetic waves in what spectral range?

16. Why do winds differ in nature near the ground (below 1 km) as compared to those aloft?

17. What is the relationship between tropical Hadley air flow patterns and high pressure systems?

18. What is the origin of most of the O_2 present in the present-day atmosphere?

19. The composition of the atmosphere is changing. What are these changes?

20. What are clouds, and what is their significance?

2   ATMOSPHERIC POLLUTANTS

In the previous chapter, the atmosphere was characterized as a mixture of gases. This gaseous mixture becomes polluted when it is changed by the addition (or, theoretically, subtraction) of particles, gases, or energy forms (e.g., heat, radiation, or noise) so that the altered atmosphere poses some harm because of its impact on weather, climate, human health, animals, vegetation, or materials. The concept of pollution entails a sense of degradation, a loss of quality, and adverse environmental effects whether it is applied to the air, water, or land.

Historically, air pollution concerns have been associated with ambient air, i.e., the freely moving air of the outdoor environment. As a consequence, control programs and texts on air pollution focus on ambient air pollution. However, air pollution is not just limited to the outdoor environment. Significant pollution can occur in occupational environments, in the built environments of our homes, offices, institutional buildings, and as a consequence of personal habits such as smoking. Although the author places primary emphasis on ambient air pollution, other air pollution problems exist which have significant implications for human health and welfare.

NATURAL AIR POLLUTION

Contamination or pollution of the atmosphere occurs as a consequence of natural processes as well as human activity (anthropogenic). Though control programs focus exclusively on anthropogenic air pollution, it is important to understand that nature too contributes to atmospheric pollution and in some instances causes significant air quality problems. As evidenced by past volcanic activity in the 19th century (e.g., Tambora and Krakatoa) and the more recent eruptions of Mt. St. Helens in Washington State, El Chinchon in Mexico, and

Mt. Pinatubo in the Philippines, as well as the 1988 forest fires in Yellowstone National Park and the western U.S., pollutants produced naturally can significantly affect regional and global air quality. In addition to volcanos and forest fires, natural air pollution results from a variety of sources: soil erosion and mineral weathering; plant and animal decomposition processes; emission of gas-phase substances from soil and water surfaces; volatile hydrocarbons (HCs) emitted by vegetation; pollen and mold spores; ocean spray; ozone (O_3) and nitrogen oxides (NO_x) from electrical storms; and O_3 from stratospheric intrusion and photochemical reactions. Emissions to the atmosphere associated with geochemical processes are described as geogenic; those produced biologically, biogenic.

Natural pollutants may pose serious air quality problems when they are generated in significant quantities near human settlements. With the exception of a few major events such as dust storms, forest fires, and volcanos, natural air pollution has not been a major societal concern. Nature does pollute more than man, as contrarion scientists and vocal opponents of contemporary air pollution control efforts maintain. However, such pollution has relatively low significance in causing health and welfare effects because (1) levels of contaminants associated with natural air pollution are typically very low, (2) large distances often separate sources of natural pollution and large human populations, and (3) major sources of natural pollution, such as forest fires, dust storms, and volcanos, are episodic and transient.

A major potential exception to our relegation of natural air pollution to a status of relative insignificance is biogenic emission of photochemically active HCs such as isoprene and α-pinene from vegetation. There is increasing evidence that natural HCs may play a significant role in photochemical oxidant production in some urban and nonurban areas (see Chapter 8).

ANTHROPOGENIC AIR POLLUTION

Anthropogenic air pollution has been and continues to be a serious problem. Its seriousness lies in the fact that elevated pollutant levels are produced in environments where harm to human health and welfare is more likely. It is this potential that makes anthropogenic air pollution a significant environmental concern.

In a historical sense air pollution became a serious problem when humans discovered the utility of fire. The smoke generated in the incomplete combustion of wood plagued cavemen and generations to come. For civilizations in Western Europe, the decimation of forests and the use of soft coal as a fuel source in the 14th century added a new dimension to the problem of air pollution. Emissions of pollutants from the combustion of coal were intense, and coal smoke became a major atmospheric pollution problem. It was the Industrial Revolution, however, in the early 19th century which gave rise to the atmospheric pollution characteristic of modern times.

Smog

In London and other British industrial cities, smoke from coal combustion mixed with fog from the North Sea. These occurrences were marked by an extreme reduction in visibility, the stench of sulfurous emissions, and, in severe episodes, illness and death. Pollution fogs in London have been reported for more than 100 years. The term smog, used to describe severe ambient air pollution conditions, was derived from the words smoke and fog. Such smog is gray in color and is often referred to as a "gray smog" or a "London-type" smog.

The term smog as it is used today is applied broadly to atmospheric pollution conditions characterized by a significant reduction in visibility. The term is applied without consideration of pollutant types, sources, or smog-forming processes.

The best-known smog problem in the U.S. is associated with the Los Angeles Basin of southern California. Although this "Los Angeles-type" smog reduces visibility, conditions which produce smog in southern California are very different from those of London and other northern European cities. The climate of the Los Angeles Basin is semiarid with numerous sunny days. Additionally, Los Angeles is situated at approximately 30 degrees North latitude, where the subsidence of air from the tropical Hadley cell (see Chapter 1) produces a semipermanent marine high pressure system with a cell center over Hawaii. The subsidence of air produces elevated inversions which are closest to the surface at the eastern edge or continental side of the marine high, i.e., over the south coast of California. These relatively low subsidence inversions effectively reduce pollution dispersion in the vertical dimension. This problem is further exacerbated by mountains, which act as barriers to horizontal air movement in the easterly and northerly directions.

The poor dispersion conditions in the Los Angeles region play a significant role in smog formation. Other factors also play a role, however. The major source of pollutants is light-duty motor vehicles, an estimated 5+ million of which release exhaust and evaporative gases into the atmosphere. Motor vehicle-generated pollutants, abundant sunlight, poor dispersion conditions resulting from inversions, and topographical barriers produce smog conditions with enormous chemical complexity. Because elevated levels of NO_2 are produced, this smog tends to be brown in color.

Smog produced over the south coast of California can be characterized as being a photochemical smog. The visibility-reducing particulate pollutants, as well as hundreds of gas-phase substances, are produced as a result of photochemical reactions in the atmosphere. Abundant solar radiation characteristic of the region is a key element in smog formation.

Los Angeles and London represent relatively unique conditions of smog development. Smogs which more rarely form over larger cities in the eastern U.S. cannot be characterized as either type. The occasional smogs over Chicago, Pittsburgh, or New York represent conditions which are somewhat intermediate between the two types. These cities have or have had significant emissions of pollutants from both automobiles and coal combustion. They do not, however, have the climatic conditions which are critical to the formation of the heavy smogs in either Los Angeles or London.

Haze

The term haze is often used in the nomenclature of air pollution. But what is haze, and how does it differ from smog? In actuality, haze and smog are closely related. Both conditions represent a reduction in visibility. They differ, however, in both intensity and geography. In general, haze refers to a more limited degree of visibility reduction. The term smog is often used to describe the marked visibility reduction over cities or large metropolitan areas. Haze, on the other hand, typically refers to the large-scale, low-level pollution-caused visibility reduction observed during the summer months over much of the Midwest, Northeast, and Southeast United States.

Nontraditional Air Pollutants

In traditional discussions of air pollution, pollutants are usually characterized as being particulate or gaseous-phase substances. Air quality can, however, be degraded by other factors, such as noise, heat, ionizing radiation, and electromagnetic fields associated with the transmission and use of electricity. With the exception of noise (which is discussed extensively in Chapter 12), nontraditional air pollutants will receive relatively little discussion in this book.

Gases and Particles

Because of their impact on the atmosphere, vegetation, human health, and materials, gases and particles have received considerable research and regulatory attention. Although the terms gases and particles are used, they actually represent all three phases of matter. Particles represent both solid and liquid phases. When solid or liquid particles are dispersed in the atmosphere, the resulting suspension is characterized as being an aerosol. Aerosols may reduce visibility, soil materials, and affect human health.

Aerosols may be generated in several ways. For example, fume aerosols are produced by the condensation of metal vapors, dust aerosols from the fragmentation of matter, mists from the atomization of liquids or condensation of vapors, and smokes from incomplete combustion of organic materials. Smoke, as we know it, is a mixture of gases, solid particles, and droplets of liquid. The photochemical aerosol of Los Angeles-type smog is produced by the condensation of vapors which result from photochemical reactions in the atmosphere. Most of the acid aerosol which contributes to the formation of haze and to the acidification of precipitation is also produced photochemically.

Although particles are the most visible manifestation of air pollution, they are only a relatively small part of the emissions problem, because on a weight basis approximately 90% of the anthropogenic emissions to the atmosphere are gaseous. Gaseous pollutants may be produced by the combustion of fuels and other materials, the smelting of mineral ores, the vaporization of volatile liquids, etc.

Pollutant Sources

Pollutants in the atmosphere may be emitted from identifiable sources, or they may be produced in the atmosphere as a result of chemical reactions. The former are classified as primary pollutants, the latter secondary.

Sources of primary pollutants may be classified as mobile or stationary, combustion or noncombustion, area or point, direct or indirect. These classifications reflect both regulatory and administrative approaches toward implementing air pollution control programs at federal, state, and local levels.

Mobile sources include automobiles, trains, airplanes, etc.; others are stationary. A point source is a stationary source whose emissions may significantly contribute to air-quality degradation. An area source, on the other hand, comprises a number of stationary and mobile sources which individually do not have a significant effect on air quality. Collectively, their impact on air quality may be enormous. Area sources include emissions from motor vehicles, aircraft, trains, open burning, and a variety of small miscellaneous sources. Facilities which themselves do not emit pollutants may be classified as indirect sources. These may include shopping centers and athletic stadia which contribute to elevated pollutant levels as a result of attracting motor vehicle traffic.

National Emission Estimates

National emission estimates for five major primary pollutants and broad source categories are presented in Tables 2.1 and 2.2. Emission estimates in Table 2.1 are for 1970, when the U.S. began its major regulatory effort to control ambient air pollution. Emission estimates, reflecting the status of this nation relative to the same pollutants and source categories more than two decades later, are summarized in Table 2.2.

Transportation sources include motor vehicles, aircraft, trains, ships, boats, and a variety of off-road vehicles. Major stationary fuel combustion sources include fossil fuel-fired electrical generating plants, industrial and institutional boilers, and home space heaters. Industrial process losses include pollutants produced in a broad range of industrial activities including mineral ore smelting, petroleum refining, oil and gas production and marketing, chemical production, paint application, industrial organic solvent use, food processing, mineral rock crushing, etc. Emissions from solid waste disposal results from onsite and municipal incineration and open burning. Miscellaneous sources include forest fires, agricultural burning, coal refuse and structural fires, and a variety of organic solvent uses. In addition to the five pollutant categories indicated in Tables 2.1 and 2.2, the U.S. has major control programs for lead and O_3. Ozone is not inventoried, as it is a secondary pollutant produced as a result of atmospheric

Table 2.1 Estimates of 1970 National Emissions of Five Primary Pollutants (10^6 short ton/year)

Source	Carbon monoxide	Particulate matter (PM_{10})	Hydrocarbons	Nitrogen oxides	Sulfur oxides	Total	Percent
Transportation	98.64	0.46	14.51	9.02	0.43	123.06	54.7
Stationary source fuel combustion	4.63	2.87	0.72	11.69	23.46	43.37	19.3
Industrial processes	9.84	7.67	12.33	0.91	7.09	37.84	16.8
Solid waste disposal	7.06	0.99	1.98	0.08	0.01	10.12	4.5
Miscellaneous	7.91	0.84	1.10	0.30	0.10	10.25	4.5
Total	128.08	12.84	30.65	22.00	31.09	224.66	
Percent	57.0	5.7	13.6	9.8	13.8		100.0

From USEPA. 1994. EPA/454-R-94-027.

Table 2.2 Estimates of 1993 National Emissions of Five Primary Pollutants (10^6 short ton/year)

Source	Carbon monoxide	Particulate matter (PM_{10})	Hydrocarbons	Nitrogen oxides	Sulfur oxides	Total	Percent
Transportation	75.26	0.59	8.30	10.42	0.72	95.29	56.2
Stationary source fuel combustion	5.43	1.21	0.65	11.69	19.27	38.25	22.5
Industrial processes	5.28	0.61	11.20	0.91	1.86	14.86	11.7
Solid waste disposal	1.73	0.25	2.27	0.08	0.04	4.37	2.6
Miscellaneous	9.51	1.03	0.89	0.30	0.01	11.74	6.9
Total	97.21	3.69	23.31	23.40	21.89	169.50	
Percent	57.4	2.2	13.8	13.8	12.9		100.0

From USEPA. 1994. EPA/454/R-94-027.

Table 2.3   Estimates of National Lead Emissions, 1970 and 1993 (10^3 short ton/year)

Source	1970	1993	Percent reduction
Transportation	180.30	1.59	99.2
Stationary source fuel combustion	10.62	0.50	95.3
Industrial processes	26.40	2.28	91.4
Solid waste disposal	2.20	0.52	76.4
Miscellaneous	0.00	0.00	
Total	219.47	4.89	97.8

From USEPA. 1994. EPA/454/R-94-027.

reactions. Emissions of lead to the atmosphere from 1970 to 1993 are summarized in Table 2.3.

The National Emissions Picture

Table 2.1 provides a variety of insights into the nature of the air pollution problem that the U.S. faced in 1970 just as it began to enact and implement tough air pollution control legislation. As can be seen, (1) transportation was responsible for more than 50% of emissions of five primary pollutants, (2) transportation was the major source of CO and second largest source of NO_x, (3) stationary source fuel combustion was the major source of sulfur and nitrogen oxides, and (4) industrial processes were the largest source of particulate matter (PM) and the second largest source of SO_x.

The impact of transportation on the nation's air quality, as evidenced in Table 2.1, has been considerable. In the transportation category, motor vehicles, which include the personal automobile, accounted for 75% of all emissions. Motor vehicles have also been the major source of atmospheric lead. Of the seven major pollutant categories for which major control programs have been initiated, transportation, particularly the motor vehicle, has been implicated as the major direct or indirect source for CO, HCs, lead, and O_3, and the second largest source of NO_x. It is no wonder that motor vehicles have received so much regulatory attention and investment in pollution control.

Significant regulatory requirements have also been placed on fossil fuel-fired electrical power generating facilities — the major sources of emissions under the stationary source fuel combustion category. Stationary source fuel combustion is the largest source of SO_x and NO_x. Sulfur oxides emitted by fossil fuel (especially coal)-fired electrical power generating stations have been of particular concern.

The relative effectiveness of pollution control efforts over the past two decades can be seen from a comparison of Tables 2.1 and 2.2. A relatively modest 25% reduction has been achieved in emissions of the five primary pollutant categories. Reductions based on source categories include: transportation, 23%; stationary source fuel combustion, 12%; industrial process losses, 48%; and solid waste disposal, 57%. On a pollutant basis, reductions in emissions have included CO, 24%; PM, 71%; HCs, 24%, and SO_x, 30%. There has been an apparent increase of NO_x emissions of approximately 6%. [Readers of *Air Quality*, 2nd ed., may note disparities in the emission reductions for 1970 to 1987 and 1970 to 1993 described here. These disparities are apparently due to changes in emission estimating practices of the U.S. Environmental Protection Agency (USEPA).]

Emission reductions from industrial processes and solid waste disposal have been significant. These reflect the regulatory attention given these source categories. Though transportation has been regulated intensively, emission reductions have been relatively modest. This has been due, in part, to an increase in the motor vehicle population and to about 10% of the motor vehicle (primarily older) population which is a very significant source of emissions. The relatively modest reductions in emissions associated with stationary source fuel combustion reflect a combination of growth in this area and regulatory policies which focused primarily on ground-level concentrations rather than emissions.

On a pollutant basis, significant emission reductions have been achieved for PM with more modest reductions for CO, HCs, and SO_x and a slight increase in NO_x. The modest reductions in HC emissions reflect the growth in the population of motor vehicles and belated regulatory efforts to control HC emissions from industrial and commercial sources. The modest reduction in SO_x emissions reflects control policies which focused on ground level concentrations. The apparent lack of progress associated with NO_x emissions reflects a variety of factors. These include (1) technical and cost problems in controlling emissions from motor vehicles, (2) policies which focused on ground-level concentrations, and (3) an O_3 control strategy which focused exclusively on reducing HC emissions.

Decreases in CO and HC emissions are a result of the application of catalytic converters and HC control technology on light-duty motor vehicles. Decreases in PM can be attributed to the installation of control equipment on industrial sources, changes in fuel use practices in the direction of cleaner-burning fuels, the banning of open burning of solid wastes, and limitations on combustion practices used for solid-waste incineration. The observed decrease in SO_x emissions has been primarily due to the use of lower-sulfur fuels and the installation of flue gas desulfurization systems on coal-fired power plants.

Reductions in lead emissions to the atmosphere (Table 2.3) have been enormous (approximately 98%), since the nation began to phase lead out of gasoline in 1974.

Looking back over the past several decades, the U.S. appears to have made moderate progress in enhancing the quality of ambient air. Yet, as we approach the 21st century, it is apparent that continued efforts to limit pollutant emissions is essential to achieve air-quality goals. Many areas in the country still do not have acceptably clean air. Significant reductions in emissions will be required to mitigate the problem of acidic deposition, to control a variety of toxic pollutants, to limit emissions of stratospheric O_3-destroying chemicals, to reduce elevated ground-level O_3 levels produced by atmospheric photochemistry, and to reduce the potential for anthropogenically forced global warming.

GASEOUS POLLUTANTS

Concentration Expression

In the U.S., concentrations of atmospheric gases and vapors have been historically expressed as mixing ratios and reported as parts per million volume (ppmv). One ppmv is equal to a volume of a gas mixture in 1 million volumes of air.

$$1\,ppmv = \frac{1\,gas\,volume}{10^6\,air\,volumes} \tag{2.1}$$

A microliter volume of gas mixed in a liter of air would therefore be equal to 1 ppmv.

$$1\,ppmv = \frac{1\,\mu L\,gas}{1\,L\,air} \tag{2.2}$$

Mixing ratios based on volume–volume ratios may also be expressed as parts per hundred million (pphmv), parts per billion (ppbv), or parts per trillion (pptv). Mixing ratios used to express air concentrations should not be confused with those used for water which are weight–volume ratios (mg/L) and for solids which are weight–weight ($\mu g/gm$, mg/km) ratios. Though all are expressed as ppm, they are not equivalent concentrations.

In the 1970s there was some movement in the U.S. to express gas concentrations in metric units as mass per unit volume ($\mu g/m^3$, mg/m^3) in conformance with international practice. This movement ran out of steam, and the expression of contaminant gas concentrations as mixing ratios continues. When the USEPA promulgated air-quality standards for five gas-phase contaminants in the 1970s, concentrations were expressed as both ppmv and μg or mg/m^3. Outside the U.S., metric expression continues to be the rule although there is some tendency to express concentrations in mixing ratios.

Nature of the Concern

A wide variety of gaseous pollutants are produced by the myriad of pollutant sources extant in the U.S. It is likely that hundreds may exist in the polluted air environment over some of our major cities. More than 400 different gaseous species have been identified in automobile exhaust alone, and many new pollutants are produced as a result of atmospheric chemistry.

Although the atmosphere may be contaminated by hundreds of pollutants, only a relatively small number have been identified as being at levels significant enough to pose a threat to human health and welfare. Major gaseous pollutants identified as having potentially significant public health and/or welfare effects are discussed here in the context of their chemical properties, reactions, sources, sinks, background levels, and urban/nonurban concentrations.

Carbon Oxides

Hundreds of millions of tons of the carbon oxides, carbon monoxide (CO), and carbon dioxide (CO_2), are produced/emitted to the atmosphere by both natural and anthropogenic sources. Carbon monoxide, because of its human toxicity, has been a major pollution concern for many decades. Carbon dioxide, on the other hand, is relatively nontoxic and is a major raw material in plant photosynthesis. A few decades ago, little significance was given to increased atmospheric levels associated with human activities. That, of course, is no longer the case as its thermal absorption properties and ever-increasing emissions and atmospheric levels raise major questions and scientific and regulatory concern that CO_2 may contribute to global warming.

Carbon Dioxide. Carbon dioxide is a relatively abundant and variable constituent of the atmosphere (Table 1.2). It is produced and emitted naturally to the atmosphere in the biological decomposition, combustion, and weathering of organic matter, and the weathering of carbonates in rock, soil, and water. The major direct anthropogenic source of CO_2 is combustion of fossil fuels and biomass burning; agriculture represents an indirect source.

Enormous changes in atmospheric CO_2 levels have occurred through geological time. Carbon dioxide has, over hundreds of millions of years, been removed from the atmosphere by plants and stored in the earth's crust as coal, oil, and natural gas. Because of their importance as energy sources, these materials have been intensively used over the past century or so, resulting in significant increases in atmospheric concentrations, from approximately 280 ppmv in the mid-18th century to approximately 360 ppmv at the present time (1996). Figure 2.1 shows changes in atmospheric levels over the past two centuries from ice core data and measurements at Hawaii's Mauna Loa Observatory. Note the rapid increase in atmospheric CO_2 levels at Mauna Loa since 1958. Both monthly and annual changes in CO_2 levels measured at Mauna Loa observatory can be seen in Figure 2.2. Since 1958 when the annual average concentration was 315 ppmv, CO_2 levels have increased by approximately 45 ppmv. Figure 2.2 also illustrates the pronounced seasonal variation of approximately 7 ppmv associated with high photosynthetic consumption of CO_2 during the spring and summer with an excess of respiratory CO_2 emissions (over photosynthesis) during autumn and winter. This seasonal cycle varies from 1.6 ppmv at the South Pole to 15 ppmv at Point Barrow, AK.

Carbon dioxide is very soluble in water, and the earth's oceans serve as an enormous CO_2 sink, removing an estimated 50% of anthropogenic emissions. The world's forests, particularly tropical forests, have historically served as major sinks for CO_2 as well. In recent decades the relentless clearing and burning of high carbon density forests and replacement by low carbon density agricultural

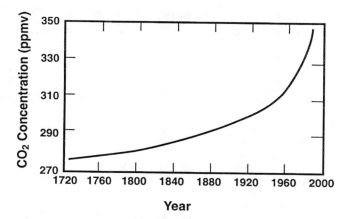

Figure 2.1   Changes in CO_2 levels in the atmosphere over the past two centuries. (From Harrison, R.M. (Ed.). 1990. Royal Society of Chemistry. Thomas Graham House, Cambridge, UK. With permission.)

or grazing lands has significantly decreased the capacity of these biological sinks. Indeed, such land-use changes may have served as a major source of CO_2 emissions to the atmosphere.

Carbon Monoxide. Carbon monoxide is a colorless, odorless, and tasteless gas. Emissions to the atmosphere result from a number of natural and anthropogenic sources, with the former being dominant in the global budget of CO in the atmosphere (Table 2.4). Carbon monoxide is produced in large quantities as a result of the incomplete combustion of fossil fuels and biomass (Equations 2.3 and 2.4).

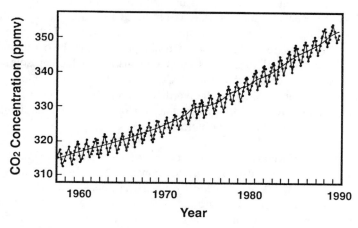

Figure 2.2   Seasonal and annual trends in CO_2 levels over Mauna Loa Observatory since 1958. (From Elsom, D.M. 1992. Blackwell Publishers, Oxford, UK. With permission.)

Table 2.4   Sources and Sinks for Carbon Monoxide

Sources	10^6 ton C/year
General	
Oceans	20
Methane oxidation	260
Oxidation of anthropogenic HCs	40
Wood fuel combustion	20
Fossil fuel combustion	190
Temperate zone	
Oxidation of natural HCs	100
Forest fires	10
Agricultural burning	10
Tropics	
Oxidation of natural HCs	150
Forest clearing	160
Burning of grasslands, agricultural land	100
Total	1060
Sinks and accumulation	
Reaction with OH	820 ± 300
Soil uptake	110
Accumulation	10
Total	940 ± 300

From Bridgman, H. A. 1994. John Wiley & Sons, Ltd. London.
With permission.

$$2C + O_2 \rightarrow 2CO \tag{2.3}$$

$$2CO + O_2 \rightarrow 2CO_2 \tag{2.4}$$

Carbon monoxide is an intermediate product in combustion-oxidation processes with significant production rates when insufficient O_2 is present.

As can be seen in Table 2.4, CO is produced naturally from a variety of sources, including the photolytic decomposition of methane (CH_4) and biogenic HCs such as terpenes, forest fires, and microbial processes in the ocean and soils.

Direct anthropogenic emissions from combustion and other sources and photolytic oxidation of anthropogenic HCs account for approximately 30% of CO emitted to the atmosphere of the Northern Hemisphere. CO emissions over North America are substantially greater than for other areas of the hemisphere because of emissions from transportation sources.

Concentrations. Carbon monoxide levels vary significantly in the atmosphere, both spatially and temporally. Background concentrations increase along a gradient from a low of 50 ppbv in high southern latitudes, with average global concentrations of 110 ppbv. Background concentrations have been increasing in the past decade at an approximate rate of 1% per year with most of this increase in the middle latitudes of the Northern Hemisphere. Highest concentrations are observed during winter months when scavenging substances such as hydroxyl radical (OH) are at relatively low concentrations.

Ambient CO levels in urban areas associated with anthropogenic sources such as motor vehicles, iron foundries, incinerators, and steel mills are orders of magnitude higher than background levels. Concentrations in urban areas are closely associated with traffic density and meteorological conditions. Peak levels occur along traffic corridors during morning and evening rush hours and decrease rapidly with distance from highways and streets. Though CO levels in urban areas are decreasing as a consequence of motor vehicle emissions controls in the U.S. over the past 25 years, CO levels in urban areas averaged over an hour range from a few ppmv to 15 ppmv or more, with historical highs in the range of 40 to 60 ppmv. Urban areas of particular significance have been the high-population, high-traffic density Los Angeles Basin and high-altitude cities such as Denver. Though CO levels in urban areas in the U.S. continue to decline, increasingly higher concentrations are being observed in developing countries, most notably in high-altitude Mexico City with its dense population and increasing motor vehicle traffic.

Sinks. With an increasing knowledge of CO sink processes, we are beginning to recognize that CO is a far more important contaminant of the atmosphere than our regulatory concerns associated with its health effects would suggest. Carbon monoxide sink processes affect both tropospheric concentrations of OH and O_3, and thus the oxidizing power of the atmosphere. They also indirectly affect tropospheric concentrations of CH_4, a major greenhouse gas, and stratospheric concentrations of H_2O vapor derived from CH_4 oxidation in the stratosphere.

Major sink processes include atmospheric photochemical processes and uptake in the soil accounting for an estimated 3.9×10^9 and 2×10^8 ton/year, respectively. The oxidation of CO in the atmosphere is initiated by reaction with OH:

$$CO + OH \rightarrow CO_2 + H \qquad (2.5)$$

followed by

$$H + O_2 + M \rightarrow HO_2 + M \qquad (2.6)$$

In the presence of significant quantities of NO_x:

$$HO_2 + NO \rightarrow NO_2 + OH \qquad (2.7)$$

NO_2 is subsequently photolyzed to produce O_3.

$$NO_2 + \xrightarrow{\quad h\nu \quad} O(^3P) + NO \qquad (2.8)$$

$$O(^3P) + O_2 + M \rightarrow O_3 + M \qquad (2.9)$$

where $O(^3P)$ is ground-state atomic oxygen. As a result of these reactions, CO is oxidized to CO_2.

$$CO + 2O_2 + hv \rightarrow CO_2 + O_3 \tag{2.10}$$

with the production of one O_3 molecule for each CO oxidized. These reactions therefore are an important source of O_3 in the troposphere.

As will be seen in the section on HCs, CO competes with CH_4 for OH. The oxidation of CH_4 is also a major source of CO, accounting for approximately 25% of its atmospheric concentration. The atmospheric residence time of CO is estimated to be approximately 2 months, with the atmospheric burden of CO replaced 5 to 6 times/year.

Sulfur Compounds

Sulfur compounds are emitted to the atmosphere from a variety of natural and anthropogenic sources. These include most importantly SO_x and a variety of species of reduced sulfur (e.g., hydrogen sulfide (H_2S), carbonyl sulfide (COS), carbon disulfide (CS_2), and dimethyl sulfide [$(CH_3)_2S$]).

Sulfur Oxides. Sulfur oxides are emitted to the atmosphere naturally from volcanoes and in the atmospheric oxidation of reduced sulfur compounds. Sulfur oxides are produced anthropogenically when metal sulfide ores are roasted and sulfur-containing fuels are combusted. Of the four known monomeric SO_xs, only SO_2 is found in appreciable quantities in the atmosphere. Sulfur trioxide (SO_3) is emitted directly into the atmosphere in metal smelting and fossil fuel combustion and is produced by the oxidation of SO_2 in the atmosphere. Because it has a high affinity for H_2O, SO_3 is rapidly converted to sulfuric acid (H_2SO_4). The formation of SO_2, SO_3, and H_2SO_4 by direct oxidation processes is summarized in the following equations:

$$S + O_2 \rightarrow SO_2 \tag{2.11}$$

$$2SO_2 + O_2 \rightarrow 2SO_3 \tag{2.12}$$

$$SO_3 + H_2O \rightarrow H_2SO_4 \tag{2.13}$$

Sulfur dioxide is a colorless gas with a sulfurous odor that can be easily detected in the concentration range of 0.38 to 1.15 ppmv. Above 3.0 ppmv, it has a pungent, irritating odor.

Concentrations. Background levels of SO_2 are very low with typical concentrations in the range of 24 to 90 pptv. In remote areas relatively unaffected by pollutant sources, concentrations are typically <5 ppbv. Average annual concentrations ranging from 1.5 ppbv in Auckland, NZ to 9 ppbv in Chicago to 82 ppbv in Shenjang, China with corresponding 98% percentile 24-hour concentrations of 4, 32, and 317 ppbv, respectively, have been reported. In the recent past, cities such as Chicago and Pittsburgh observed 1-hour maximum concentrations in the range of 100 to 500 ppbv. Maximum 1-hour concentrations in the range of 1.5 to 2.3 ppmv have been reported around large nonferrous metal smelters. With the implementation of significant SO_x reductions in the early 1970s, hourly SO_2 levels in urban areas in the U.S. above 100 ppbv have become relatively rare.

Atmospheric Reactions and Sink Processes. After SO_2 is emitted to the atmosphere, it is oxidized by various gas and liquid phase reactions. The oxidation of SO_2 may be direct (Equations 2.11 to 2.13), photochemical, or catalytic. Because direct oxidation is slow, it is a relatively insignificant SO_2 conversion process. Gas-phase oxidation mechanisms include photo-oxidation, reaction with OH, O_3, biradical, ground-state atomic oxygen [$O(^3P)$] and peroxy radicals (HO_2, RO_2). Reaction with OH is the most important atmospheric sink process. It appears to proceed as follows.

$$SO_2 + OH \rightarrow HOSO_2 \tag{2.14}$$

$$HOSO_2 + O_2 \rightarrow HO_2 + SO_3 \tag{2.15}$$

$$SO_3 + H_2O \rightarrow H_2SO_4 \tag{2.16}$$

Subsequent aerosol formation occurs by nucleation and condensation processes. Sulfuric acid will react with ammonia (NH_3) to form sulfate salts.

Sulfur dioxide can dissolve in fog, cloud, and rain droplets, as well as hygroscopic aerosols, to form a dilute solution of sulfurous acid.

$$SO_2 + H_2O \rightarrow SO_2 \cdot H_2O \tag{2.17}$$

$$SO_2 \cdot H_2O \rightarrow HSO_3^- + H^+ \tag{2.18}$$

In the aqueous phase, SO_2 can be oxidized by a variety of mechanisms to form H_2SO_4. These include oxidation by nitrous acid (HNO_2), O_3, hydrogen peroxide (H_2O_2), organic peroxides, and catalysis by iron and manganese.

Sulfur dioxide and its oxidation products can be removed from the atmosphere by wet and dry deposition processes. In the latter case, reaction with plant surfaces appears to be a major gas-phase removal process. Particulate-phase sulfates are removed from the atmosphere by wet and dry deposition processes, as described later in this chapter. The atmospheric lifetime of SO_2 is believed to be 2 to 4 days.

Reduced Sulfur Compounds. A variety of reduced sulfur compounds are emitted to the atmosphere from both natural and anthropogenic sources. Major reduced sulfur compounds emitted or formed in the atmosphere include H_2S,

COS, CS_2, and $(CH_3)_2S$. Carbonyl sulfide is the most abundant sulfur species in the atmosphere. Though it is produced in combustion processes, most COS enters the atmosphere from biogenic sources. Background concentrations are approximately 500 pptv with an atmospheric lifetime of 44 years. Because of its limited reactivity, it contributes little to pollution problems associated with atmospheric sulfur. Carbon disulfide is produced from sources similar to COS but is more photochemically reactive, with a lifetime of 12 days. As a consequence, global concentrations are relatively variable, ranging from 15 to 190 pptv. Dimethyl sulfide is released from oceans in large quantities. It has a very short lifetime (0.6 days) and is rapidly oxidized to SO_2.

The reduced sulfur compound of major human concern is H_2S. It is primarily produced by anaerobic decomposition processes. It has a relatively short (4.4 days) lifetime in the atmosphere, being scavenged by OH and COS to produce approximately 50% of the background levels of SO_2. Background concentrations of H_2S are in the range of 30 to 100 pptv.

Anthropogenic sources include oil and gas extraction, petroleum refining, coke ovens, and kraft paper mills. Anthropogenic sources account for less than 5% of emissions to the atmosphere.

Hydrogen sulfide has a characteristic "rotten egg" odor which is detectable at about 500 pptv. Though it is a relatively toxic gas, its low atmospheric concentrations appear to pose no threat to human health. The major air-quality concern associated with it and other anthropogenically emitted reduced sulfur compounds such as methyl and ethyl mercaptan are malodors and, for H_2S, reaction with and discoloration of lead-based paints.

Nitrogen Compounds

A variety of gas- and particulate-phase nitrogen compounds are commonly found in the atmosphere. Gas-phase substances include nitrogen gas (N_2), nitrous oxide (N_2O), nitric oxide (NO), nitrogen dioxide (NO_2), nitrate radical (NO_3), dinitrogen pentoxide (N_2O_5), nitrous acid (HNO_2), nitric acid (HNO_3), peroxyacyl nitrate ($CH_3COO_2NO_2$), other organic nitrates, ammonia (NH_3), and hydrogen cyanide (HCN). The ionic species NO_3^-, NO_2^-, and NH_4^+ are found in the aqueous phase.

Nitrogen gas is the major gaseous constituent of the atmosphere, making up approximately 78% of the atmosphere's mass. Other nitrogen compounds are found in only trace quantities. Despite their relatively low concentrations, substances such as NO, NO_2, N_2O, organic nitrates, gas- and particulate-phase nitrogen acids, and ammonia (NH_3) play very significant roles in atmospheric chemistry and especially the chemistry of polluted environments. Nitric oxide and NO_2 are of major environmental concern because their concentrations increase significantly in the atmosphere as a result of human activity, and they serve as precursor molecules for a large variety of atmospheric reactions. Because of their rapid chemical interconvertibility, atmospheric concentrations are usually described by the collective term, NO_x.

Nitrous Oxide. Nitrous oxide is a colorless, slightly sweet, relatively nontoxic gas. It is widely used as an anesthetic in medicine and dentistry. It is called laughing gas because exposures to elevated concentrations produce a kind of hysteria.

Though a natural constituent of the atmosphere, its concentration has been steadily increasing (since about 1940) at a rate of 0.8 ppbv/year from preindustrial concentrations (based on glacial ice core measurements) of 288 ppbv to an approximate 1995 atmospheric average of 316 ppbv. It is produced by both natural (microbiological processes in the world's oceans) and anthropogenic sources (soil disturbance, application of industrially produced nitrogen fertilizers, and burning of fossil fuels). It has no known tendency to react with other substances in the atmosphere, and as a result, has an atmospheric lifetime of approximately 150 years. Its photolysis and subsequent oxidation by singlet oxygen, $O(^1D)$ in the stratosphere is the only known sink process.

Nitrous oxide is an important chemical factor in stratospheric O_3 chemistry. Nitric oxide produced from N_2O photolysis in the stratosphere is transported down into the troposphere where it may affect NO_x levels, especially in unpolluted areas.

Increased N_2O levels in the atmosphere pose two major concerns to humans: (1) stratospheric O_3 depletion, and, because of its thermal absorptivity, (2) global warming. (See Chapter 4 for a discussion of these two phenomena and their relationship to N_2O.)

Nitric Oxide. Nitric oxide is a colorless, odorless, tasteless, relatively nontoxic gas. It is produced naturally by anaerobic biological processes in soil and water, by combustion processes, and by photochemical destruction of nitrogen compounds in the stratosphere. On a global basis, natural emissions of NO are estimated to be approximately 5×10^8 ton/year.

Major anthropogenic sources include automobile exhaust and stationary sources such as fossil fuel-fired electric generating stations, industrial boilers, incinerators, and home space heaters. Nitric oxide is a product of high-temperature combustion:

$$N_2 + O_2 \rightarrow 2NO \qquad (2.19)$$

As this reaction is endothermic, the equilibrium moves to the right at high temperatures. At lower temperatures, it shifts completely to the left. If the cooling rate is rapid, equilibrium is not maintained and high NO emissions result. High combustion temperatures, rapid cooling, and instantaneous dilution promote high NO emissions.

In 1970, combined worldwide emissions of NO and NO_2 were estimated to be about 5.3×10^7 ton/year. This was about 10% of that estimated to have been produced by natural sources. Anthropogenic NO_x emissions for the U.S. in 1970 and 1993 are summarized in Tables 2.1 and 2.2.

Nitrogen Dioxide. Nitrogen dioxide is a colored gas which is light-yellowish-orange to reddish-brown at relatively low and high concentrations, respectively.

It has a pungent, irritating odor. It is also relatively toxic and is extremely corrosive because of its high oxidation rate.

Nitrogen dioxide can be produced by the direct oxidation of NO:

$$2NO + O_2 \rightarrow 2NO_2 \tag{2.20}$$

At low atmospheric NO levels, Equation 2.20 is slow, accounting for less than 25% of all NO conversion. Photochemical reactions involving O_3, RO_2, and odd hydrogen species (OH, HO_2, H_2O_2, etc.) are the primary means by which NO is converted to NO_2 in the atmosphere. Typical reactions include:

$$NO + O_3 \rightarrow NO_2 + O_2 \tag{2.21}$$

$$RO_2 + NO \rightarrow NO_2 + RO \tag{2.22}$$

$$HO_2 + NO \rightarrow NO_2 + OH \tag{2.23}$$

NO_x Concentrations. Average NO_x (combined NO and NO_2) concentrations measured in remote locations have been reported to range from 0.02 to 0.04 ppbv for marine environments and 0.02 to 0.08 ppbv for tropical forests. Rural locations are typically reported to range from 0.02 to 10 ppbv and urban/suburban areas in the U.S. from 10 to 1000 ppbv. Urban/suburban concentrations during peak morning hours (6 to 9 a.m.) in U.S. cities are reported to have maximum concentrations as high as 200 ppbv with average concentrations less than 100 ppbv. Maximum 1-hour concentrations of NO were observed to exceed 1 ppmv with maximum NO_2 levels as high as 0.5 ppmv in monitoring studies conducted in the U.S. in the 1960s and early 1970s.

Ambient levels of NO in urban areas show considerable variation. In cities such as Los Angeles, peak concentrations are associated with morning rush hour traffic with a small peak very late in the day (Figure 2.3). Peak morning NO concentrations are followed several hours later by peak NO_2 levels which result from photochemical conversion of NO to NO_2.

Atmospheric levels of NO and NO_2 appear to show seasonal trends depending on location. In Los Angeles, the quarterly means of hourly average concentrations of NO are significantly higher in the cooler, lower solar insolation months of January–March and October–December than in spring and summer. This pattern appears to be similar but less pronounced for NO_2. These seasonal differences appear to be associated with the increased use of heating fuels during cooler months and more intense photochemical activity during warmer months.

NO_x Sink Process. The nitrogen oxides NO and NO_2 are removed from the atmosphere in chemical reactions in which NO is converted to NO_2 and then to HNO_3. The major NO_x sink process involves the reaction of NO_2 with OH radical:

$$NO_2 + OH + M \rightarrow HNO_3 + M \tag{2.24}$$

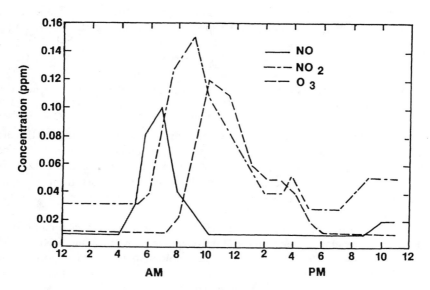

Figure 2.3 Variation in levels of NO, NO_2, and O_3 on a smoggy day in Los Angeles. (From NAPCA. USDHEW 1969. Publication No. AP-63.)

M is an energy absorbing chemical species such as O_2 or N_2. Nitrogen dioxide is also converted to HNO_3 by nighttime reactions involving O_3:

$$NO_2 + O_3 \rightarrow NO_3 + O_2 \tag{2.25}$$

$$NO_2 + NO_3 \rightarrow N_2O_5 \tag{2.26}$$

$$N_2O_5 + H_2O \rightarrow 2HNO_3 \tag{2.27}$$

In Equations 2.25 and 2.26, NO_3 is nitrate radical. It is a key reactant in nighttime chemistry. Reactions of NO_3 with NO_2 produce dinitrogen pentoxide (N_2O_5) which reacts rapidly and irreversibly with H_2O to produce HNO_3. Nitric acid can also be produced by hydrogen abstraction reactions involving NO_3, formaldehyde (HCHO), and hydrocarbon radicals (RH):

$$NO_3 + HCHO \rightarrow HNO_3 + CHO \tag{2.28}$$

$$NO_3 + RH \rightarrow HNO_3 + R \tag{2.29}$$

Nitrate radical also has a significant role in the nighttime oxidation of naturally produced organic compounds such as isoprene and the pinenes — compounds which may play a significant part in tropospheric O_3 chemistry. It is rapidly photolyzed at daybreak, being depleted along with N_2O_5 by a shift in the equilibrium between NO_2, NO_3, and N_2O_5 toward NO_2.

Nitric acid in the atmosphere tends to remain in the gas phase. Some HNO_3 will react with NH_3 and other alkaline species to form salts such as NH_4NO_3:

$$NH_3 + HNO_3 \rightarrow NH_4NO_3 \tag{2.30}$$

Nitrate aerosol particles can be removed from the atmosphere by wet and dry deposition processes.

Ammonia. Ammonia and its particulate-phase anion ammonium (NH_4^+) are important nitrogen species in the atmosphere. Major sources include animals and their wastes, soil, and biomass burning. Minor sources include coal combustion, human excreta, and fertilizers. Background concentrations vary from 0.1 ppbv over remote oceans to 6 to 10 ppbv in continental background air. Atmospheric residence time is approximately 6 days.

Ammonia is rapidly converted to NH_4^+ which is a major constituent of two of the most prevalent aerosols in the atmosphere: ammonium sulfate [$(NH_4)_2SO_4$] and ammonium nitrate (NH_4NO_3). As such, ammonia plays a very important role in sink processes involving sulfur and nitrogen compounds. Ammonia itself can be oxidized by chemical reactions initiated by OH.

Hydrogen Cyanide. Hydrogen cyanide (HCN) has only recently been recognized in the atmosphere. With a background concentration of 160 pptv, its oxidation comprises a relatively small source of tropospheric NO_x.

Organic Nitrate Compounds. Photochemical reactions involving HCs and NO_x produce a variety of organic nitrate compounds including peroxyacyl nitrate (PAN), peroxypropronyl nitrate (PPN), and peroxybutryl nitrate (PBN). PAN is a potent eye irritant and at elevated concentrations found in the Los Angeles basin causes smog injury on sensitive vegetation.

PAN is formed from the reaction of NO_2 with acetylperoxy radicals:

$$CH_3C = OOO\cdot + NO_2 + M \rightarrow CH_3C = OOONO_2 \tag{2.31}$$

The sink process for PAN is thermal degradation to reform acetylperoxy radicals and NO_2. PAN is permanently removed when the acetylperoxy radical reacts with NO.

If PAN is formed or transported to the upper troposphere, it may have an atmospheric lifetime of several months or longer. As a result, it may serve as an important reservoir and carrier of NO_x in long-range transport. Because its decomposition increases significantly with temperature, it can be formed in colder regions, transported, and then decomposed to release NO_2 in warmer regions. Concentrations in urban areas have reportedly ranged from 10 to 50 ppbv.

Other nitrated organic compounds have also been observed to form in the atmosphere. These include polycyclic aromatic hydrocarbons such as 2-nitrofluorothene, identified in the organic extract of ambient particulate matter.

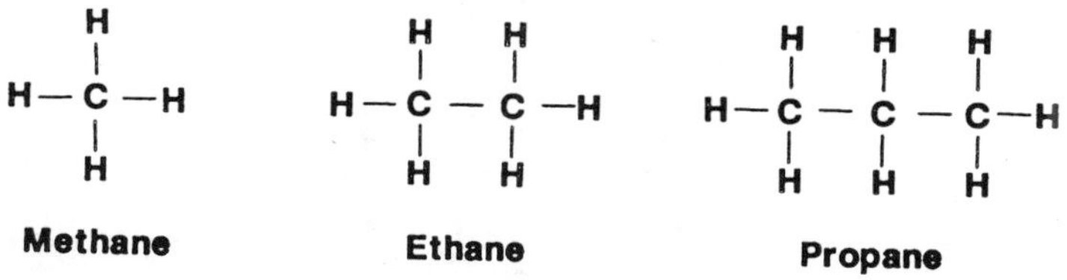

Figure 2.4 Paraffinic hydrocarbons.

Hydrocarbons and Hydrocarbon Derivatives

Hydrocarbons include a broad range of organic compounds whose chemical structure consists of carbon and hydrogen atoms covalently bonded to each other. Based on carbon–carbon bonding patterns, HCs may have straight or branched chains and cyclic or combined cyclic structures. Based on the number of electron pairs shared by covalently bonded carbon atoms, HCs may be saturated or unsaturated. In the former case, carbon atoms share a single electron pair, thus possessing a single bond and described as being saturated. When two or more electron pairs are shared by carbons in the molecular structure of a molecule, it has double bonds and, in the case of three electron pairs, a triple bond. Compounds with double or triple bonds are described as being unsaturated. Because compounds with double bonds have high chemical reactivities, such HC species play a significant role in atmospheric photochemical processes.

Hydrocarbons and their derivatives in the atmosphere may be present in gas, liquid, or solid phases. Hydrocarbons which contain one to four carbons are gases under normal conditions of atmospheric temperature and pressure. Hydrocarbons in the 5 to 12 carbon range tend to be volatile liquids. Those with higher carbon numbers and molecular weights are solids and are often found condensed on atmospheric aerosols.

Hydrocarbons are classified on the basis of their being saturated (paraffins or alkanes), unsaturated (olefins or alkenes), and unsaturated with ring structures (aromatics). Paraffinic and olefinic HCs are also described as being aliphatic HCs.

The paraffins or alkanes include a series of straight-chained, branched or cyclic compounds which contain only single covalent bonds between carbon atoms. Several straight-chained paraffins are illustrated in Figure 2.4. Olefins include straight or branch-chained HCs which contain one or more double bonds. Several straight-chained olefinic HCs are illustrated in Figure 2.5.

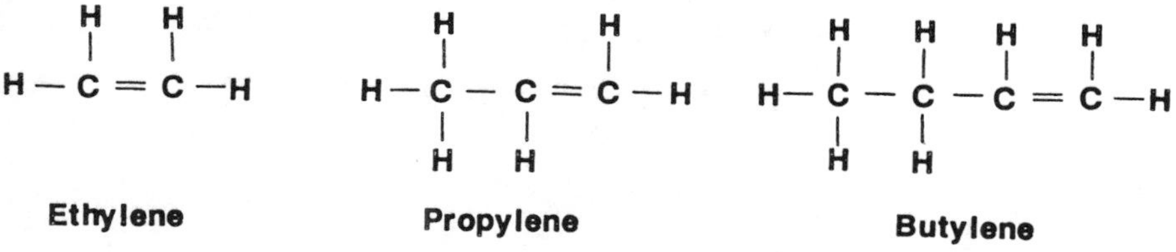

Figure 2.5 Olefinic hydrocarbons.

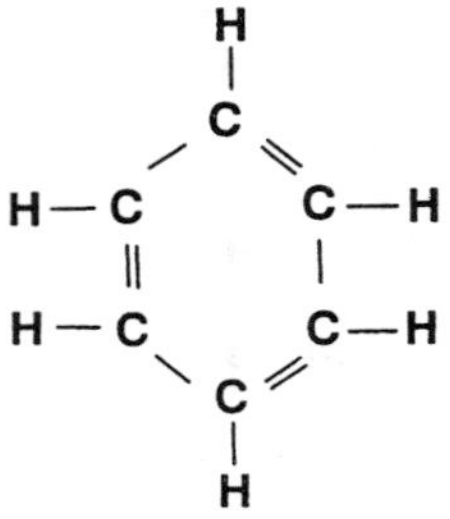

Figure 2.6 Structure of benzene.

Aromatic HCs are compounds which are based on the ring structure of benzene (Figure 2.6). Benzene and similar substances such as toluene and xylene are volatile liquids and are common contaminants of urban atmospheres. Benzene rings serve as the basic structural unit for naphthalene (two benzene rings) and a family of polycyclic aromatic HCs that are solids under normal atmospheric temperature conditions. The best known of this group (referred to as PAH or POM) is benzo(α)pyrene (Figure 2.7). The PAHs are produced in the combustion of a variety of organic fuels and are a common component of atmospheric aerosol. PAHs are known to be carcinogens (cancer-producing substances).

Within the context of photochemical processes in the atmosphere, unsaturated compounds such as the olefins and aromatics are of special concern because of their reactivity, particularly in urban and suburban areas. The paraffins are relatively unreactive and have a very limited role in urban and suburban photochemistry. Their importance increases as olefinic and aromatic HCs are depleted as a result of photochemical reactions and as polluted air masses move downwind of source regions.

A wide variety of HC derivatives can be formed as a result of reactions with oxygen, nitrogen, and sulfur compounds, and halogens (such as chlorine, bromine, and fluorine). Of particular importance as atmospheric pollutants are oxygenated and halogenated HCs. Because of unique atmospheric concerns associated with the latter, halogenated HCs are discussed separately in this chapter.

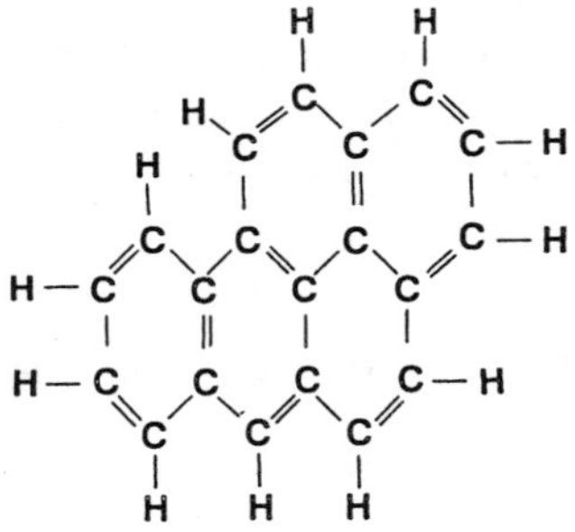

Figure 2.7 Structure of benzo(α)pyrene.

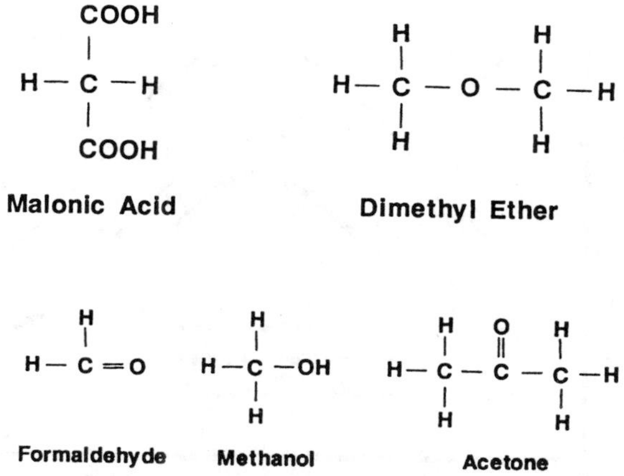

Figure 2.8 Representative hydrocarbons.

Hydrocarbon derivatives produced on reaction with O_2 are major atmospheric contaminants. These oxygenated HCs include aldehydes, acids, alcohols, ethers, ketones, and esters. Oxygenated HCs are emitted directly into the atmosphere from a variety of combustion sources (notably, motor vehicles) and industrial processes. Chemical structures of various oxyhydrocarbon compounds are illustrated in Figure 2.8. Oxyhydrocarbons participate in photochemical reactions and may themselves be produced photochemically.

Formaldehyde, acetaldehyde, acrolein, and other aldehydes are major byproducts of combustion processes, with significant emissions in motor vehicle exhaust. Aldehydes are also produced as a consequence of atmospheric photochemistry with peak levels occurring at solar noon (Figure 2.9). Both HCHO and acrolein are potent mucous membrane irritants.

Malonic acid is one of a series of dicarboxylic acids produced in polluted atmospheres. Dicarboxylic acids are common components of photochemical aerosol, formed as a result of condensation processes.

Sources and Emissions. Nonmethane hydrocarbons (NMHCs) are the focus of air-quality concerns in many countries because of their role in atmospheric photochemistry. They are produced and emitted from a variety of anthropogenic and biogenic sources.

Major anthropogenic sources of NMHCs include mobile and stationary source fuel usage and combustion, petroleum refining and petrochemical manufacturing, industrial, commercial, and individual solvent use, gas and oil production, and biomass burning. Emissions have been of particular concern in urban areas. In a source apportionment of NMHC emissions conducted in Los Angeles in 1976, the weight percentage of emissions (not including industrial emissions and solvent use) was estimated to be 49% motor vehicle exhaust, 16% gasoline spillage, 13% gasoline evaporation, 15% natural gas and oil fuel production, and 5% natural gas distribution and use. A more inclusive source apportionment for

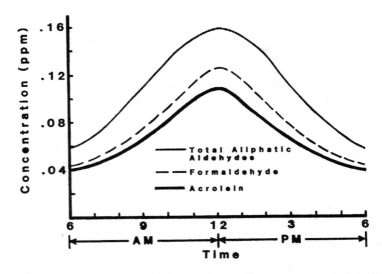

Figure 2.9 Variation in aldehyde levels associated with emissions and photochemical activity. (From NAPCA. USDHEW. 1970. Publication No. AP-64.)

Sydney, Australia conducted in 1983 estimated NMHC emissions as 36% motor vehicle exhausts, 16% gasoline spillage, 16% gasoline evaporation, 23% solvents, 4% commercial use of natural gas, and 5% industrial processes. In both cases, source apportionment revealed that motor vehicle exhaust and gasoline usage were responsible for more than 70% of NMHC emissions in these two cities. In a recent study of Mexico City, the major source of NMHCs appeared to be emissions of unburned liquefied petroleum gas (LPG) used for heating and cooking, rather than motor vehicles and gasoline, as previously thought.

Emissions associated with gasoline usage and petroleum-related industries include a variety of paraffinic, olefinic, and aromatic HCs; with natural gas they include light paraffins; with solvents they include higher paraffinic and aromatic HCs. Total global emissions from these anthropogenic sources are estimated to be 8.9×10^7 ton/year. This compares to an approximately similar quantity associated with biomass burning in the tropics. NMHC emissions from biomass burning consists primarily of light paraffins and olefins.

Emissions from biogenic sources which include foliar emissions from forest trees and grasslands and emissions from soils and ocean water are approximately an order of magnitude higher on a global basis than anthropogenic emissions. Foliar emissions from forest trees are comprised mainly of isoprene and mono-terpenes with some paraffins and olefins; grasslands, light paraffins and higher HCs; soils, mainly ethane; and ocean water, light paraffins, olefins, and C_9-C_{28} paraffins.

On a broader scale over the U.S., USEPA has estimated that 40% of NMHC anthropogenic emissions result from transportation (with light-duty cars and trucks comprising the largest fraction); 32% result from solvent usage, and 28% from industrial manufacturing activities and fuel combustion.

Identification. A number of investigators have attempted to identify the many HC species and their derivatives which may be present in the atmosphere. Because of the complexities and expense associated with HC sampling and analysis, the atmosphere has not been well characterized, particularly in rural and remote areas. Many HC species produced under intense photochemical activities may be present in subtrace amounts and may, in many cases, be below the limit of detection of sampling systems utilized. As a result, the probable presence of many compounds must be inferred from smog chamber studies and from analysis of motor vehicle exhaust. In the latter case, investigators have identified more than 400 HC and oxyhydrocarbon derivatives. Oxyhydrocarbons, which include aldehydes, ketones, organic acids, alcohols, ethers, esters, and phenol, have been reported to comprise 5 to 10% of the total NMHC concentration in auto exhaust. The aldehyde fraction is dominated by HCHO and acetaldehyde. In addition to emissions from automobiles, HCHO and other aldehydes are produced by photochemical reactions reaching peak concentrations around solar noon (Figure 2.9).

A variety of HC species, in addition to oxyhydrocarbons, have been found or are believed to be present in the atmosphere. These include straight, branch-chained, and cyclic paraffinic compounds, olefinic compounds with one or two double bonds, acetylene-type compounds, and benzene and its derivatives. Compounds with multiple benzene rings (PAHs) are found condensed on aerosol particles. Biogenic compounds commonly found in the atmosphere include isoprene and monoterpenes such as λ and β-pinene. The most abundant HC species sampled over Atlanta, GA are summarized in Table 2.5. As can be seen, air samples included a large number of short- and longer-chained paraffins, a relatively limited number of olefins, and a variety of aromatic HCs. P-cymene and isoprene are biogenic in origin. Acetylene is produced in combustion and has a relatively long atmospheric lifetime. It is an indicator of motor vehicle-related emissions.

Hydrocarbon Concentrations. Most HC data are based on the measurement of total NMHCs, usually averaged from 6 to 9 a.m., which corresponds to early morning commuting traffic in urban areas. Concentrations are determined by flame ionization and are reported in CH_4 or carbon equivalents. The air-quality standard for this 3-hour period is 0.24 ppmv, with most urban concentrations at levels less than this. Substantially higher levels were reported in the Los Angeles Basin in the 1950s and 1960s prior to emission control requirements on motor vehicles.

Concentrations of individual HC species based on carbon equivalents are presented in Table 2.5 for 35 major HC species identified in samples collected over Atlanta, GA. Similar data are not readily available for oxyhydrocarbon species. However, concentrations of HCHO in both remote and urban locations have been well documented. Concentrations range from a few ppbv in remote areas to between 70 and 100 ppbv under severe smog conditions.

Sink Processes. The primary sink process for HCs and their oxygenated derivatives is the oxidative production of alkylperoxy radicals (ROO·) on reaction with OH or O_3. In the presence of NO, ROO· is converted to alkoxy radical (RO·)

Table 2.5   Major Hydrocarbon Compounds Measured in Atlanta, GA in 1981

Species	Concentration (ppbC)	Species	Concentration (ppbC)
i-pentane	19.8	3 methyl pentane	3.4
n-butane	16.9	ethylene	3.0
toluene	14.7	3 methyl-cyclopentane	2.9
p-cymene	11.0	ethylbenzene	2.8
n-pentane	9.4	o-xylene	2.8
benzene	8.8	3 methyl-hexane	2.6
m- and p-xylene	7.6	2,3 dimethyl-pentane	2.5
2-me-pentane	5.9	1,4 diethyl-benzene	2.4
cyclohexane	5.4	isobutene	2.2
2-me-hexane	5.2	2,2,4 trimethyl-pentane	2.2
ethane	5.0	1,2,4 trimethyl-pentane	2.2
undecane	4.9	i-butyl-benzene	1.8
propane	4.8	2-methyl-2-butene	1.8
i-butane	4.8	1,3,5 trimethylbenzene	1.8
isoprene	4.6	cyclopentane	1.6
acetylene	4.3	propene	1.5
n-hexane	3.8	i-propyl-benzene	1.5
m and p ethyl-toluene	3.6		
Total ppbC			197

From NRC. 1991. National Academy Press, Washington, DC. With permission.

which reacts with O_2 to produce aldehydes or, in the case of higher HCs, butane, aldehydes, and ketones. The oxidation of ethane, a relatively unreactive HC, is summarized in the following equations.

$$C_2H_6 + OH \rightarrow H_2O + C_2H_5 \cdot \tag{2.32}$$

$$C_2H_5 \cdot + O_2 \rightarrow C_2H_5OO \cdot \tag{2.33}$$

$$C_2H_5OO \cdot + NO \rightarrow C_2H_5O \cdot + NO_2 \tag{2.34}$$

$$C_2H_5O \cdot + O_2 \rightarrow HO_2 + CH_3CHO \tag{2.35}$$

Ethane oxidation produces acetaldehyde, which, like other aldehydes, is considerably more reactive toward OH than its HC precursor. As a result, acetaldehyde will react with OH and, by a series of reactions, produce HCHO, which can undergo further oxidation or photodecomposition.

Formaldehyde, acetaldehyde, and acetone can photodecompose on absorption of ultraviolet light in the wavelength range of 330 to 350 nm. The photodecomposition of HCHO proceeds via either of two pathways. In both cases, photodecomposition produces CO. These pathways are summarized below.

$$HCHO + hv \rightarrow HCO + H \tag{2.36}$$

$$H + O_2 + M \rightarrow HO_2 + M \tag{2.37}$$

$$HCO + O_2 \rightarrow HO_2 + CO \qquad\qquad (2.38)$$

$$HO_2 + NO \rightarrow NO_2 + OH \qquad\qquad (2.39)$$

It can be seen in the first pathway that the photodecomposition of HCHO leads to the production of HO_2 which, on reaction with NO, generates OH which then becomes available for HC oxidation. In the second photodecomposition pathway (Equation 2.40), H and HCO are rapidly converted to CO and H_2.

$$H + HCO \rightarrow H_2 + CO \qquad\qquad (2.40)$$

Aldehydes and ketones produced from HC oxidation can be removed from the atmosphere by scavenging processes involving dry and wet deposition. The oxidation of longer-chain aliphatic and aromatic HCs may lead to the formation of condensible products such as dicarboxylic acids which enter the particulate phase and are removed from the atmosphere by wet and dry deposition processes.

All HCs, with the exception of CH_4, react relatively rapidly with OH. However, HCs in the atmosphere vary considerably in their OH reactivity, with unsaturated hydrocarbons including olefinic and aromatic HCs being the most reactive. Many unsaturated HCs react with O_3 at rates that are competitive with OH. In the troposphere, the lifetime of olefins and biogenically produced terpenes is on the order of hours, whereas for less reactive paraffinic species the atmospheric lifetime may be days. Because the most reactive compounds are removed at a faster rate, the abundance spectrum of HCs changes in the direction of less reactive species such as ethane in both urban areas as well as downwind.

Photochemical Precursors. Oxidation products produced in HC sink processes such as $ROO\cdot$, $RO\cdot$, HO_2, and CO serve as major reactants in the production of photochemical smog and associated elevated tropospheric levels of O_3 and other oxidants (the formation of which will be discussed later). In addition to oxidants, photochemical reactions produce a large variety of HC species which may comprise 95% of atmospheric HC during a severe smog episode.

Methane. Because of its low reactivity and relative abundance compared to other HC species, CH_4 levels are subtracted from total HC values measured in urban atmospheres. As previously indicated, the air-quality standard for HCs (Chapter 8) is a nonmethane HC standard.

Methane has low reactivity with OH relative to olefins, aromatics, and even other paraffins, and, as a result, it is of little significance in urban photochemistry producing elevated O_3 levels. Methane and other low-reactivity paraffins such as ethane become important when reactive HCs are depleted over urban areas as polluted air masses travel downwind of urban sources. Methane is also a thermal absorber and, as a result of its increasing atmospheric levels, may play a significant role in global warming (see Chapter 4).

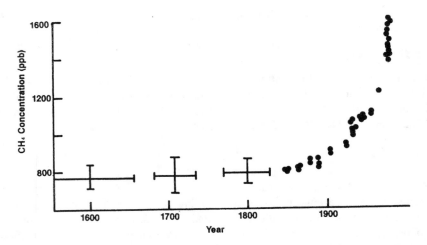

Figure 2.10 Trends in atmospheric methane levels. (From NRC. 1989. National Academy Press, Washington, DC. With permission.)

Global CH_4 levels in 1995 averaged about 1.72 ppmv. In the 1980s and early 1990s, CH_4 levels were increasing at a rate of approximately 20 ppbv/year. Surprisingly, in 1994, no apparent increase in atmospheric CH_4 was observed. Increases of CH_4 over the last 300 years as measured in air trapped in Antarctic and Greenland ice as well as recent atmospheric measurements, can be seen in Figure 2.10. Note the significant inflection of the curve during this century.

Emission of CH_4 to the atmosphere results from natural sources, natural sources influenced by human activity, and anthropogenic sources. Biogenic emissions result from anaerobic decomposition of organic matter in the sediments of swamps, lakes, rice paddies, and sewage wastes, and in the digestion of cellulose by livestock (and other ruminants) and termites. Methane emissions also occur from coal and lignite mining, oil and natural gas extraction, petroleum refining, leakage from natural gas transmission and usage, and automobile exhaust. In the last case, CH_4 comprises 15% of motor vehicle HC emissions and 70% of urban concentrations after correcting for background levels.

There are two major sink mechanisms for CH_4. The most important is reaction with OH:

$$OH + CH_4 \rightarrow C\dot{H}_3 + H_2O \qquad (2.41)$$

Subsequent reactions produce HCHO, CO, and ultimately CO_2. The second sink is the stratosphere where CH_4 photodecomposition produces H_2O vapor. This is the major source of H_2O vapor in the stratosphere. One major factor contributing to increased atmospheric CH_4 levels may be CO depletion of OH (Equation 2.5). The atmospheric lifetime of CH_4 is estimated to be 7 years.

Photochemical Oxidants

Photochemical oxidants are produced in the atmosphere as a result of chemical reactions involving sunlight, NO_x, O_2, and a variety of HCs. Photochemical oxidants produced from such reactions include O_3, NO_2, PAN, odd hydrogen compounds (HO, HO_2, H_2O_2, etc.), and RO_2. Because of its significant environmental effects, elevated tropospheric O_3 has received major scientific and regulatory attention.

As indicated in Chapter 1, O_3 is a normal constituent of the atmosphere with peak concentrations in the middle stratosphere. As a consequence of anthropogenic influences, O_3 represents two major and distinct environmental concerns. In the stratosphere it is an issue of trace gas-related depletion (Chapter 4), whereas in the troposphere it is one of the levels significantly elevated above background. It is the tropospheric problem that is the focus of this discussion.

Ozone is formed in the atmosphere when molecular O_2 reacts with ground-state $O(^3P)$ atomic oxygen:

$$O_2 + O(^3P) + M \rightarrow O_3 + M \tag{2.42}$$

In the troposphere, photodissociation of NO_2 at wavelengths of 280 to 430 nm is the only significant source of atomic oxygen:

$$NO_2 + h\nu \rightarrow NO + O(^3P) \tag{2.43}$$

The reaction of $O(^3P)$ with O_2 produces O_3, which reacts immediately with NO to regenerate NO_2.

$$NO + O_3 \rightarrow NO_2 + O_2 \tag{2.44}$$

This O_3-producing process is presented diagrammatically in Figure 2.11.

Equations 2.42 to 2.44 proceed rapidly producing a small steady-state concentration of 20 ppbv under solar noon conditions in mid-latitudes at atmospheric NO_2/NO concentration ratios equal to 1.

In both urban and nonurban atmospheres, O_3 concentrations are often much higher than those that occur from NO_2 photolysis. The key to elevated tropospheric O_3 levels is chemical reactions that convert NO to NO_2 without consuming O_3. In very polluted and even lightly polluted atmospheres, such shifts in O_3 chemistry occur in the presence of RO_2 produced by the oxidation of HCs:

$$RO_2 \cdot + NO \rightarrow NO_2 + RO \cdot \tag{2.45}$$

$$NO_2 + h\nu \rightarrow NO + O(^3P) \tag{2.46}$$

$$O(^3P) + O_2 + M \rightarrow O_3 + M \tag{2.47}$$

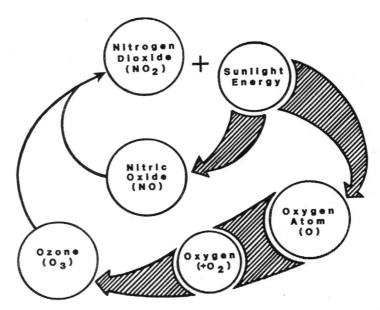

Figure 2.11 NO_2 photolysis and O_3 formation. (From NAPCA. USDHEW. 1969. Publication No. AP-63.)

$$\text{Net: } RO_2 \cdot + O_2 + h\nu \rightarrow RO \cdot + O_3 \tag{2.48}$$

This process is presented diagrammatically in Figure 2.12.

The rate of O_3 formation is closely related to the concentration of RO_2. Peroxy radicals are produced when OH and HO_x (odd hydrogen species) react with vapor-phase HCs. As indicated previously, odd hydrogen species such as OH are produced by reactions involving the photodissociation of O_3, carbonyl compounds (aldehydes), and N_2O. Ozone photodissociation and formation of OH are indicated in Equations 2.49 and 2.50.

$$O_3 + h\nu \rightarrow O(^1D) + O_2 \tag{2.49}$$

$$O(^1D) + H_2O \rightarrow 2OH \tag{2.50}$$

where $O(^1D)$ is singlet (excited) atomic oxygen.

In very polluted atmospheres, O_3 concentrations will be a function of the intensity of sunlight, NO_2/NO ratios, reactive HC type and concentrations, and other pollutants, such as aldehydes and CO, that react photochemically to produce RO_2. The increase in NO_2/NO ratios caused by atmospheric reactions involving RO_2 may result in significant increases in tropospheric O_3 levels.

Tropospheric O_3 Concentrations. In the relatively clean atmospheres of remote sites, ground-level O_3 concentrations are estimated to range from 20 to

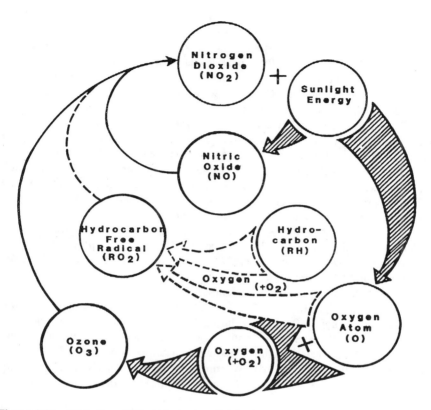

Figure 2.12 Interaction of NO_x, HCs, and sunlight in the production of elevated tropospheric O_3 levels. (From NAPCA. USDHEW. 1969. Publication No. AP-63.)

50 ppbv (0.02 to 0.05 ppmv) during the warmer months of the year. Major sources of background O_3 are photochemical processes and the movement of stratospheric O_3 in the region of so-called "tropopause folds" into the lower atmosphere, particularly during spring months. There is general agreement among atmospheric scientists that the former is the major source of background O_3. Depending on location, O_3 associated with processes uninfluenced by human activities may account for 20 to 50% of observed rural monthly average concentrations, particularly during the warmer months.

Ozone concentrations over large, economically developed land masses such as North America, Europe, Eastern Australia, and Southeast Asia are often significantly higher than those found in remote and rural areas. These relatively high O_3 levels are associated with the anthropogenic emissions of NMHCs and NO_x and the biogenic emissions of relatively reactive NMHCs and subsequent photochemical reactions summarized in Equations 2.45 to 2.48. In urban areas, O_3 levels reach their peak near solar noon, while in rural areas peak levels may occur at night.

Our understanding of atmospheric chemistry associated with the formation of elevated tropospheric O_3 levels has evolved considerably from the 1950s and early 1960s when it was understood to be an urban problem associated with such cities as Los Angeles. At that time, peak levels were observed at solar noon with a subsequent rapid destruction associated with NO_x scavenging processes (Figure 2.3). In the 1970s, elevated O_3 levels were observed in many rural areas far removed from urban centers that were the focus of earlier O_3 concerns. Rural O_3 levels elevated above background were observed to be associated with (1) the long-range transport of O_3 which was protected from the normal scavenging processes by ground-based inversions, and (2) the transport of O_3 precursors such as NO_x and less reactive HCs. In the mountain areas of the eastern U.S., elevated O_3 was observed to be both persistent and to occur at night, contradicting the previous paradigm of O_3 production and destruction which was characteristic of urban areas with large motor vehicle populations.

Our understanding of O_3 chemistry continues to evolve as we come to understand the significant role that HCs emitted by forest trees play in producing elevated O_3 levels on a regional basis (see additional discussion in the public policy section of Chapter 8).

In the U.S., noontime peak concentrations have been at their highest in the precursor and sunlight-rich Los Angeles Basin with one-hour average concentrations ranging as high as 0.40 ppmv during severe smog episodes and concentrations of 0.15 to 0.30 ppmv commonly reported. Second daily maximum concentrations (one-hour averaging time) are summarized for a number of U.S. cities in Table 2.6 for the year 1993. All levels are above the ambient air quality standard of 0.12 ppmv (one hour average). The number of U.S. cities which exceed the

Table 2.6   Ozone Levels Measured in
             Selected U.S. Metropolitan
             Areas in 1993

Metropolitan area	O_3 (ppmv)[a]
Danbury, CT	0.14
New London, CT	0.13
Bridgeport, CT	0.17
Oakland, CA	0.13
Dallas, TX	0.14
Louisville, KY	0.14
Baltimore, MD	0.15
Birmingham, AL	0.13
Orange County, CA	0.17
Phoenix, AZ	0.13
St. Louis, MO	0.13
Los Angeles, CA	0.25
Riverside, CA	0.23
Philadelphia, PA	0.14

[a] Second highest daily maximum one-hour concentration.

From USEPA. 1995. EPA/600/AP-95/001c.

0.12 ppmv air quality standard varies from year to year, with over a hundred cities exceeding it in the very hot summer of 1988.

Though the focus of concern relative to elevated urban O_3 levels has historically been in the U.S., the problem is much more widespread. In Mexico City, O_3 levels are a significant problem with the Mexican one hour standard of 0.11 ppmv exceeded on 71% of the days in 1991 and 98% in 1992, with values in the 0.20 to 0.30 ppmv range common and values as high as 0.48 ppmv recorded. During episodes of elevated O_3 levels in the more limited solar radiation Europe concentrations of 0.10 ppmv are common, with one severe episode in southern England exceeding 0.25 ppmv.

Ozone Sinks. The two major sink processes for O_3 are surface destruction or deposition and photochemical reactions. Surface deposition includes reaction with plants, bare land, ice and snow, and man-made structures. Deposition of O_3 is of its greatest magnitude over daytime forests and croplands.

The primary sink for O_3 is its photodissociation on absorption of UV light and subsequent formation of OH radical (Equations 2.49–2.50)

In polluted atmospheres, O_3 reacts with NO to produce $NO_2 + O_2$ (Equation 2.44). During nighttime hours, O_3 reacts with NO_2 to subsequently produce HNO_3 (Equations 2.25 to 2.27).

Halogenated Hydrocarbons

A variety of halogenated HCs are emitted to the atmosphere naturally and from anthropogenic sources. These include both volatile and semivolatile substances which contain one or more atoms of chlorine, bromine, and/or fluorine. Halogenated compounds are relatively unique contaminants of the atmosphere because of their persistence. They include volatile substances such as methyl chloride (CH_3Cl), methyl bromide (CH_3Br), methyl chloroform (CH_3CCl_3), trichloroethylene, perchloroethylene, and carbon tetrachloride (CCl_4). Methyl chloride and CH_3Br are produced naturally in the environment; the other volatile halogens indicated above have been primarily used as solvents. A variety of semivolatile halogenated HCs are land, air, water, and biological contaminants. These include chlorinated HCs such as DDT, chlordane, dieldrin, aldrin, etc., used as pesticides; polychlorinated biphenyls (PCBs) used as solvents and electrical transformer insulators; and polybrominated biphenyls (PPBs) used as fire retardants.

Halogenated HCs which include chlorine and fluorine in their chemical structure are called chlorofluorocarbons (CFCs). They represent a number of chemical species. The most commonly used and emitted of these are trichlorofluoromethane ($CFCl_3$) dichlorodifluoromethane (CF_2Cl_2), and trichlorotrifluoroethane ($C_2Cl_3F_3$). These are generally described as CFC-11, CFC-12, and CFC-113. The CFCs, because of their desirable properties including low reactivity, low toxicity, thermal absorption, and solvent properties have been used as aerosol propellants, refrigerants, degreasers, and foaming agents.

Because of their low chemical reactivity, halogenated HCs break down very slowly and, as a result, are characterized by their long atmospheric lifetimes, estimated at 6.3 years for CH_3CCl_3, 40 years for CCl_4, 75 years for CCl_3F, and 111 to 170 years for CF_2Cl_2.

Though halogenated compounds such as CH_3Cl and CH_3CCl_3 have tropospheric sinks, the CFCs apparently do not. As a result, concentrations in both the troposphere and stratosphere have been increasing rapidly since the mid-1970s. Tropospheric concentrations of CFCs and other halocarbons are reported in the pptv range. Average global concentrations reported in the early 1980s for CF_2Cl_2 were $\approx$300 pptv; CH_3Cl, $\approx$600 pptv; CCl_3F, $\approx$180 pptv; CH_3CCl_3, $\approx$120 pptv; CCl_4, $\approx$120 pptv; $C_2Cl_3F_3$, $\approx$70 pptv; and CH_3Br, $\approx$10 pptv. Concentrations of CFCs and other halogenated HCs vary spatially over the surface of the earth with highest concentrations corresponding to source regions over the Northern Hemisphere.

Methyl chloride is produced naturally and is reported to be second to CF_2Cl_2 as a source of stratospheric chlorine. In the Northern Hemisphere, stratospheric chlorine concentrations have increased from 1.8 ppbv in 1974 to 3.5 ppbv in 1986 and are projected to increase to 5.0 ppbv in the year 2000 despite the implementation of the Montreal Protocol (see Chapter 8).

Brominated HCs also are of interest in the atmosphere because bromine atoms have a significant potential to deplete stratospheric O_3. Methyl bromide is the most abundant bromine compound in the atmosphere with average atmospheric concentrations in the range of 9 to 13 pptv. Its major sources include marine phytoplankton, agricultural pesticide use, and biomass burning. Other sources of bromine in the atmosphere are CF_3Br (Halon 1301) and CF_2BrCl (Halon 1211).

PARTICULATE MATTER

Particulate matter is a collective term used to describe small solid and liquid particles that are present in the atmosphere over relatively brief (minutes) to extended periods of time (days to weeks). Individual particles vary in size, geometry, mass, concentration, chemical composition, and physical properties. They may be produced naturally or as a direct or indirect result of human activities.

Of major concern are particles <20 μm because they can remain suspended in the atmosphere where (depending on actual particle size) they can settle out relatively slowly. Atmospheric aerosol is characterized by these relatively small particles.

The formation and increase of atmospheric aerosol by both natural and anthropogenic sources is a major air-quality concern because aerosol particles may (1) scatter light, reducing visibility, (2) pose an inhalation hazard to humans and animals, (3) affect climate on a regional and global scale, and, along with settlable particles, (4) be a nuisance because of their soiling potential.

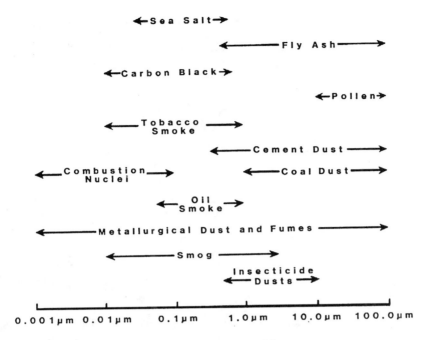

Figure 2.13 Size ranges of common atmospheric particles.

Size

Particles in the atmosphere range in size over many orders of magnitude from molecular clusters to those which are visible with the unaided eye, from approximately 0.005 to >100 μm in diameter. Small particles (<1 μm aerodynamic diameter) behave as gases; that is, they are subject to Brownian motion (the random motion of molecules), follow fluid stream lines around obstacles, are capable of coagulation, and settle out very slowly. Particles >1 μm are strongly affected by gravity and rarely coalesce. Size ranges for a variety of natural and anthropogenically derived particles are illustrated in Figure 2.13.

The expression of particle size is based on particle behavior in the earth's gravitational field. The aerodynamic equivalent diameter refers to a spherical particle of unit density (1 g/cm^3) that falls at a standard velocity. Particle size is also expressed as the Stokes equivalent diameter or Stokes number. This value refers to the diameter of a spherical particle of the same average density and settling velocity.

Size is a very important characteristic of particles because it determines atmospheric lifetime, effects on light scattering, and deposition in human lungs. If one looks at particle size relative to the number of particles present, it is evident that most particles are very small (<0.1 μm), whereas most of the particle volume and mass is associated with particles >0.1 μm.

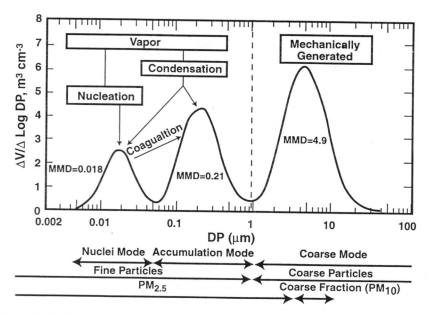

Figure 2.14 Size characteristics of atmospheric particles. (From USEPA. 1995. EPA/600/AP-95/001c.)

The size distribution of particles in the atmosphere has been characterized as multimodal. Particle volume distributions with respect to size in ambient air are almost always bimodal with minimum values between 1 and 3 μm. Particles in the larger mode are described as being "coarse," those smaller "fine." Particle volume distributions associated with motor vehicle traffic are trimodal (Figure 2.14), with the fine particle fraction consisting of two modes. The smallest, the nuclei mode, consists of particles smaller than 0.08 μm. Nuclei mode particles are formed from the nucleation and condensation of low vapor pressure substances generated in the gas phase by high temperature vaporization or chemical reactions in the atmosphere. Nuclei mode particles grow in size by coagulation and condensation (gases/vapors condense onto existing particles). The rate of growth depends on particle number, velocity, and surface area. These decrease rapidly as particle size reaches 1 μm. As a result, particles tend not to grow above 1 μm. Particles in the size range of 0.1 to 1 μm comprise the accumulation mode. These particles do not grow into coarse particles.

Coarse particles are usually produced by mechanical processes such as the fragmentation of matter and atomization of liquids. The majority of their mass is in the particle size range of 2 to 100 μm. Though there is some overlap in size ranges, coarse particles differ from fine particles in atmospheric behavior, deposition, and the sources that produce them.

In urban areas, particulate aerosols are about evenly divided on a mass basis between fine and coarse particles. Under calm atmospheric conditions, the fine particle mass begins to exceed the coarse particle mass with increasing downwind distance from urban areas. Because of their low settling velocities, fine particles may be transported relatively long distances from source regions.

Classes and Sources

Particulate matter can be described as being primary or secondary, based on its origin and processes of formation. Primary particles are emitted directly into the atmosphere from a variety of natural and anthropogenic sources. In the former case, these include volcanos, forest fires, ocean spray, biologic sources (mold, pollen, bacteria, plant parts) and meteoric debris; the latter includes transportation, fuel combustion in stationary sources, a variety of industrial processes, solid waste disposal and miscellaneous sources such as agricultural activities and fugitive emissions from roadways. Secondary particles are formed in the atmosphere as a result of chemical processes involving gases, aerosol particles, and moisture.

Emission estimates for five major source categories have been made annually by USEPA for the past several decades. In 1970 when pollution control efforts were just beginning, the major sources of particles with a cutoff diameter of ≤ 10 μm (PM_{10}) were industrial processes and stationary source fuel combustion, accounting for 60 and 22% of PM_{10} emissions, respectively (Table 2.1). By 1993 (Table 2.2), PM_{10} emissions declined overall by 72%, with reductions of 92% for industrial processes and 58% for stationary source fuel combustion. Major industrial sources have included chemical and related product manufacturing, metals processing, petroleum and related industries, minerals handling and processing, agricultural and wood products handling and processing, and a variety of other manufacturing activities.

In 1970, mobile sources (primarily light-duty motor vehicles) accounted for approximately 4% of PM_{10} emissions to the atmosphere. Since 1970, PM_{10} emissions from the transportation sector have increased by about 30%. In the recent past motor vehicle-related PM_{10} emissions have included particles dominated by lead halides, sulfate, and elemental carbon with a variety of adsorbed organic compounds. With the lead phaseout in gasoline beginning in 1974, lead emissions associated with motor vehicle PM_{10} have declined dramatically (Table 2.3).

In 1990, USEPA began to estimate PM_{10} emissions associated with fugitive dust from paved and unpaved roadways as well as emissions from agricultural and forestry activities. Because Tables 2.1 and 2.2 are presented here for comparative purposes, these data were not included in the 1993 emissions estimates in Table 2.2. They are, however, quite significant accounting for 4.12×10^7 short ton/year in 1993, dwarfing emissions from major anthropogenic sources.

Secondary particles as indicated previously are produced both naturally and anthropogenically. In both cases, secondary aerosol particles include sulfates, nitrates, and oxyhydrocarbons which are produced by direct, catalytic, and photochemical oxidation of sulfur, nitrogen, and volatile HCs. Approximately 30% of atmospheric aerosol in the particle size range <5 μm is estimated to have been produced as a result of chemical reactions in the atmosphere involving gases/vapors from natural sources.

Chemical Composition

Particles that make up ambient aerosol vary in their chemical composition. The composition of an individual particle depends on its source or origin and its subsequent atmospheric history. Because of the large variety of sources, the atmospheric formation of secondary particles, and atmospheric behavior, particles may contain hundreds of different chemical species. Most notable because of their significant concentrations in fine particles ($PM_{2.5}$) are sulfates, nitrates, elemental carbon, condensed organic compounds, and a variety of trace metals.

Sulfate compounds comprise the major portion of the fine particle ($PM_{2.5}$) mass collected in the ambient air of many urban areas in North America, accounting for 40 to 50% of $PM_{2.5}$ (cutoff diameter ≤ 2.5 μm) in the eastern U.S. These include $(NH_4)_2$ SO_4, ammonium bisulfate (NH_4HSO_4), sulfuric acid (H_2SO_4), calcium sulfate ($CaSO_4$), and a variety of metallic salts. Of these, $(NH_4)_2SO_4$ is the most common and abundant sulfate species in ambient aerosol samples. Sulfuric acid also comprises a significant fraction of atmospheric sulfate. Although sulfates are considerably lower in western states, they nevertheless are found at significant levels. In the global context, aerosol sulfate concentrations in remote areas of the planet range from 1 to 2 μg/m^3, <10 μg/m^3 in nonurban continental areas, and >10 μg/m^3 in areas under urban and anthropogenic influence.

Another major component of the fine particle mass of atmospheric aerosols are nitrates. Nitrate concentrations are typically low (approximately 1% of fine particle mass) in the eastern U.S. (as can been seen in Table 2.7) with much higher concentrations in western U.S. cities such as Los Angeles, Phoenix, and Denver. Indeed, under severe smog conditions in Los Angeles, nitrate concentrations as high as 22.64 μg/m^3 have been reported, accounting for 25% of the sample mass. Ammonia nitrate is the major nitrate in urban aerosol. It is relatively unstable compared to $(NH_4)_2SO_4$, with concentrations depending on atmospheric temperature and the relative abundance of NH_4^+, NO_3^-, and SO_4^{-2}.

Table 2.7   Major Components of PM$_{2.5}$ Particle Mass in Selected U.S. Cities/Regions

	Los Angeles Summer 1987		Los Angeles Fall 1987		Phoenix Fall 1989/ Winter 1990		Houston September 1980	
	μg/m^3	% mass	μg/m^3	% mass	μg/m^3	% mass	μg/m^3	% mass
Mass	26.30	—	18.50	—	33.09	—	24.80	—
OC	3.34	12.6	4.89	26.4	4.46	13.4	3.10	12.5
EC	0.22	>1.0	1.21	6.5	0.84	2.5	—	—
Nitrate	5.13	19.5	4.86	26.3	0.86	2.6	1.63	6.5
Sulfate	1.87	7.1	1.01	5.6	0.37	1.1	0.91	3.6

Note: OC, organic carbon; EC, elemental carbon.

From USEPA. 1995. EPA/600/AP-95/001a.

The fine particle fraction contains significant concentrations of both elemental and organic carbon. The major source of elemental carbon is fuel combustion. Particles containing elemental carbon are relatively small, with average aerodynamic diameters of approximately 0.01 μm. Significant concentrations of elemental carbon occur in $PM_{2.5}$ samples collected in urban areas and even in some rural sites. As seen in Table 2.7, elemental carbon concentrations as a percentage of collected $PM_{2.5}$ were as low as 3.7% in Houston and as high as 25.4% in Phoenix.

The organic carbon component of atmospheric particulate mass includes primary HCs emitted in combustion and condensed vapors associated with volatile liquids, and secondary products produced by photochemical reactions. Organic compounds produced photochemically include aliphatic organic nitrates, dicarboxylic acids, benzoic and phenylacetic acids, and terpene products. A variety of organic compounds may be sorbed on the surfaces of elemental carbon. Specific classes of organic compounds identified have included aliphatic and aromatic HCs, PAHs, aliphatic and aromatic aldehydes and ketones, ozareanes, phenols, quinones, polyols, phthalic acid esters, sulfur heterocyclics, aryl and alkyl halides, chlorophenols, and alkylating agents. Organic compounds are a significant component of $PM_{2.5}$ aerosols (Table 2.7) collected in western U.S. cities such as Los Angeles (20 to 45%), Phoenix (≈34%) and Denver (≈34%).

A variety of trace metals have been identified in fine particles. Elements with more than 75% of their mass associated with $PM_{2.5}$ include sodium (Na), cesium (Cs), chloride (Cl), bromide (Br), copper (Cu), arsenic (As), silver (Ag), cadmium (Cd), lead (Pb), indium (In), tin (Sn), and antimony (Sb). Concentrations of trace elements such as As, Cd, nickel (Ni), Pb, vanadium (V), zinc (Zn), chromium (Cr), iron (Fe), mercury (Hg), and manganese (Mn) are typically orders of magnitude higher in urban samples compared to those in remote areas in the U.S., and several times to orders of magnitude higher than those in rural areas.

The chemical composition of coarse particles (>2.5 ≤ 10 μm) has not been as well characterized as it has been in fine particles. Coarse particles may originate from a large variety of sources, including mineral/soil erosion, mineral processing, road aggregate, wood ashes, pollen, mold spores, plant debris, etc. A large percentage of coarse particles collected in samples in the eastern U.S. are crystalline (mineral in origin) comprising >50% of particle mass. Common elements include silicon (Si), aluminum (Al), Fe, potassium (K), calcium (Ca), and other alkaline and transition elements. In general, the coarse fraction is basic; the fine fraction acidic. The coarse fraction appears to comprise about 35% of PM_{10} mass in the eastern U.S., about 50% in the central part of the country, and approximately 60% in the west. Component concentrations are summarized in Table 2.8.

Atmospheric Behavior

Aerosol particles have their own unique atmospheric history. It may begin with the emission of primary particles from various natural or anthropogenic sources or by the nucleation of condensable vapors produced as a result of chemical reactions between low molecular weight gas-phase chemical species.

Table 2.8 Major Components of Coarse (2.5 to 10 μm) Particle Mass in Selected U.S. Cities/Regions

Component	Los Angeles Summer 1987		Los Angeles Fall 1987		Phoenix Fall 1989/Winter 1990		Denver January 1982		Houston September 1980		Camden, NJ Summer 1982		Smoky Mts. September 1978	
	μg/m³	% mass	μg/m³	% mass	μg/m³	% mass	μg/m³	% mass	μg/m³	% mass	μg/m³	% mass	μg/m³	% mass
Mass	41.10	—	90.20	—	29.37	—	20.73	—	38.60	—	28.70	—	24.00	—
OC	8.27	20.0	18.46	44.9	10.10	34.4	7.11	34.3	5.68	14.7	2.05	7.1	2.22	9.3
EC	2.37	5.7	7.28	8.0	7.47	25.4	2.15	10.4	1.42	3.7	1.87	6.5	1.10	4.6
Nitrate	4.34	10.5	22.64	25.1	3.60	12.3	2.22	10.7	0.54	1.5	<0.48	<1.6	0.30	1.3
Sulfate	9.41	22.8	4.38	4.8	1.33	4.5	2.06	9.9	14.61	37.8	11.20	39.0	12.00	50.0

Note: OC, organic carbon; EC, elemental carbon.

From USEPA. 1995. EPA/600/AP-95/001a.

Depending on various chemical and physical characteristics, as well as the environment surrounding them, particles may change size, sorb and release vapor-phase molecules, collide, coalesce or adhere to each other, sorb and condense H_2O vapor, undergo deliquescence (rapid uptake of H_2O), undergo changes in electrical condition, and be removed from the atmosphere by depositional processes.

Fine particles produced by gas-to-particle conversion processes commonly undergo changes in size. Homogeneous gas-phase chemical reactions produce condensible species such as H_2SO_4 which can nucleate to form a new particle (nucleation) or condense on the surface of an existing particle (condensation). The relative amount of nucleation as compared to condensation that occurs depends on the rate of formation of the condensable species and on the surface area of existing particles. In urban areas, new particle formation appears only to occur near major sources of nuclei such as freeways. In many cases, the available surface area associated with aerosol particles rapidly scavenges newly formed condensable species. New particle formation is inversely related to available surface area of aerosol particles.

Particle growth may occur as a result of the condensation of homogenous-phase reaction products on particle surfaces, condensation of H_2O vapor, collisions between particles and subsequent coalescence or adherence, surface adsorption, and heterogenous chemical reactions. In the last case, both gas- and particulate-phase substances participate in chemical reactions (as when SO_2 is oxidized to sulfate in an aqueous droplet). The rate of particle growth by this mechanism is limited by chemical reactions at the surface or within the particle. Collisions with other particles which result in coalescence/adherence also result in particle growth. Associated with this phenomenon is a decrease in particle number and total surface area while the average volume per particle increases. Coagulation produces chain agglomerates in soot and some metal-based particles.

Though fine particles are usually thought of as having the potential to increase in size as a result of the processes described above, they, in theory, also have the potential for decreasing in size. Aerosol species such as HNO_3 and certain organic compounds such as PAHs are semivolatile and thus can exist in the gas and particulate phase depending on vapor pressure, particle surface area and composition, and atmospheric temperature. Daily temperature fluctuations can result in desorption and absorption of these semivolatile species with subsequent changes in particle volume and mass as well as chemical composition.

Water is a very important component of many atmospheric particles. The behavior of atmospheric particles in association with changes in ambient humidity are of considerable importance in the global hydrologic cycle and energy budget, and in atmospheric chemistry and optical phenomena. Atmospheric sulfates are both hygroscopic and water soluble. Their hygroscopicity results in physical and chemical changes which affect their size, shape, pH, reactivity, and refractive index. Crystalline solid particles such as $(NH_4)_2SO_4$ undergo a phase change with increasing relative humidity to become aqueous solution particles. This phase transition is abrupt, with a sudden uptake of H_2O taking place at a relative humidity above the deliquescence point.

Particles may undergo a change in their electrical state as they change size. A particle's charge depends on the state of its surface, the value of its dielectric constant, and/or its size. In general, particles greater than 3 μm carry a negative charge and particles less than 0.01 μm carry a positive charge. Electrical charges can affect coagulation and the rate of dry deposition.

Particle behavior also includes the ability to scatter visible light and absorb and reradiate thermal energy. This behavior is affected by particle size, geometry, and chemical composition.

In the history of aerosol particles in the atmosphere, there is a time to be born, a time to grow, and in a metaphorical sense, a time to die. The life of aerosol particles is relatively short. In the planetary boundary layer (defined by the mixing height; see Chapter 3), the residence time for particles is usually less than 3 days. In the upper troposphere the residence time may be one week.

The removal of atmospheric particles occurs by both dry and wet depositional processes. In the former, particles are brought to the earth's surface in the absence of precipitation by a variety of processes including Brownian motion, gravitational settling, and impaction. Diffusion by Brownian motion is of major importance for the deposition of small particles, sedimentation, and impaction for larger particles. In wet deposition, particles are removed from the atmosphere by "in cloud" processes in which they participate in rain formation (rainout) and by "below cloud" processes in which they are struck and brought to the earth's surface by rain droplets or snow and ice crystals (washout).

Atmospheric Concentrations

As described in Chapter 7, ambient PM concentrations can be determined by a variety of techniques. Until the mid to late 1980s, PM was sampled at hundreds of sites in the U.S. by high-volume samplers with concentrations reported as μg/m^3 total suspended particulates (TSP). With the development of fractionating instruments, it became possible to sample only those particles which were relevant to human health, those that were inhalable, aerodynamic diameters ≤10 μm (PM$_{10}$) or respirable, aerodynamic diameters ≤2.5 μm (PM$_{2.5}$). As a result, USEPA promulgated a new PM standard in 1987 based on the collected mass of particles ≤10 μm and has continued to study the possibility of a PM$_{2.5}$ standard as well. A considerable quantity of both PM$_{10}$ and PM$_{2.5}$ data have been collected across the U.S. since 1985.

Urban PM$_{10}$ monitoring is conducted by state and local agencies as a result of federal regulatory requirements to determine compliance with the PM$_{10}$ standard (50 μg/m^3 annual arithmetic average, 150 μg/m^3 24-hour concentration not to be exceeded more than once per year). PM$_{10}$ concentrations for selected U.S. cities in 1993 are summarized in Table 2.9. Not surprisingly, highest annual average concentrations (>30 μg/m^3) are reported for high population or heavily industrialized urban areas such as Chicago, IL, Denver, CO, Gary, IN, Los Angeles, CA, New Haven, CT, Provo, UT, Philadelphia, PA, St. Louis, MO, with the second highest 24-hour maximum concentration in excess of 100 μg/m^3.

Table 2.9   PM$_{10}$ Levels (μg/m^3) in Selected U.S. Urban Areas
in 1993

Urban area	Annual average	Second highest 24-hour concentration
Asheville, NC	22	58
Baltimore, MD	35	70
Birmingham, AL	36	85
Casper, WY	18	41
Chicago, IL	47	147
Dallas, TX	30	74
Danbury, CT	19	46
Denver, CO	41	142
Gary, IN	34	122
Honolulu, HI	24	58
Los Angeles, CA	47	101
Louisville, KY	37	43
New Haven, CT	52	178
New York, NY	47	86
Philadelphia, PA	34	531
Phoenix, AZ	44	92
Provo, UT	40	209
Santa Fe, NM	15	35
San Francisco, CA	29	72
St. Louis, MO	44	101
Steubenville, OH	40	177

From USEPA. 1995. EPA/600/AP-95/001c.

Both regional and seasonal differences in PM$_{10}$ concentrations are evident in aerometric data for urban areas. In the eastern states (east of the Rocky Mountains), highest PM$_{10}$ concentrations exceeding 40 μg/m^3 are observed during the summer months in the Ohio Valley from Pittsburgh to St. Louis as well as in cities such as Nashville, TN, Birmingham, AL, Atlanta, GA, Philadelphia, PA, and Chicago, IL. In the western states, high (>50 μg/m^3) PM$_{10}$ concentrations occur in localized areas during the cold season over large urban areas such as Salt Lake City, UT, and in the smaller mountain valleys of Colorado, Idaho, Montana, Oregon, Washington, and Wyoming. In California, high concentrations occur in the Los Angeles Basin and San Joaquin Valley.

Significant downward trends in national and regional annual average PM$_{10}$ concentrations have been observed in urban areas from 1985 to 1993. The national average has decreased by 39% from 48 to 25 μg/m^3; the eastern states by 29% from 35 to 25 μg/m^3; the western states by 55% from 57 to 26 μg/m^3.

READINGS

1. Blake, D.R. and S. Rowland. 1996. "Urban Leakage of Liquefied Petroleum Gas and Its Impact on Mexico City Air Quality." *Science* 264:953–956.
2. Bridgman, H.A. 1994. *Global Air Pollution*. John Wiley & Sons, New York.
3. Elsom, D.M. 1992. *Atmospheric Pollution — A Global Problem*. 2nd ed. Blackwell Publishers, Oxford, UK.

4. Harrison, R.M. (Ed.). 1990. *Pollution — Causes, Effects and Control.* Royal Society of Chemistry. Thomas Graham House, Cambridge, UK.

5. Lamb, B. 1984. "Gaseous Pollutant Characteristics." pp. 65–96. In: S. Calvert and H. Englund (Eds.). *Handbook of Air Pollution Technology.* 3rd ed. John Wiley & Sons, New York.

6. NAPCA. USDHEW. 1969. *Air Quality Criteria for Photochemical Oxidants.* Publication No. AP-63.

7. NAPCA. USDHEW. 1970. *Air Quality Criteria for Hydrocarbons.* Publication No. AP-64.

8. National Research Council. 1976. *Vapor-phase Organic Pollutants.* National Academy of Sciences, Washington, DC.

9. National Research Council, Subcommittee on Nitrogen Oxides, Committee on Medical Biological Effects of Environmental Pollutants. 1977. *Nitrogen Oxides.* National Academy of Sciences, Washington, DC.

10. National Research Council, Subcommittee on Carbon Monoxide, Committee on Medical and Biological Effects of Environmental Pollutants. 1977. *Carbon Monoxide.* National Academy of Sciences, Washington, DC.

11. National Research Council, Subcommittee on Airborne Particles. 1979. *Airborne Particles.* University Park Press, Baltimore, MD.

12. National Research Council. 1984. *Global Tropospheric Chemistry: A Plan for Action.* National Academy Press, Washington, DC.

13. National Research Council. 1989. *Ozone Depletion, Greenhouse Gases and Climate Change.* National Academy Press, Washington, DC.

14. National Research Council. 1991. *Rethinking the Ozone Problem in Regional and Urban Air Pollution.* National Academy Press, Washington, DC.

15. Patterson, R.G. 1984. "Particulate Pollutant Characteristics." pp. 97–134. In: S. Calvert and H. Englund (Eds.). *Handbook of Air Pollution Technology.* 3rd ed. John Wiley & Sons, New York.

16. Robinson, E. and R.C. Robbins. 1970. "Gaseous Atmospheric Pollutants from Urban and Natural Sources." pp. 50–64. In: S.F. Singer (Ed.). *Global Effects of Environmental Pollution.* Springer-Verlag, New York.

17. Rowland, F.S. and I.S.A. Isaken. 1987. *The Changing Atmosphere.* John Wiley & Sons, Chichester, UK.

18. Rowland, F.S. 1989. "Chlorofluorocarbons and the Depletion of Stratospheric Ozone." *American Scientist* 77:36–45.

19. Seinfeld, J.H. 1986. *Atmospheric Chemistry and Physics of Air Pollution.* John Wiley & Sons, New York.

20. Seinfeld, J.H. 1989. "Urban Air Pollution: State of the Science." *Science* 243:745–752.

21. Urone, P. 1976. "The Primary Air Pollutants — Gaseous. Their Occurrence, Sources and Effects." pp. 24–71. In: A.C. Stern (Ed.). *Air Pollution. Vol. I. Air Pollutants, Their Transformation and Transport.* 3rd ed. Academic Press, New York.

22. USEPA. 1982. *Air Quality Criteria for Particulate Matter and Sulfur Oxides. Vol. II.* EPA/600/8-82-029a-c.

23. USEPA. 1994. *National Air Pollutant Emission Trends, 1900-1993.* EPA/454/R-94-027.

24. USEPA. 1995. *Air Quality Criteria for Particulate Matter. Vol. I.* EPA/600/AP-95/001a.

25. USEPA. 1995. *Air Quality Criteria for Particulate Matter. Vol III.* EPA/600/AP-95/001c.

26. Warneck, P. 1988. *Chemistry of the Natural Atmosphere.* Academic Press, San Diego, CA.

27. Watson, A.Y., R.R. Bates, and D. Kennedy (Eds.). 1988. *Air Pollution: The Automobile and Public Health.* Health Effects Institute. National Academy Press, Washington, DC.

28. Whelpdale, D.M. and R.E. Munn. 1976. "Global Sources, Sinks, and Transport of Air Pollution." pp. 290–322. In: A.C. Stern (Ed.). *Air Pollution. Vol. I. Air Pollutants, Their Transformation and Transport.* 3rd ed. Academic Press, New York.

QUESTIONS

1. How is the rural O_3 problem uniquely different from that which occurs in urban areas?

2. What is the significance of aldehydes in photochemical smog? Your answer should include chemical considerations as well as potential health effects.

3. Indicate the final chemical form to which the following pollutants are converted by sink processes: SO_2, NO_2, HCs, CO, and O_3.

4. By means of your senses, how can you recognize the following pollutants: H_2S, PAN, CO, and NO_2?

5. What are the principal processes by which nuclei mode, accumulation mode, and coarse particles are produced?

6. When can peak levels of NO, NO_2, CO, HCs, and aldehydes be expected in an urban area? Indicate reasons for each of your answers.

7. In the removal of aerosol particles from the atmosphere, indicate the nature and significance of dry deposition, washout, and rainout.

8. Describe the differences between smoke, fume, dust, and mist aerosols.

9. If nature pollutes more than humans, why should we be concerned about anthropogenic air pollution?

10. What is the atmospheric significance of anthropogenic CO emissions?

11. Why do CFCs and other halogenated compounds accumulate in the atmosphere?

12. Describe the significant role that OH plays in atmospheric chemistry. How does CO affect OH, CH_4, and O_3 concentrations?

13. How do haze and smog differ?

14. How do elemental concentrations differ in fine and coarse particles? What are the reasons for these differences?

15. Why are biogenically produced hydrocarbons important in the atmospheric chemistry of O_3?

16. What is PAN? How is it produced?

17. What processes lead to the formation of sulfuric acid and other sulfate compounds in the atmosphere?

18. Why is NO_2 such an important atmospheric pollutant?

3

DISPERSION

Since man first discovered the utility of fire, he has used the atmosphere as a sink to carry away the airborne waste products of his activities. For thousands of years the atmosphere with its enormous volume and constant motion has easily accepted and dispersed the relatively small pollutant burdens imposed on it.

Until the modern era of large cities and intense industrialization, it was natural to depend on the atmosphere to carry away air pollutants. However, as is the case of any sink whose capacity is exceeded, the atmosphere, too, has its limits. Poor dispersion of pollutants, resulting from localized overloading of the atmosphere and the vagaries of meteorological phenomena, allows the buildup of pollutant levels that degrade the quality of the environment and endanger public health.

The awareness that the atmosphere is not limitless, and that pollutants threaten our health and reduce the quality of the environment, has been responsible for recent vigorous efforts to control air pollutants before they enter the atmospheric sink. Although considerable progress has been made, we have neither the technological resources nor the economic commitment to eliminate all emissions. As a result, the atmosphere, to a more limited extent, must continue to serve as a disposal medium for these airborne by-products of our civilization.

Our constantly moving atmosphere provides an enormous sink for the dispersal of air pollutants. Given sufficient time, the whole troposphere with its 5×10^8-m^3 volume and average 10-km thickness is available for this purpose. However, when pollutants are first released from a source, their concentrations are usually high, and effects on environmental quality may be immediate. Dispersion must take place quickly in a limited portion of the atmosphere. Initially, dispersion is influenced by micro- and mesoscale motion. Macroscale motions disperse substances with long atmospheric residence time, such as carbon dioxide (CO_2) and chlorofluorocarbons (CFCs), throughout the troposphere. Scales of air motion and characteristic phenomena associated with them are summarized in Table 3.1.

Table 3.1   Scales of Air Motion

Scale	Geographical area (mi)	Time period	Meteorological phenomena
Macroscale	>100	weeks-months	Atmospheric circulation Weather fronts High and low pressure systems Hurricanes
Mesoscale	10–100	hrs-days	Land/sea breeze Valley winds Urban heat island
Microscale	<10	min	Plume behavior Building downwash and eddy currents

MICRO- AND MESOSCALE DISPERSION OF POLLUTANTS

In the initial dispersion process from point or area sources, pollutants are released into the ambient environment where their transport and subsequent dilution depend on local meteorological phenomena and the influence of topography. In these local dispersion processes important meteorological phenomena include wind (speed and direction), turbulence, and atmospheric stability.

Wind

Horizontal winds play a significant role in the transport and dilution of pollutants. As the wind speed increases, the volume of air moving by a source in a given period of time also increases. If the emission rate is relatively constant, a doubling of the wind speed will halve the pollutant concentration, as the concentration is an inverse function of the wind speed.

The speed of the horizontal wind is affected by friction, which is proportional to surface roughness, which is determined by topographical features such as mountains, valleys, rivers, lakes, forests, cultivated fields, and buildings. Wind speeds over smooth surfaces, e.g., cultivated fields and lakes, tend to be on the average higher than those over rougher surfaces, e.g., mountains and buildings. The effect of surface roughness on wind speed as a function of elevation over urban areas, suburbs, and level country is illustrated in Figure 3.1. Note that as the surface roughness associated with urban areas increases, wind speeds near the surface are significantly reduced.

Pollutant dispersion is also significantly affected by variability in wind direction. If wind direction is relatively constant the same area will be continuously exposed to high pollutant levels. If, on the other hand, wind direction is constantly shifting, pollutants will be dispersed over a larger area and concentrations over any given exposed area are lower. Large changes in wind direction may occur over short periods of time. For example, a change of wind direction of 30° or so over an hour is common, and over a 24-hour period wind direction may shift 180°. Seasonal changes may result in wind direction variations as much as 360°.

Wind direction frequency and wind speed for a given time period may be summarized by constructing a wind rose (Figure 3.2). Data are given for eight

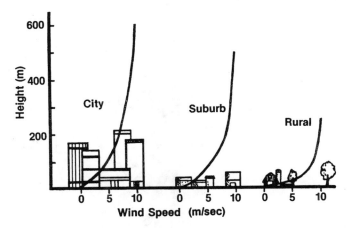

Figure 3.1 Relationships between surface roughness and wind speed changes with height. (From Turner, D.B. 1969. *Workbook for Atmospheric Dispersion Estimates.* EPA Publication No. AP-26.)

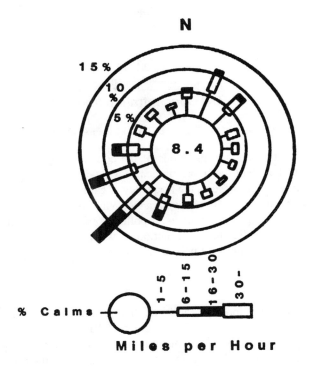

Figure 3.2 Wind rose. (From Slade, D.H. (Ed.). 1968. *Meteorology and Atomic Energy.* AEC, Washington, D.C.)

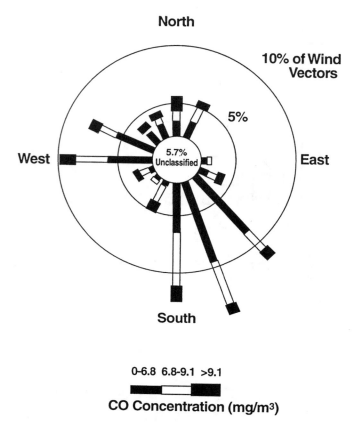

Figure 3.3 Pollution rose for CO. (From USEPA. 1991. EPA/600/8-90/045f.)

primary and eight secondary directions of the compass. Wind speed is divided into ranges. The length of the spokes indicates wind direction frequency. Starting from the center, which represents the frequency of calms, each segment represents the frequency of wind speed in the given range. Wind roses may be used to predict dispersion from point and area sources. A pollution rose indicating the frequency of given levels of a pollutant as a function of wind direction can also be constructed. Figure 3.3 illustrates a pollution rose for CO.

Turbulence

Air does not flow smoothly near the earth's surface; rather, it follows patterns of three-dimensional movement called turbulence. Turbulent eddies are produced by two specific processes: (1) thermal turbulence, resulting from atmospheric heating and (2) mechanical turbulence caused by the movement of air past an obstruction. Usually both types of turbulence occur in any given atmospheric situation, although in some situations one or the other may prevail. Thermal turbulence is dominant on clear, sunny days with light winds. Although mechanical turbulence occurs under a variety of atmospheric conditions, it is dominant on windy nights with neutral atmospheric stability. The effect of turbulence is to enhance the dispersion process, although in the case of mechanical turbulence, downwash from the pollution source may result in high pollution levels immediately downwind.

Lapse Rates and Atmospheric Stability

In the troposphere, temperature decreases with height up to an elevation of approximately 10 km. This decrease is due to the reduction of heating processes with height and radiative cooling of air, and reaches its maximum in the upper levels of the troposphere. Temperature decrease with height is described by the lapse rate. On the average, temperature decreases $-0.65°C/100$ m or $-6.5°C/km$. This is the *normal lapse rate*.

If a parcel of warm dry air were lifted in a dry environment, it would undergo adiabatic expansion and cool. This adiabatic cooling would result in a lapse rate of $-1°C/100$ m or $-10°C/km$, the *dry adiabatic lapse rate*.

Individual temperature measurements in the vertical may vary considerably from either the normal or dry adiabatic lapse rate. This change of temperature with height for any given measurement is the *environmental lapse rate*. Values for environmental lapse rates characterize the stability of the atmosphere and profoundly affect vertical air motion and the dispersion of pollutants.

The relationship of environmental lapse rates to stability is illustrated in Figure 3.4. When there are no sources or sinks of thermal energy the temperature of a rising air parcel will decrease with height at a rate of $-1°C/100$ m, or the dry adiabatic lapse rate. If a parcel of warm air is released into an environment where the temperature decrease with height is greater than the adiabatic lapse rate, say $-2°C/100$ m, the parcel will rise rapidly. The atmosphere will be unstable and conditions for the vertical dispersion of pollutants will be excellent. This is a superadiabatic lapse rate (Figure 3.4A).

If a warm parcel of air is placed in an environment where the lapse rate approaches the adiabatic lapse rate (Figure 3.4B), the decrease of temperature with height would be the same as the cooling rate of the parcel and the atmosphere would be neutrally stable. The parcel then would always be warmer than its surroundings and, as a result, it would rise completely through the neutral layer. Good dispersion of air pollutants is achieved in a neutrally stable environment.

If a warm parcel of air were released into an environment in which the temperature remained constant with height, i.e., were isothermal (Figure 3.4C), it would rise until it reached equilibrium with the temperature of its surroundings. Under isothermal conditions the atmosphere tends to be more stable and the dispersion of pollutants becomes somewhat limited or moderate.

If a warm parcel of air is released into an environment where the temperature is inverted (Figure 3.4D), that is, as temperature increases with height, it will rise until the temperature and density are in equilibrium with the environment. If the

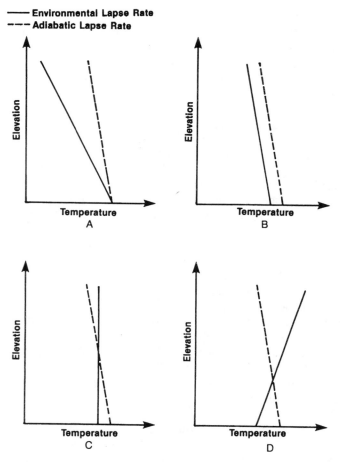

Figure 3.4 Comparison of environmental lapse rates with the adiabatic lapse rate with relationships to stability conditions.

parcel temperature is cooler than its surroundings, the parcel will sink. In an inversion, atmospheric conditions are very stable, that is, they tend to resist change, and vertical motions are restricted. Dispersion potential under inversion conditions is very poor.

Using the neutral or dry adiabatic lapse rate with its good dispersion characteristics as a reference point, we can characterize the dispersion characteristics of other environmental lapse rates. If the slope is to the left of the dry adiabatic lapse rate, dispersion characteristics are good to excellent. The greater the deviation to the left, the more unstable the atmosphere and the more enhanced dispersion. Moving to the right of the dry adiabatic lapse rate, the atmosphere becomes stable and dispersion becomes more limited. The greater the deviation to the right from the adiabatic lapse rate, the more stable the atmosphere and the poorer the dispersion potential.

Figure 3.5 Average summertime mixing heights for U.S. cities.

Mixing Height

Movement of air in the vertical dimension is enhanced by vertical temperature differences. The steeper the temperature gradient, the more vigorous the convective and turbulent mixing of the atmosphere. The larger the vertical column in which turbulent mixing occurs, the more effective the dispersion process.

The *maximum mixing height* (MMH), or *maximum mixing depth* (MMD), is the depth of the convective layer associated with the maximum surface temperature. Average MMHs for the U.S. cities are summarized in Figure 3.5. Note the relatively low mixing height for Los Angeles.

The MMH shows diurnal and seasonal variations. In addition, it is markedly affected by topography and macroscale air circulation (anticyclones). During the day, minimum MMH values occur just before sunrise. As solar energy heats up the ground and warms the air, it expands, becomes less dense, and rises by convection. Cooler air from aloft descends to take its place (Figure 3.6). As solar

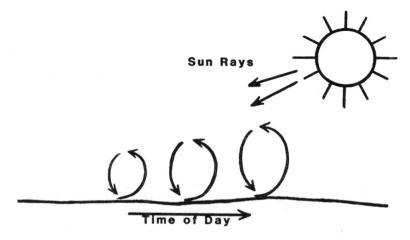

Figure 3.6 Convective air flows in the atmosphere with reference to the time of day.

heating continues, vertical motions become intense and the MMH is reached in early afternoon, with values of several thousand meters. The minimum values of MMH, observed at night and in the early morning hours, may be a result of surface level inversions produced by the radiative cooling of the ground and the air on clear nights. In these situations the MMH may be near zero. These surface level inversions are particularly intense in mountain valleys. The MMH shows seasonal variability with maximum values in the summer and minimum values in late fall and winter. Fortunately, however, the lower average MMHs of winter are associated with higher average surface wind speeds. As a result dispersion is not as adversely affected.

Topographical features such as water surfaces with their large heat absorbing capacities result in lower MMHs. Land surfaces with little vegetative cover — e.g., deserts — may have mixing heights of 5000 m or more.

The MMH is affected by macroscale air motions as well. Subsidence inversions produced in semipermanent marine high pressure systems and stagnating migratory anticyclones can significantly reduce the MMH.

Topography

Topography may affect micro- and mesoscale air motion near point and area sources. Most of the large urban centers in the U.S. are located along sea (New York, Los Angeles) and lake coastal (Chicago, Detroit) areas and much heavy industry is located in river valleys, e.g., the Valley of the Ohio. The local air flow patterns in these regions have a significant impact on pollution dispersion processes.

Lake, Sea, and Land Breezes. Land/water mesoscale air circulation patterns develop as a result of the differential heating and cooling of land and water surfaces. During the summer, when skies are clear and prevailing winds are light, land surfaces heat more rapidly than water. The warm air rises and moves toward water. Because of differences of temperature and pressure, air flows in from the water and a sea or lake breeze is formed (Figure 3.7). Over water the warm air from the land cools and subsides to produce a weak circulation cell. At night the more rapid radiational cooling of the land surfaces results in a horizontal flow toward water and a land breeze is formed (Figure 3.8). In general, land breezes are weaker than lake and sea breezes. These circulation patterns form under light prevailing winds. If prevailing winds are strong, however, land/water airflows may be overridden. The weak circulation cells of land/water breezes may cause pollutants to be recirculated and carried over from one day to the next.

Because of seasonal differences in heating and cooling of land and water surfaces, inversions may be produced near the sea/lake shore, resulting in reduced vertical mixing and pollutant dispersal. This occurs in late spring, when large water bodies are still cold relative to warmer adjacent land areas and air moves over a relatively large area of cool water before coming ashore. As the air moves inland it warms and the inversion is replaced by superadiabatic lapse rate conditions.

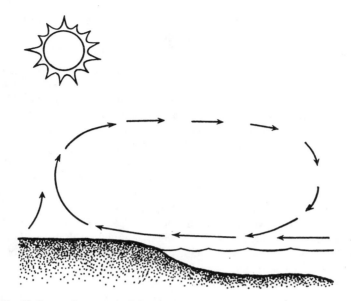

Figure 3.7 Air flow patterns associated with sea and lake breezes.

Valleys. Because of the availability of water for the transportation of raw materials and for process use, river valleys have historically been ideal locations for the growth of heavy metallurgical industries such as the smelting and processing of iron ore and iron and steel products. Because of terrain effects on local

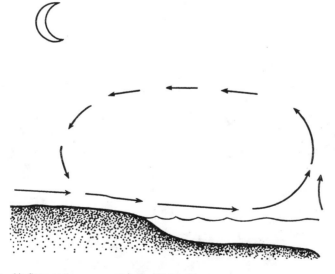

Figure 3.8 Air flow patterns associated with land breezes.

meteorology and the intensely polluting nature of such industries, these valleys have suffered severe air pollution problems.

River valleys were formed by the erosive force of water following a path of least resistance, moving from higher elevations toward the sea. In many ways the micro- and mesoscale movement of air in valleys is similar to that of water. This is particularly true for movement of cool dense air on clear nights that results from the radiative cooling of ridgetops and the air immediately above them.

Like water, air flows downhill into the valley floor. Winds produced are called slope winds. As the air reaches the valley floor it flows with the path of the river. This air movement is called the valley wind. The formation of the valley wind lags several hours after that of the slope winds. Because of a smaller vertical gradient, downriver valley winds are lighter, and because of the large volume, cool dense air tends to accumulate, flooding the valley floor and intensifying the surface inversion that would normally be produced by radiative cooling. The inversion deepens over the course of the night and may reach its maximum depth just before sunrise. The height of the inversion layer depends on the depth of the valley and the intensity of the radiative cooling process.

In some mountain valleys the inversion layer may be several hundred meters thick. A valley surface inversion and resulting pollutant buildup are illustrated in Figure 3.9. Note that pollutants do rise somewhat but do not break through the inverted lapse rate boundary. It is along this boundary that the highest pollutant levels occur. In the Appalachian mountains nocturnal inversions occur, on the average, one-third of the days of the year.

Nocturnal surface inversions are relatively nonpersistent, that is, they are readily destroyed by early morning heating of the ground and resulting convective

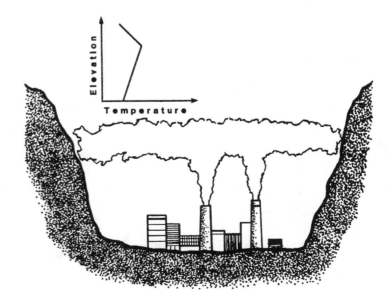

Figure 3.9 Nocturnal inversion conditions in a river valley.

and turbulent air mixing. As convection cells reach the top of the inversion, the layer of elevated pollutant concentrations is brought to the ground (fumigation) and maximum ground level concentrations occur about two hours after sunrise.

In some cases valley ground-based inversions may be associated with fogs produced by the radiative cooling of humid air. Because the fog droplets have a high albedo, solar heating of the ground is reduced and inversion breakup is delayed. The increased life of the inversion increases the pollution burden of the valley. Such a situation occurred in the Monongahela Valley during the Donora disaster of 1948.

Hills and Mountains. As previously discussed, mountains can affect local air flow by increasing surface roughness and thereby decreasing wind speed. In addition, mountains and hills can form physical barriers to the movement of air. The severe smog problem in the Los Angeles Basin is due in part to the mountain barriers which rise from 400 to 2000 m to restrict advective flow of air eastward. Air flow patterns in the basin are dominated by local seaward drainage at night and landward movement during the day. An air flow pattern develops which tends to move polluted air back and forth over land and sea surfaces. As a result, polluted air may be carried over from one day to another. The reduced ventilation of the basin is exacerbated by the subsidence inversions associated with semi-permanent marine high pressure systems off the California coast. This aspect will be discussed in another section.

Dispersion from a Point Source

A visible plume usually forms as pollutants are being emitted from a smoke-stack into the atmosphere. The subsequent history of this plume will depend on the interplay of a number of factors, which include: (1) the physical and chemical nature of the pollutants, (2) meteorological parameters occurring at any period of time, (3) the location of the source relative to obstructions, and (4) downwind topography. As these factors affect the plume, primary pollutant concentrations in most cases will tend to be reduced. In general, maximum ground level concentrations will occur in a range from the vicinity of the smokestack to several kilometers downwind.

Nature of the Pollutants. Plumes are a mixture of both gases and particulate matter. Large particles have appreciable settling velocities and settle near the source. Smaller particles tend to remain suspended in the atmosphere for longer periods of time and their dispersion behavior may be similar to that of gases. The gaseous nature of the plume would, given sufficient time, allow for its dispersion by simple diffusion, whereby gaseous molecules randomly move from an area of high to an area of low concentration. Although diffusion is an important parameter in the dispersion of a plume, it cannot by itself account for the dispersion observed.

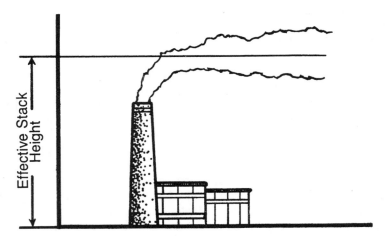

Figure 3.10 Plume rise and effective stack height.

At this point it would be desirable to discuss the fate of the plume as a two-part dispersion process: (1) the rise of the plume from the source, and (2) downwind transport and dilution.

Plume Rise and Transport. Initial plume rise is important, as the height of the plume will determine subsequent pollutant concentrations measured near the ground. The larger the rise, the greater the downwind distance it will be carried before it reaches the ground. As a consequence, ground level concentrations will be reduced. The height of the rise will depend on the temperature of the emissions, the cross-sectional area of the stack, the emission velocity, the horizontal wind speed, and the vertical temperature gradient. For a given set of these conditions the plume will rise to some height (Δh). The height of the stack plus the plume rise (Δh) is called effective stack height (Figure 3.10). The higher the effective stack height, the greater dispersion will be. Effective stack height can be increased by building taller stacks and emitting pollutants at higher temperatures. Many large fossil-fuel power plants disperse their effluents in this way. Stack heights of 250 to 300 m are commonly used. A few are 400 m tall.

Horizontal wind speed also affects plume rise — the higher the wind speed, the more quickly the plume bends over and becomes transported downwind. Although higher wind speeds decrease plume rise, dispersion may not be affected adversely because of the greater volume of air moving past the source. High winds usually enhance the dispersion process.

Plume rise is adversely affected by atmospheric stability. As the atmosphere becomes stable, that is, under inversion conditions, plume rise can be markedly reduced. Dispersion is decreased, resulting in higher ground level concentrations downwind.

As the plume moves downwind, it expands. Mean concentrations of pollutants in this plume show a Gaussian, that is, a bell-shaped, distribution, with

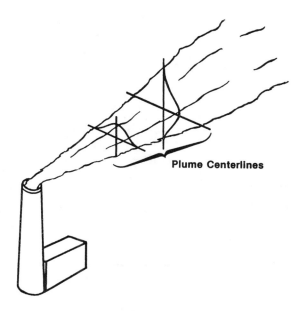

Figure 3.11 Gaussian plume spread associated with downwind plume movement. (From USEPA. 1981. EPA/450/2-81-0113.)

highest concentrations in the center line and decreasing as a function of distance from the center line (Figure 3.11).

Plume Characteristics. Because of differences in atmospheric stability, plumes may take on characteristic shapes or profiles under certain atmospheric conditions. Six such plumes are illustrated in Figure 3.12. The looping plume of Figure 3.12A occurs under unstable atmospheric conditions, with a superadiabatic lapse rate and considerable convective mixing. Looping plumes are usually produced on clear sunny days with light winds. The looping plume may be drawn to groundlevel near the source. A coning plume is illustrated in Figure 3.12B. It is produced when the lapse rate is neutral to isothermal and the atmosphere is slightly unstable. Coning occurs on cloudy or windy days or nights. A coning plume may be brought to the surface at a greater distance downwind than a looping plume. A fanning plume (Figure 3.12C) is produced when the lapse rate is inverted near the surface and the atmosphere is stable. Although the plume can spread horizontally, little vertical mixing results. Such inversions are produced on clear nights with little or light wind. A lofting plume (Figure 3.12D) may be produced when there is a superadiabatic lapse rate above a surface inversion through which a smokestack protrudes. The plume rises upward but downwind movement is restricted by the inversion beneath. A lofting plume may be produced at sunset on a clear evening in open country. As the inversion deepens it is replaced by a fanning plume. A fumigating plume (Figure 3.12.E) is produced when a

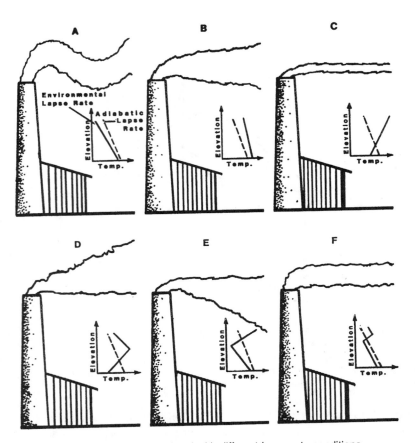

Figure 3.12 Plume geometry associated with different lapse rate conditions.

fanning plume and its associated surface inversion are broken up by the thermal turbulence associated with the solar heating of the ground surface. As convective cells grow, the inversion is destroyed and the polluted air mass near the top of the inversion layer is drawn to the surface, resulting in high ground-level concentrations. Fumigating plumes are formed on clear sunny days with light winds. When an inversion occurs above and below the plume, trapping results (Figure 3.12F). A trapping plume may be produced on clear sunny days and clear nights with light winds and a stagnating anticyclone.

As a plume from an isolated point source moves downwind, it mixes with relatively unpolluted air by diffusion, convection, advective displacement, and mechanical turbulence, resulting in a significant reduction (dilution) in pollutant levels. At some distance downwind sufficient dilution occurs, so that the plume appears to lose its physical integrity and is no longer visible. For some large sources, however, such as large fossil fuel-fired power stations, plumes may be visible for 100 km downwind.

Figure 3.13 Plume downwash associated with mechanical turbulence. (From Briggs, G.A. 1969. AEC Critical Review Series, Washington, D.C.)

Effect of Terrain and Source Configuration. Terrain and the location of buildings relative to a smokestack may have a significant effect on plume behavior. In mountain valleys under stable conditions a plume may strike the mountainside downwind of the source, with elevated pollutant levels occurring in the area of contact. If a source is located downwind of a ridge, the resulting mechanical turbulence and downwash may also result in elevated ground level concentrations. In the case of a source with a short stack, the flow characteristics of air moving over the building may induce a downwash on the lee side, drawing the plume to the ground near the source (Figure 3.13). This downwash problem can be mitigated by the use of a taller stack. As an operating rule, the stack height should be at least twice the height of adjacent buildings.

MACROSCALE DISPERSION

Urban Plume

Large urban centers represent a multitude of individual point and mobile sources. These collectively form an urban plume whose dispersion represents one of the most critical air-quality problems facing modern society. For it is in the initial stages of urban plume development that high pollutant concentrations may affect human health and welfare.

The normal air motions that disperse plumes from isolated point sources and emissions from low-volume motor vehicle traffic may be overwhelmed by the emissions from the large number of sources which form the urban plume. Although processes similar to those which disperse emissions from point sources occur for individual or mobile sources within the urban plume, they are usually less effective because dilution must occur in an already contaminated environment.

For an urban plume, prime consideration must be given to the ventilation rate of the whole region, as dispersion may be affected by both meso- and macroscale air motions. Conditions that restrict or inhibit vertical and horizontal movements, such as stagnating anticyclones, surface inversions and topographical barriers, will intensify the urban plume and result in high pollutant concentrations. On the other hand, rapidly moving weather systems such as migrating cyclones and associated high winds are particularly effective in dispersing urban plumes.

The urban plume moves downwind of large metropolitan areas and can affect air quality hundreds of kilometers away. It is responsible for elevated O_3- levels at remote sites and may be the partial or major cause of elevated background levels of contaminants such as SO_2 or CO in urban areas downwind and relatively nearby.

Long-Range Transport

Without evidence to the contrary it was widely believed that once polluted air was transported for some distance downwind, the enormity of the atmosphere and dilution processes associated with it reduced pollutants to background levels. However, elevated levels of pollutants may occur hundreds to thousands of kilometers downwind of large point sources and areas producing urban plumes. This phenomenon is now known as long-range transport. As we shall see in Chapter 4, long-range transport is a significant factor in producing Arctic haze and exacerbating problems of acidic deposition. It also is responsible for the rural O_3 problem discussed in Chapter 2. Because of long-range transport, there are very few if any locations east of the Mississippi River that are uncontaminated by anthropogenic air pollution.

Jet Streams

Jet streams have considerable influence on the dispersion and removal of pollutants from the atmosphere. Associated with the jet stream are areas of discontinuities between the troposphere and the stratosphere. It is through these discontinuities that stratospheric particulate matter from volcanoes is removed.

The dispersion and removal of pollutants from the troposphere are strongly influenced by migrating precipitation-producing storm systems or cyclones. Jet streams intensify these surface cyclones and influence their speed and direction.

High-Pressure Systems

Anticyclones may significantly affect the dispersion of pollutants over large regions. Near 30° north and south latitude, cool air descends from aloft to form a tropical Hadley cell. This descending or subsiding air produces semipermanent marine high-pressure systems centered over the world's major oceans. These subtropical marine highs shift their position toward the poles in summer and toward the equator in winter in response to the earth's changing inclination to the sun.

As the cool air descends it is warmed adiabatically (by compression) so that air aloft has a higher temperature than that below. This results in a subsidence inversion (Figure 3.14). On the easterly side of marine high-pressure systems, the inversion layer approaches closer to the surface with increasing distance from

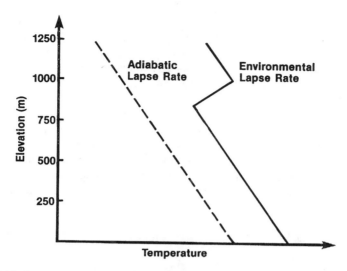

Figure 3.14 Lapse rate conditions associated with a subsidence inversion.

the cell center. As a result, west coasts of continents have relatively low subsidence inversions. Inversions below 800 m occur over southern California approximately 90% of the time during the summer months. Subsidence inversions reduce vertical dispersion significantly. As a result, regions such as southern California have severe air pollution problems. The base of the subsidence inversion does not reach the ground during the day because of convective mixing associated with solar heating of the ground surface. As warm air rises the base of the inversion is eroded. At night the base of the inversion may approach close to the ground.

In the temperate zone between 30 and 60° north latitude, migrating anticyclones may have a significant effect on the ventilation of major cities. The air stagnation associated with migrating anticyclones may result in elevated pollutant levels, with potentially dangerous implications for public health. Air pollution disasters in the Meuse Valley in 1931, Donora in 1948, and London in 1952 and 1956 were caused by such stagnating anticyclones.

The base of the inversion layer varies but is often found at an altitude of 500 m, extending upward another 500 to 1000 m. The strongest and most persistent inversions in the central and northeastern parts of the U.S. occur in the fall of the year and are associated with the beautiful days and light winds of Indian summer.

Cleansing Processes — Precipitation

If pollutants were continuously emitted to the atmosphere without being removed, background levels would increase over time. With the exception of CO_2, CFCs, and a few other chemically stable compounds, no appreciable accumulation of pollutants occurs in the atmosphere. This is primarily due to wet and dry removal processes. In dry deposition, particles are removed by gravity or impaction and gases diffuse to surfaces where they are absorbed or adsorbed. Wet deposition is the major removal process for most particles and may to some degree be a factor in the removal of gaseous contaminants as well. Wet removal may involve in-cloud capture of gases or particles (rainout) or below-cloud capture (washout). In washout, raindrops or snowflakes strike particles and carry them to the surface; gases are removed by absorption.

Meteorological phenomena that increase the frequency and intensity of precipitation in a region will significantly reduce the pollutant burden and enhance air quality. Precipitation associated with migratory cyclones is particularly effective in cleansing the atmosphere. On the other hand, pollution problems are exacerbated in regions of low precipitation, e.g., the Los Angeles Basin.

AIR POLLUTION CONTROL — APPLICATIONS OF METEOROLOGY

Industrial Siting

Historically, siting decisions for large industrial facilities did not take into account the air pollution consequences of site selection. As a result, problems in many areas of the world are more severe than they need be. The heavily industrialized river valleys are a prime example of the exacerbated pollution problems caused by topography and local meteorological phenomena. One cannot, however, condemn industrial planners of the past for the poor location of facilities in respect to air pollution since there existed neither the meteorological understanding nor societal concern for pollution problems. Today the circumstances are very different and the use of meteorological air pollution information is an integral part of the site selection process, particularly for large point sources such as coal-fired plants. After considerations of resource availability, the dispersion characteristics of the atmosphere at a proposed large industrial site are the principal determinants of site selection by industry. Dispersion characteristics of the site are also of prime importance in approval of a proposed facility by regulatory agencies.

Episode Planning

The major air pollution disasters in the Meuse Valley in 1931, Donora in 1948, and London in 1952 and 1956 resulted from the buildup of high air pollutant concentrations associated with inversion conditions produced by stagnating migratory anticyclones. In the late fifties and early sixties the U.S. Weather Bureau, now the National Weather Service, observed a statistical correlation between high air pollution levels and certain weather patterns, namely, slow-moving anticyclones. As a result, "Air Pollution Potential" forecasts have been issued since 1963 when meteorological conditions indicate the presence of large slow-moving anticyclones expected to persist for thirty-six hours. These forecasts are applied to affected areas of at least 90,000 km^2. This limitation is meant to exclude those local inversion conditions which may be associated with meso-scale phenomena.

Although there has been considerable reduction of emissions from pollution sources in the past decade and a half, the occurrence of stagnating anticyclones and associated subsidence inversions still poses a significant threat to the health and welfare of individuals dwelling in industrial/urban centers. Because of our ability to forecast these conditions, it is possible to reduce the consequences of these "episodes." When air stagnation is forecast, air pollution control tactics developed by federal, state, and local air pollution regulatory agencies are implemented. These tactics, called episode plans, include health warnings to those most likely to be adversely affected and a phased reduction of emissions from sources in the region (see Chapter 8).

Dispersion Modeling

Present air pollution control strategies in the U.S. are based on allowing some degree of air quality degradation. In general, this degradation is not to exceed air quality standards. Such standards, discussed in Chapter 8, are limits on pollutant levels in the ambient environment that theoretically do not pose a threat to human health. Attainment of air quality standards requires a knowledge of the kind and quantity of pollutants emitted by sources in a region and the resultant reduction of pollutant levels by atmospheric dilution.

Ground level concentrations of pollutants downwind of an existing source or a planned source or sources can be predicted by the use of complex dispersion models. Meteorological variables, notably windspeed and atmospheric stability, are major variables in these models. The quality of model predictions depends in great measure on the accuracy of meteorological and source emissions data.

Dispersion models are used extensively by regulatory agencies as a tool in their programs of atmospheric surveillance. Within this context, dispersion modeling is treated in greater detail in Chapter 7.

READINGS

1. Bowne, N.E. 1984. "Atmospheric Dispersion." 859–893. In: S. Calvert and H. Englund (Eds.). *Handbook of Air Pollution Technology.* John Wiley & Sons, Inc., New York.
2. Briggs, G.A. 1969. *Plume Rise.* AEC Critical Review Series, Washington, DC.
3. Dobbins, R.A. 1979. *Atmospheric Motion and Air Pollution.* John Wiley & Sons, Inc., New York.
4. Scorer, R.S. 1990. *Meteorology of Air Pollution Implications for the Environment and Its Future.* Ellis Horwood, New York.
5. Singer, I.A. and P.C. Freudenthal. 1972. "State of the Art of Air Pollution Meteorology." *Bull. Amer. Meteor. Soc.* 53:545–547.
6. Slade, D.H. (Ed.). 1968. *Meteorology and Atomic Energy.* AEC, Washington, D.C.

7. Strom, G.H. 1977. "Transport and Diffusion of Stack Effluents." 401–448. In: A.C. Stern (Ed.). *Air Pollution. Vol. I. Air Pollutants, Their Transformation and Transport.* 3rd ed. Academic Press, New York.

8. Turner, D.B. 1969. *Workbook for Atmospheric Dispersion Estimates.* EPA Publication No. AP-26.

9. Turner, D.B. 1994. *Workbook of Atmospheric Dispersion Estimates: An Introduction to Dispersion Modeling.* CRC/Lewis Publishers, Boca Raton, FL.

10. USEPA. 1981. *Air Course 411. Meteorology — Instructor's Guide.* EPA/450/2-81-0113.

11. USEPA. 1991. *Air Quality Criteria for Carbon Monoxide.* EPA/600/8-90/045f.

QUESTIONS

1. What meteorological factors affect the dispersion of pollutants from a point source?

2. A parcel of polluted air is released into the ambient environment from a smokestack. Describe the parcel's behavior (a) if its temperature is higher than that of the surrounding environment, and (b) if its temperature is lower than the surrounding environment.

3. Describe the physical basis for the formation of (a) nocturnal ground-based inversions, and (b) elevated inversions occurring over a broad geographical area.

4. Describe two ways by which precipitation removal of contaminants from the atmosphere occurs.

5. Specifically describe how meteorological information is used on a day-to-day basis in air pollution control programs.

6. Describe dispersion characteristics of the atmosphere under the following lapse rate conditions: $-2°C/100$ m, $0°C/100$ m, and $1°C/100$ m.

7. The speed of air moving past a source with a constant emission rate increases from 1 m/sec to 3 m/sec. For a given pollutant, x, what is the quantitative effect of this change of wind speed on its concentration?

8. What atmospheric conditions contribute to long-range transport?

9. Why cannot (in most cases) polluted air penetrate an inversion layer?

10. Under what lapse rate conditions are the following plume geometries observed: looping, coning, fanning?

11. What is the effect of mesoscale air motions such as sea/land breezes on the dispersion of pollutants from a coastal city?

12. How does turbulence affect the behavior of a plume?

13. What atmospheric conditions produce a fumigating plume or a fumigation?

14. What is an urban plume? How is it formed?

15. What factors cause plume downwash?

16. What effect do jet streams have on pollutant dispersion?

17. What meteorological factors affect plume rise from a smokestack?

18. What microscale phenomena affect pollutant dispersion?

19. What is mixing height? How does it affect pollutant dispersion?

20. How do wind speed and direction affect dispersion?

4   ATMOSPHERIC EFFECTS

Atmospheric processes in pollution dispersion and removal were discussed in Chapter 3. It was noted in that discussion that the atmosphere, serving as a sink for pollutants, has a considerable capacity for self-renewal, that is, for cleansing itself. Although it significantly affects the fate of pollutants, the atmosphere may also experience significant short- or long-term pollution-induced changes. These changes may be local, regional, or global in scale, depending on such factors as sources, long-range transport, movement into the stratosphere, and the accumulation of substances that are only but slowly removed by sink processes. Atmospheric effects may include changes in (1) visibility, (2) the urban climate, (3) quantity and frequency of rainfall and associated meteorological phenomena, (4) precipitation chemistry, (5) stratospheric ozone (O_3) depletion, and (6) global warming or cooling. The significance of these changes may be slight (visibility impairment) to potentially very serious (stratospheric O_3 depletion and changes in climate). Changes in stratospheric O_3 levels and/or global climate have the potential for adversely affecting the earth's ecological systems and the well-being of billions of humans.

VISIBILITY

Visibility reduction associated with smokestack plumes and urban smog is the most obvious effect of atmospheric pollution perceived by those affected. Regional hazes, which occur primarily in the warm months of the year, have a less obvious connection with the polluting activities of humans.

The term visibility implies that an object can be seen by an observer, but that the ability to see an object clearly is relative. As humans, we perceive how visible an object or scene is to us. This perception is influenced by a variety of physical factors in the atmosphere and factors that are characteristic of the vision system of humans and the unique psychology of each individual. These include

(1) illumination of the observed area by the sun as mediated by clouds, ground reflection, and the atmosphere, (2) reflection, absorption, and scattering of incoming light by target objects and the sky, (3) scattering and absorption of light from a target object and the source of illumination by the atmosphere and contaminants, (4) psychological response of the eye–brain system to the resulting light distribution, and (5) a subjective judgment of the perceived images by the observer.

Fundamental to the concept of visibility is the ability to detect contrast between an object and its surroundings. Without contrast, or where contrast is very slight, an object cannot be "seen"; that is, it is not visible. The threshold contrast is that atmospheric condition when an object begins to be perceptible against its surroundings. Contrast is significant because it determines the maximum distance at which an object or the various components of an observed scene can be discerned. Visibility is usually described in the context of light intensity, but it can also include color. Color is a sensation produced by the human visual system in response to the spectral distribution of incoming light.

The ability to see an object clearly is described as a function of distance. The terms visual range, prevailing visibility, and meteorological range are used to describe this distance. Visibility measurements are routinely recorded by human observers at commercial and military airfields. These measurements are conducted in conjunction with, or in addition to, the National Weather Service, which has been recording visibility observations since 1939. Visibility measurements are reported as "prevailing visibility," defined as the greatest visibility attained or surpassed around at least one-half of a horizon circle. Prevailing visibility measurements are made by observers viewing the horizon and noting whether objects (daytime) or lights (nighttime) can be seen at known distances.

Reported prevailing visibility values are expressed in miles or kilometers. In theory there is no standard upper limit, but in most cases visibility observations are not reported beyond 15 mi (25 km) because suitable markers are usually unavailable beyond this distance.

Visibility Reduction — Natural

Our ability to see objects clearly in an "unpolluted" atmosphere is limited by blue sky scattering, the curvature of the earth's surface, and suspended natural aerosols. At sea level a particle-free atmosphere scatters light and limits visual range to 200 mi (330 km). The scattering of light by air molecules is called Rayleigh scattering and is responsible for the blue color of the sky on "clear days." Because Rayleigh scattering is a function of atmospheric density, visual range increases with altitude. At about 12,000 ft (4000 m), the potential visual range is about 300 mi (500 km).

When dark objects such as distant mountains are viewed through a relatively particle-free atmosphere, they appear blue because blue light is scattered preferentially into the line of sight. Clouds on the horizon can appear yellow-pink because the atmosphere scatters more of the blue light from bright targets out of the line of sight, leaving longer-wavelength colors.

In addition to Rayleigh scattering, which in conjunction with the earth's curvature serves to place a theoretical limit on how far one can see, there are other natural causes of visibility reduction. These include fog, rain, snow, wind-blown dust, and natural hazes. Natural hazes are formed from aerosols produced from volcanic emissions of sulfur dioxide (SO_2) and from chemical reactions involving HCs and various sulfur species produced bio- and geogenically. Reactions involving terpenes produce particles with diameters of less than 0.1 μm. Such particles scatter blue light preferentially and are likely to be the cause of blue hazes over heavily forested areas such as the Blue Ridge and Great Smoky Mountains.

Visibility Reduction — Anthropogenic

Visibility reduction associated with anthropogenic activities is due primarily to light scattering by particles and to a lesser extent to the absorption of light by atmospheric gases such as NO_2. Nitrogen dioxide absorbs short-wavelength blue light, with longer wavelengths reaching the eye. Its color therefore appears yellow to reddish-brown. Light can also be absorbed by particles. Such absorption is important when dark soot or large amounts of windblown dust are present. In general, most atmospheric particles are relatively inefficient in absorbing light.

Fine fraction solid or liquid particle aerosols are responsible for most atmospheric light scattering. Particles with diameters similar to the wavelengths of visible light (0.4 to 0.7 μm) are particularly effective in light scattering (Figure 4.1). This wavelength-dependent interaction of light with the atmosphere can be described mathematically by radiative transfer equations.

$$-dI = BeId_x \qquad (4.1)$$

where $-dI$ = decrease in intensity (extinction), Be = extinction coefficient, I = initial beam intensity, and d_x = length of light path.

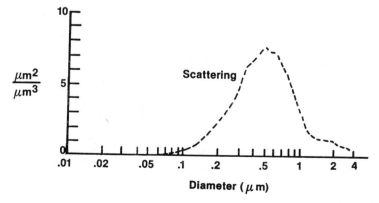

Figure 4.1 Relationship between light scattering and particle size. (From USEPA. 1982. EPA/600/8-82-029.)

Equation 4.1 can be used to describe the effect of the atmosphere on a light beam as it passes from a source to an observer. *Be* is the extinction or attenuation coefficient. Its value is determined by the scattering and absorption of light by particles and gases and varies as a function of wavelength and contaminant concentration.

The intensity of a beam in the direction of an observer decreases with distance as light is scattered or absorbed. Over a short path length, this decrease is proportional to the path length and the intensity of a beam at that point. Similarly, when an observer is looking at a distant object such as a mountain, the light from the mountain is diminished by scattering and absorption. This reduced brightness is not usually the major factor limiting visibility, however. In addition to the light originating from an object, the observer receives light scattered into the line of sight by the intervening atmosphere. It is this "air light" that forms the optical phenomenon we recognize as haze.

Air light intensity scattered into the sight path of an observer depends on the distribution of light intensity from all directions, including direct sunlight, diffuse sky light, surface reflection, and the light-scattering characteristics of air molecules and aerosols. This air light for a given short distance is described by the following equation:

$$dI = Be \ [W_a Q_v(\theta_v) I_v d\Omega] d_x \qquad (4.2)$$

where bracketed parameters [ ] = sum of light intensity from all directions scattered into the line of sight, dI = added air light, Be = extinction coefficient, and d_x = path length.

The total amount of light scattered into the sight path is a function of aerosol and air-scattering parameters ($W_a Q_v$) and illumination intensity (I_v) and angle (θ_v) summed over all directions (Ω).

The change in light intensity from an object to an observer is a direct function of the extinction of transmitted light and the addition of air light. For a short distance:

$$dI = -dI \ (\text{extinction}) + dI \ (\text{air light}) \qquad (4.3)$$

$$= -Be \ [I d_x + W_a Q_v(\theta_v) I_v d\Omega d_x] \qquad (4.4)$$

This radiative transfer equation can be used to determine the effects of pollutants on visibility. Its use is difficult, however, because a number of the parameters must be approximated.

The effect of the atmosphere on the perceived brightness of target objects can be seen in Figure 4.2. With increasing distance, extinction and added air light cause both dark and bright objects to be washed out and approach the brightness of the horizon. As a result, the contrast of an object relative to the horizon and other objects decreases.

The apparent contrast and the visual range of a large dark target object can be calculated from the following series of exponential equations.

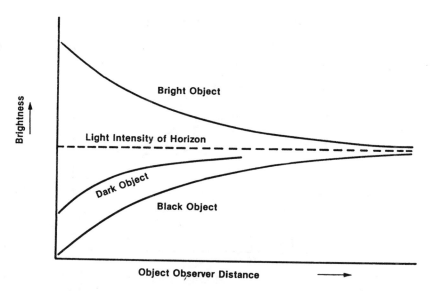

Figure 4.2 Effect of the atmosphere on perceived brightness of target objects. (From USEPA. 1982. EPA/600/8-82-029.)

$$C = C_o e^{-Bex} \qquad (4.5)$$

where C = apparent contrast at observer distance, C_o = initial contrast at the object, e = natural log base, 2.718, Be = extinction coefficient, and x = observer/object distance.

For a black object, the initial contrast is −1, and Equation 4.5 becomes:

$$C = (-1)e^{-Bex} \qquad (4.6)$$

By assuming a standard threshold of perception of 0.02, we can proceed to calculate the visual range (V_r), which is equal to the observer/object distance.

$$0.02 = -e^{-BeV_r} \qquad (4.7)$$

$$V_r = 3.92/Be \qquad (4.8)$$

Equation 4.8 is called the Koschmieder formula. Under certain limitations it can be used to determine the visual range of large dark objects when measurements of the extinction coefficient are available.

The extinction coefficient (Be) consists of both an absorption and a scattering component (Bs). The scattering coefficient is easier to measure than the extinction coefficient. For nonabsorbing atmospheric media Be and Bs have the same value.

The visual range correlates inversely with Bs (Figure 4.3). The scattered points in Figure 4.3 indicate measurements, and the straight line indicates those predicted by the Koschmieder formula.

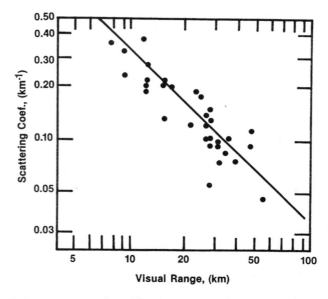

Figure 4.3   Relationship between atmospheric scattering and visual range. (From USEPA. 1982. EPA/600/8-82-029.)

Light scattering by aerosol particles includes a number of physical phenomena: (1) reflection, which causes backscatter, and (2) diffraction and refraction, which cause scattering in the forward direction (Figure 4.4). In diffractive scattering, a light ray is bent by the edge effect of an aerosol particle into its shadow. In refraction, two effects are produced. For particles with a refractive index $n > 1$, the speed of the light wave is reduced, causing a reduction in the wavelength

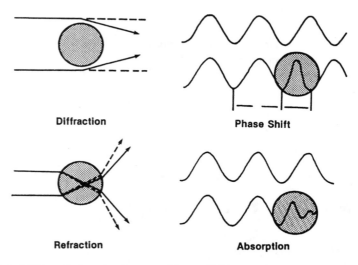

Figure 4.4   Light scattering phenomenon. (From USEPA. 1982. EPA/600/8-82-029.)

inside the particle. This results in a phase shift, which produces positive and negative interferences. Refraction also bends light waves, causing them to pass through and out of a particle at a different angle from that at which they enter. This refractive bending is responsible for the separation of the color components of white light by a prism. Refractive bending of light by aerosol particles can also have chromatic effects. Coarse particles (>2 μm) scatter light in the forward direction primarily by diffraction and refraction. For particles in the accumulation mode (0.1 to 1.0 μm), physical interactions with light are complex and enhanced. Though coarse particles are efficient in scattering light, their contribution to visibility reduction is limited due to their relatively small numbers.

Light scattering may be affected by the chemical composition of aerosol particles. As a rule, particles of differing chemical composition will have different light scattering efficiencies. In addition, hygroscopic compounds will cause particles to grow; as a result, they will typically scatter light more efficiently. This is particularly true of aerosols that contain sulfates [e.g., H_2SO_4, $(NH_4)_2SO_4$, NH_4HSO_4].

Meteorology

Meteorological factors may affect visibility in several ways. Sunlight, for example, significantly affects visibility by promoting secondary aerosol formation. Atmospheric photochemistry produces the major visibility-reducing aerosols, i.e., sulfates, nitrates, and oxyhydrocarbons.

Wind speed and atmospheric stability affect visibility because they determine atmospheric dispersion and therefore concentration of aerosol particles. In general, as the wind speed increases, visibility improves, as the wind-induced atmospheric mixing results in lower aerosol concentrations. During periods of atmospheric stagnation (associated with slowly moving high pressure systems), vertical mixing is suppressed, aerosol concentrations increase, and visibility is reduced. The resultant haze may cover hundreds of thousands of square kilometers. It is not uncommon to have a major portion of the Midwest, Southeast, or the East Coast covered by a "blanket of haze."

Visibility may be significantly affected by relatively humidity. Hygroscopic particles such as sulfates take up moisture and increase in size under favorable humidity conditions. These larger particles (in the accumulation mode) are very effective in scattering light and therefore reducing visibility. Marked visibility reduction has been noted when relative humidity exceeds 70%, although the importance of this effect extends over a range of relative humidities. In general, regions and periods of low relative humidity are associated with good visibility.

Visibility Problems

Pollution-induced changes in visibility that have received both scientific and regulatory interest include the intense smog conditions of cities such as Los Angeles and Denver, and to a lesser extent other cities as well; seasonal regional hazes that affect large areas in the midwestern, eastern, and southeastern U.S.;

the springtime haze over the Arctic Circle; and the plume blight and hazes produced by large point sources in the pristine air quality regions of the American southwest and the Rocky Mountains.

Smog. Smog problems in Los Angeles and other cities are long-standing. Despite three decades of pollution control efforts, smog problems persist and, in cases such as Los Angeles and Denver, even worsen over time. They have become, unfortunately, a part of the ambiance of those cities. In British cities such as London, pollution control efforts have been very effective, with the density of the so-called London smogs diminished significantly.

Regional Haze. The problem of regional haze in the midwestern, eastern, and southern U.S. appears to have grown progressively worse since the mid-1950s. This is particularly evident during the summer months. Studies of long-term visibility trends in the Northeast, for example, have shown a significant (10 to 40%) reduction in annual visibility at suburban and nonurban locations. The greatest decline has occurred during the summer months, when atmospheric extinction has increased from 50 to 150%, with corresponding visibility reductions in the range of 25 to 60%.

Regional haziness in the U.S. during the summer months is associated with slow-moving high pressure systems. These typically track up the Ohio Valley, where major point sources emit large quantities of SO_x, which, in combination with warm temperatures, abundant sunlight, high relative humidity, and reduced vertical mixing, provide conditions favorable for the production of regional haze. Such haze is visible from orbiting earth satellites. Statistical evidence indicates that sulfate aerosols are the most optically active components of regional haze, accounting for 50+% of observed light extinction. The decline in visibility observed during the summer months in the Northeast and Midwest is strongly correlated with increased use of coal over the past three decades to produce electrical power needed to serve air conditioning needs in residential and nonresidential buildings. Summertime visual range values for 1974 to 1976 in the U.S. are summarized in Figure 4.5. Lowest average visibility values have been observed for the Ohio Valley and the Southeast.

Arctic Haze. There is significant evidence to indicate that the springtime haze over the western Arctic Ocean, a phenomenon first reported by weather reconnaissance crews in the 1950s, is caused by the macroscale transport of pollutants on the order of 10,000 km from sources in northern Europe. This Arctic haze contains sulfate levels 10 to 20 times greater than would be expected from natural sources in the region. The presence of vanadium in Arctic haze strongly implicates anthropogenic sources of sulfates because it is strongly associated with industrial fuel oil use.

Pristine Air. Historically, visibility impairment has been a problem in our large urban centers and increasingly on a broad regional scale in many areas east of the Mississippi River. In the regulatory, sense, however, it has been the pristine

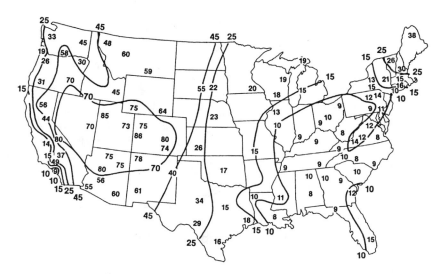

Figure 4.5   Median summertime visual range (mi) in the U.S., 1974 to 1976. (From USEPA. 1982. EPA/600/8-82-029.)

areas of the Rocky Mountain West that have received the most attention. The scenic beauty associated with our western national parks, monuments, and wilderness areas is due in great measure to the high air quality and high visibility [visibilities in the range of 60 to 100 mi (100 to 164 km) are not uncommon] and contrast characteristics of these regions. It is also due to the recognition that states such as Utah (with five national parks) have significant coal resources and an economic incentive to use these resources for the large-scale production of electricity. Major concerns include visibility-reducing and aesthetically degrading power plant plumes and regional hazes produced by photochemical conversion of SO_x. The former is commonly described as plume blight. In pristine areas, power plant plumes may maintain their distinct character for many miles or kilometers downwind. The plume and the haze that it will eventually produce cause blighted vistas, destroying the unique character of some of the U.S.'s most beautiful landforms and environments.

The problem of visibility degradation of pristine air and the protection of areas of high air quality have been specifically addressed by Congress and USEPA. Prevention of significant deterioration (PSD) provisions and programs are discussed in Chapter 8.

TURBIDITY

Visibility is a measure of horizontal light scattering. Turbidity, on the other hand, is a measure of vertical extinction. Vertical extinction is dependent on the same atmospheric light scattering characteristics of aerosols associated with visibility reduction. The principal effect of atmospheric turbidity is diminution of

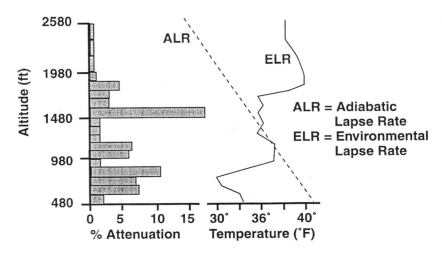

Figure 4.6 Dust dome effect on lapse rate conditions in the Cincinnati metropolitan area. (From Bach, W. 1971. *Geog. Rev.* 61:573–594. With permission.)

the intensity of sunlight received at the ground. Though intensity is reduced, most light scattering is in the forward direction.

Turbidity, or optical thickness of the atmosphere, can be directly related to aerosol concentrations, as sunlight must pass through the entire thickness of the atmosphere between the sun and the observer.

Aerosols in both the tropospheric and stratospheric layers of the atmosphere contribute to turbidity. In general, tropospheric extinction can be directly attributed to aerosols from human activity. On the other hand, stratospheric turbidity is largely attributable to aerosols produced from volcanos.

The turbid air over major cities is often described as a dust dome. In studies in the Cincinnati metropolitan area, a complex stratification of the dust dome has been observed with alternating polluted and cleaner air layers that correlate with different lapse rate conditions (Figure 4.6). High turbidity was often observed to occur between 11 and 12 a.m. on either relatively clean or polluted days. On relatively clean days, the lowest 1000 m contributed 30% of the solar attenuation; on polluted days, 65%. Other turbidity-related phenomena in the troposphere include: (1) a commonly observed turbid air mass centered over the eastern U.S., (2) maximum turbidity during the summer and minimum turbidity in winter, (3) losses of direct radiation of up to 50% in the rural midwestern U.S. during summer months, resulting in up to a 20% reduction of total radiation received at the ground, and (4) an increasing upward trend in turbidity over Washington, DC and Davos, Switzerland over the past four decades.

Stratospheric Turbidity

An aerosol layer, described as the Jange layer, forms in the stratosphere from volcanic eruptions and is maintained between interruptions by particle formation and transport in the upper tropical troposphere. This stratospheric aerosol layer varies in height from 9 to 28 km with maximum concentrations at approximately 18 km. This aerosol consists of highly concentrated liquid sulfuric acid (H_2SO_4) droplets that are typically 0.1 to 0.3 μm in diameter. Most of the mass of this aerosol layer is formed from reduced sulfur gases such as carbon disulfide (CS_2), carbonyl sulfide (COS), and SO_2. They are oxidized in the stratosphere to H_2SO_4 which condenses on pre-existing nuclei.

The stratospheric aerosol layer can have a variety of atmospheric effects. By backscattering sunlight it reduces the amount of solar radiation reaching the ground, thus cooling the atmosphere. Cooling of portions of the troposphere by such turbidity has been observed after major volcanic eruptions, particularly those which have high sulfur gas emissions. In addition, aerosol particles provide sites for heterogeneous chemical reactions which repartition nitrogen species and decrease O_3 concentrations. In polar regions reactions on these particles can result in the catalytic destruction of O_3 by chlorine (Cl) atoms.

THERMAL AIR POLLUTION

Historically, only gases and aerosol particles have been considered to be atmospheric pollutants. However, because of the effect of surfaces and waste heat on the urban heat balance and the resultant alteration of the atmospheric environment, heat or thermal energy can also be considered a form of air pollution.

In making temperature measurements on a clear night (with light winds) in an urban area and also in the surrounding countryside, it is possible to detect and characterize a phenomenon known as the urban "heat island." In this heat island, areas of elevated temperatures (Figure 4.7) can be identified, with the entire urban area being warmer than the surrounding countryside. Urban-rural differences of 1 to 2°C are common, and differences may be as great as 5°C. The principal contributors to heat island development include (1) waste heat dumped into the environment from energy-utilizing processes, (2) solar energy absorption and heat storage properties of urban surfaces, and (3) decreased urban ventilation resulting from the increased surface roughness of buildings. Urban surfaces such as buildings, sidewalks, and streets more readily absorb and store heat from sunlight during daylight hours than vegetated areas external to the city. In addition, urban surfaces cool more gradually at night.

In cities such as New York, anthropogenic heat may be 2.5 times greater than that received from the sun during winter months; during the summer it still may represent a significant portion of the total city energy budget — on the order of 17%. On an annual basis, anthropogenic heat for many cities may exceed 10% of that received naturally from the sun.

Urban heat islands may have significant effects on weather and local climate within urban centers and in some cases downwind. These may include (1) longer frost-free seasons, (2) less frequent fog, (3) decreased snow accumulation, (4) lower relative humidity, (5) decreased likelihood of surface-based nocturnal inversions, (6) increased rainfall downwind, and (7) a distinctive nocturnal metropol-

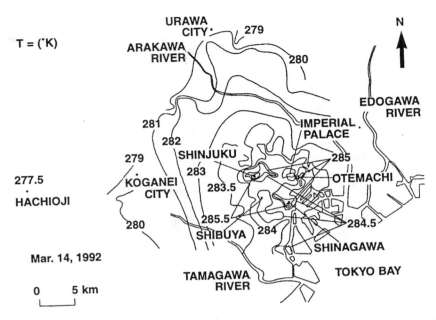

Figure 4.7 Urban heat island in Tokyo, Japan. (From Saitoh, T.S., T. Shimade, and H. Hoshi. 1996. *Atmos. Environ.* 30:3431–3442. With permission.)

itan air circulation pattern. A nocturnal air circulation pattern associated with an urban heat island is illustrated in Figure 4.8.

This pattern is much like a sea breeze on an island. As the warm air rises from the urban center it flows outward toward rural areas. Cooler air outside of the urban center sinks and flows inward toward the city. This thermally induced

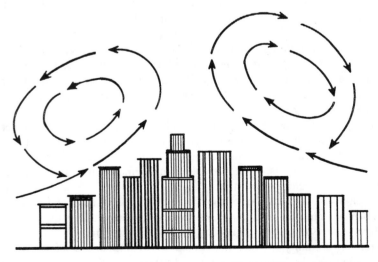

Figure 4.8 Urban air circulation pattern associated with heat island phenomenon.

air flow creates a turbid air mass over the city that tends to be highest in the middle of the heat island. The resultant "dust dome" forms best under conditions of light wind and compact city form.

Because of thermal turbulence and vertical mixing, surface-based inversions generally do not occur over urban heat islands. However, elevated nocturnal inversions of longer duration may occur frequently. These elevated inversions may result in part from the dust dome absorption of thermal energy. They may also result from the advection of stable rural air over warmer urban surfaces. As this air moves over the city it is heated from below, resulting in an inversion layer above the lower, relatively warm air.

The increased rainfall downwind of cities (discussed in a subsequent section) may be due in part to the heat island effect. Enhanced vertical convective mixing tends to make precipitation more probable.

EFFECTS ON PRECIPITATION AND PRECIPITATION PROCESSES

Pollutants emitted in urban areas may increase precipitation and affect other related meteorological phenomena. In some cases the opposite effect has been observed; that is, pollution sources have decreased downwind precipitation.

Atmospheric particles can serve as sites for the condensation of H_2O vapor. Condensation around nucleating particles is an important factor in cloud development and precipitation. Only a small fraction of the total aerosol load of the atmosphere serves as weather-active or cloud condensation nuclei (CCN).

The concentration of CCN determines the initial size and number of water droplets, which determine the frequency and amount of precipitation. An inverse relationship exists between CCN concentration and initial droplet size; that is, as CCN concentration increases, droplet size decreases. Water droplets increase in size by coalescence or agglomeration. Because larger droplets coalesce more effectively, clouds with larger droplets will have a higher probability of falling to the earth's surface as rain. The more numerous and consequently smaller cloud droplets produce less rainfall. This phenomenon can be used to explain decreased rainfall observed downwind of sources such as sugarcane fires in Australia and forest fires in the Pacific Northwest. These sources apparently contribute large numbers of CCN, which decrease droplet size.

Where pollution sources contribute significant numbers of giant condensation nuclei, small-sized CCN may be unimportant in influencing precipitation. The increase in rainfall observed downwind of some cities may be due to the contribution of giant nuclei (>1 μm) which promote coalescence and increase precipitation probability.

Pollutants may also affect precipitation in cold cloud processes by influencing the concentration of freezing nuclei. Because atmospheric levels of freezing nuclei are frequently low, the addition of freezing nuclei in the form of pollution particles may enhance precipitation under cold cloud conditions. An application of this principle is cloud seeding by silver iodide crystals.

In previous paragraphs it was noted that increased precipitation has been observed downwind of some U.S. cities. Studies of precipitation downwind of St. Louis indicate that summer rainfall is increased in a range of 6 to 15% as a result of urban effects. Precipitation increases appear to be associated with elevated pollutant levels.

ACIDIC DEPOSITION

In addition to affecting frequency and quantity, pollutants affect the chemical nature of precipitation. Changes in precipitation chemistry have been reported for over three decades. During that time, national governments in Europe and North America have come to recognize the seriousness of precipitation acidification by sulfur and nitrogen pollutants of anthropogenic origin.

When the problem of pollutant-caused acidification of precipitation was first described, it was called quite naturally acid rain. As studies of the phenomenon of rain acidification were expanded, it became apparent that snow was also being affected. Thus, it became more appropriate to describe the problem as acid precipitation. Because acidic materials are deposited on land and water surfaces in the absence of precipitation (dry deposition), the concept was expanded again. It is apparent that this phenomenon is more than acid rain and snow and dry deposition. It is also acid fog (with a pH as low as 1.8), acid clouds, acid dew, and probably acid frost as well. To take into account all of these phenomena, the term acidic deposition is used by scientists and regulators. However, from a public perspective, the term acid rain was applied first and has therefore stuck. Thus, in public discourse the terms acid rain and acidic deposition are used synonymously.

Acidic deposition as a significant environmental problem was first recognized in Sweden in the mid-1960s. Swedish ecologists observing significant changes in species composition in a number of freshwater lakes made a connection between observed lake ecological changes and the acidity of rainfall received over affected watersheds. In North America, the problem was first recognized in the mid-1970s. Studies conducted in Europe, North America, and a variety of other locations indicate that acidification of precipitation is widespread.

When water condenses after evaporation or distillation, it has a pH of 7 and is neither acidic nor alkaline. It rapidly changes to a pH of 5.7 as it comes into equilibrium with CO_2 in the atmosphere. Cloud water, rain water, and snow in the absence of pollution and in equilibrium with CO_2 would naturally be acidic with a pH of 5.7.

The term pH refers to the negative log of the hydrogen ion [H^+] concentration. A solution with a pH of 6 has an H^+ concentration of 0.000001 molar (M), and one with a pH of 5 would have an H^+ concentration of 0.00001 M. Values of pH are geometric, so that a pH of 5 is 10 times more acidic than a pH of 6, and a pH of 4 is 100 times more acidic than a pH of 6.

A pH of 5.7 has been used as a reference for assessing the significance of pollution-related precipitation acidification. Studies in remote areas in the world

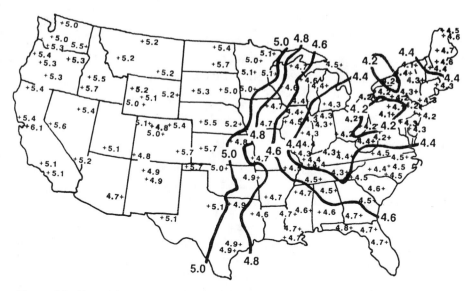

Figure 4.9 Hydrogen ion (H+) precipitation (meq/m²) in the U.S. in 1980. (From USEPA. 1979. EPA/600/8-79-028.)

where anthropogenic effects on precipitation are likely to be minimal indicate that the average pH value is closer to 5.0 than 5.7.

Average H+ precipitation deposition values (millequivalents/m²) for the U.S. in 1980 are plotted in Figure 4.9. Areas with the highest H+ precipitation deposition are clustered around the Ohio Valley, an area where industrialization is intense. In this area, pH values of rainfall average approximately 4.2 to 4.3, with individual values as low as 3.0. Using a reference value of pH = 5, precipitation in the Ohio Valley is approximately 6 to 7 times more acidic than it would normally be.

In the northeastern U.S., increased precipitation acidity appears to be due to sulfuric (65%) and nitric acids (30%). The apparent major sources of precursors (sulfur and nitrogen oxides) for these strong acids include fossil fuel-fired power plants, industrial boilers, metal smelters, and automobiles. In the western states where SO_x sources are few, precipitation acidity is predominantly due to nitric acid, with motor vehicles being the major source of precursors.

The major source region for acid precursors in the eastern U.S. is the middle Ohio Valley. Sulfur oxide emissions for Ohio (≈3 million ton/year) are the highest for any state and are twice the combined emissions of the six New England states, New York, and New Jersey. In addition, significant SO_x emissions occur in Indiana, Illinois, Kentucky, Pennsylvania, West Virginia, Tennessee, Michigan, and Missouri.

Approximately 60 to 70% of these SO_x emissions are reported to be associated with large electric power plants which burn high-sulfur coal. These plants with their tall stacks (200 to 400 m) disperse emissions at higher altitudes, providing longer residence times for the conversion of acid precursors to strong acids. Consequently, the emission of SO_x from the Midwest and from the northern Appalachian states may significantly affect precipitation acidity over large geographical areas downwind, including trans-boundary transport into Canada. It has been estimated that approximately 50% of the acid-forming pollution burden in Canada originates in the U.S. There is, of course, evidence that Canadian sources also contribute to acidic deposition in the U.S.

There is little doubt that long-range transport contributes to acidic deposition at considerable distances downwind of sources of acid precursors. In many cases, however, acidic deposition at a particular receptor site is likely to be more strongly affected by local sources as well. The relative contribution of distant sources as compared to local sources to acidic deposition is a matter of intense debate. Nevertheless, it is apparent that long-range transport of acid precursors has a profound effect on acidic deposition hundreds of kilometers downwind of sources.

Although nitric acid is responsible for only about 30 to 35% of the H^+ ion concentration in acidic precipitation and dry deposition in the eastern U.S., the importance of NO_x and their contribution to acidic deposition may be far more significant than these numbers indicate. In the northeastern U.S., sulfur emissions have not increased much in the past 25 years, although significant increases have occurred in the Southeast. What has changed substantially since 1960 has been emissions of NO_x, from automobiles and electrical power generation. It is the absorption of sunlight by NO_2 which initiates photochemical reactions that promote the atmospheric production of sulfuric and nitric acids. The problem of acidic deposition is not simply a matter of emissions of acid precursors, but atmospheric processes which produce strong acids as well.

Ecological Effects

Acidic deposition poses a threat to ecosystems which, because of local or regional geology (crystalline/metamorphic rock), soils and surface waters cannot adequately neutralize acidified rain, snow, and dry-deposited materials. Such soils and waters contain little or no calcium or magnesium carbonates and therefore have low acid neutralizing capacity (ANC). Acid-sensitive ecosystems in North America are illustrated in Figure 4.10. These include the mountainous regions of New York and New England; the Appalachian Mountain chain, including portions of Pennsylvania, West Virginia, Virginia, Kentucky, and Tennessee; upper Michigan, Wisconsin, and Minnesota; the Pre-Cambrian Shield of Canada, including Ontario, Quebec, and New Brunswick; and various mountainous regions of the North American west.

Aquatic Systems.

Chemical Changes. Acidic deposition has been linked to the chronic acidification of thousands of lakes and episodic acidification of thousands of streams in the Scandinavian peninsula and in southeastern Canada. In the U.S., surface-water acidification by acidic deposition has been significant but more limited.

Figure 4.10 Acid sensitive ecosystems in North America. (From USEPA. 1979. EPA/600/8-79-028.)

The National Surface Water Survey (NSWS), conducted in acid-sensitive regions, indicated that the chemical composition of approximately 75% (880) of acidic lakes and 47% (2200) of acidic streams was dominated by sulfate from atmospheric deposition, indicating that acidic deposition was the primary cause of acidification. Most surface waters acidified by acidic deposition were found in six regions: New England, the southwest Adirondacks, small-forested watersheds in the mid-Atlantic mountains, the mid-Atlantic coastal plain, the north-central portion of Florida, and in the upper peninsula of Michigan and northeastern Wisconsin. Virtually no acidic surface waters were observed in the western U.S. or the southeastern mountains.

Though a lake or stream may have low ANC and therefore be acid-sensitive, acidification, whether it is chronic or episodic, will depend on source factors as well as processes occurring within a watershed. As rain or snowmelt waters move through a watershed their chemistry is usually altered by processes that neutralize acids. The likelihood of neutralization increases as the distance to receiving waters increases.

Surface water chemistry changes associated with acidic deposition may include (1) increases in sulfate concentration, (2) little change in nitrate levels (because watersheds effectively retain nitrate), (3) increases in base cations such as Ca^{+2} and Mg^{+2} associated with acid neutralization processes, (4) decreases in pH and ANC, (5) increases in Al^{+3}, and (6) decreases in dissolved organic carbon.

Though chronic acidification of lakes has received the most attention, it is important to note that episodic acidification is a common phenomenon in regions that receive high levels of acidic deposition. In lakes and streams where chronic acidification has lowered ANC levels, small decreases in ANC caused by pulses of nitrate, sulfate, or organic anions can lower pH to critical levels. Pulses of nitrate commonly contribute to acidic episodes in portions of New England, the mid-Atlantic mountains, and the Adirondacks. Pulses of sulfate associated with acidic deposition have been shown to be significant contributors to episodic acidification in some streams in the mid-Atlantic mountains.

Biological Changes. Chemical constituents which have been demonstrated to cause significant biological changes include H^+, Al^{+3}, and Ca^{+2}. Increased H^+ (decreased pH) and Al^{+3} cause a variety of adverse biological effects that may be exacerbated by low Ca^{+2} concentrations. As pH decreases below 5.5, Al^{+3} concentrations increase to what are often toxic levels.

Changes in water chemistry associated with chronic or episodic acidification can affect aquatic organisms at all levels. Because of their large size and recreational value, significant scientific attention has been given to the effects of acidification on fish populations. Declines in fish populations in acidic and low ANC lakes in the Adirondacks have been directly linked to acidic deposition-induced acidification. More than 200 lakes are now known to be fishless. Fish population declines in lakes and streams in New England and mid-Atlantic mountains may also be associated with acidification. In Canada, thousands of lakes in Ontario, Quebec, and New Brunswick are fishless, apparently due to acidic deposition.

The extinction of fish populations in freshwater lakes and the decline of fish populations (and fish kills) in streams appears to be a result of chronic as well as episodic acidification. Sudden shifts in pH following spring snowmelt has reportedly caused fish kills. Long-term increases in H^+ and Al^{+3} concentrations result in recruitment failure; that is, few young fish reach maturity. Recruitment failure results from the high mortality of eggs, larvae, and young fish, which are particularly vulnerable to increases in H^+ and Al^{+3}. As a result, acidified waters often show decreased population density with a shift of size and age toward larger and older fish. As acidification continues, a lake will become fishless. Because acidification is a progressive phenomenon, the most sensitive fish species will be affected first. Based on toxicity models developed from bioassay data, about 20% of lakes in the Adirondacks, 6% in New England, and 3% in the upper Midwest have H^+, Al^{+3}, and Ca^{+2} index values which are unsuitable for the survival of sensitive fish species such as the common minnow and rainbow trout. Almost 10% of lakes in the Adirondacks have H^+, Al^{+3}, Ca^{+2} index values which make them unsuitable for the more acid-tolerant brook trout. In the mid-Atlantic coastal plain, an estimated 60% of upstream sites during spring have H^+ and Al^{+3} levels that would kill 50% of anadromous fish species such as the larval stage of the blueback herring.

As with other stressful environmental conditions, one initial significant biological change which occurs in acidified waters is a reduction in species diversity.

Shifts in species composition begin to occur at pH levels of 6.0 to 6.5. Unicellular phytoplankton species decline and certain filamentous species increase in abundance and in severely acidified lakes cover all surfaces with a thick mat. Acidification decreases the diversity and abundance of benthic invertebrates which because of their role as microdecomposers are essential to nutrient cycling in aquatic systems. Reduction in invertebrate populations results in fewer prey species for higher aquatic organisms.

Decreased decomposition by benthic organisms and bacteria results in an accumulation of organic debris with less mineralization and release of organically bound nutrients. The strong acidity binds nutrient ions to mineral particles decreasing their uptake by phytoplankton.

The effects of acidification are profound, whether they are associated with acidic deposition or natural processes. As lakes become acidified, significant species changes occur as well as decreases in biomass production.

Terrestrial Ecosystems and Agricultural Crops. Acidic deposition has the potential to affect terrestrial plants in natural ecosystems and croplands both directly and indirectly. Major scientific attention has been given to evaluating potential causal connections between acidic deposition and "declines" of Norway spruce and beech in Germany and other countries of central Europe, high elevation spruce/fir forests in the northern and southern Appalachian mountains of the eastern U.S., low-elevation spruce/fir forests in the northeastern U.S., southern pines in the southeastern U.S., and sugar maple in the northeastern U.S. and southeastern Canada.

Forests and crops may be exposed and adversely affected through several different pathways. Foliage surfaces intercept acidic substances from rain, clouds, or fog. The most likely plant responses to such exposures is the leaching of cation nutrients such as Ca^{+2}, Mg^{+2}, and K^+. In situations where cation nutrients are already limiting growth, increased foliar leaching could exacerbate further plant/soil nutritional imbalances. Wet and dry acidic deposition may also affect plants by modifying soil chemistry, thereby affecting root system functions such as nutrient uptake and symbiotic relationships between tree roots and fungi known as mycorrhizae.

Based on studies conducted under the National Acid Precipitation Assessment Program (NAPAP) conducted in the U.S. in the 1980s, there appears to be no evidence that acidic deposition occurring at current levels (pH = 4.0 to 5.0) is responsible for any widespread adverse effects on forest ecosystems. There is evidence from some locations that soil chemistry changes are occurring consistent with patterns of acidic deposition. These changes include acidification, cation exchange, aluminum mobilization, and cation leaching. Long-term changes in sensitive soils are expected. However, whether these changes result in reduced forest health is very uncertain. It is also uncertain as to how such effects may be manifested and how long they may take to occur. Model estimates suggest that, at current deposition rates, significant changes in forest soil nutrient status could occur in some soils within 50 to 60 years.

There is evidence from NAPAP studies to indicate that acidic deposition is at least in part responsible for growth reductions and mortality of red spruce at high elevations in the northern Appalachians and growth reductions in the southern Appalachians. Results from controlled-exposure studies, as well as field assessments, indicate that several different mechanisms may be involved. Most notable among these is acidic cloud water (pH 2.8 to 3.8). Exposures to acidic cloud water appears to result in alterations in plant nutrition, cold hardening, and a variety of physiological processes. Observed symptoms on red spruce appear to be due to reduced foliar nutrition due to leaching and reduced soil nutrient availability, interference with normal root function associated with increased mobilization of aluminum (Al^{+3}) and competitive inhibition of Ca^{+2}, Mg^{+2}, and P^{+2} uptake, reduced carbon uptake due to reduced photosynthetic capacity and increased respiratory carbohydrate consumption, and increased sensitivity to cold damage. The increased soil solution concentration of Al^{+3} associated with acidic deposition (whatever its form) can affect tree growth by direct toxic effects on roots or by inhibiting the uptake of other cations. The latter effects can occur at concentrations well below the threshold of Al^{+3} root toxicity.

NAPAP studies of agricultural groups indicate that, although yield reductions in some crop varieties are occurring on a regional scale (and appear to be associated with atmospheric pollution), acidic deposition is not responsible, and that direct injury by ambient levels is unlikely. Because sulfur and nitrogen are essential plant nutrients, acidic deposition may have beneficial effects on crop production. Unlike many forest soils, agricultural soils tend to be insensitive to significant soil chemistry changes that could occur from acidic deposition.

STRATOSPHERIC OZONE DEPLETION

In the late 1960s and early to mid-1970s, the specter of anthropogenically caused stratospheric O_3 depletion was raised by scientists. Initial concern was expressed about the building and subsequent extensive use of high-altitude supersonic transport (SST) planes in commercial aviation. Such aircraft would fly in the lower stratosphere (in order to reduce friction), where they would release substantial quantities of NO, which can catalytically destroy O_3. Scientific and public concern associated with potential O_3 depletion resulted in the cancellation of government support for the development of an SST commercial aviation capability in the U.S.

About this time, atmospheric scientists proposed that nuclear weapons testing in the atmosphere could cause stratospheric O_3 depletion. The high temperatures and explosive energy would produce large quantities of O_3-destroying NO which would be injected into the stratosphere.

After the cancellation of the U.S. SST program, stratospheric O_3 concerns abated somewhat. In 1974 the O_3 depletion specter rose anew when two University of California scientists, Rowland and Molina, proposed that chlorofluorocarbons (CFCs), widely used as aerosol propellants, refrigerants, and foam-blowing agents, could be broken down in the stratosphere releasing Cl atoms that could catalytically destroy O_3 molecules.

The Rowland-Molina theory of O_3 depletion received considerable scientific and public policy attention and was in a major way responsible for the banning of CFC use as aerosol propellants by the U.S., Canada, and Sweden in 1978. Most other developed countries that used CFCs industrially and commercially viewed this as an unverified theory and continued their use without any limits.

In the late 1970s, Rowland proposed that the O_3 layer may also be threatened by the increasing release of N_2O to the atmosphere from denitrification processes of ammonium and nitrate-based fertilizers. In this new theory, N_2O would migrate to the stratosphere where it would be photolytically destroyed to produce NO, an O_3-destroying chemical associated with previous concerns such as SSTs and atmospheric nuclear testing.

From 1985 to 1988, the potential for anthropogenically caused stratospheric O_3 depletion went from the realm of unverified theory to the reality that significant O_3 depletion was occurring over the South Pole in the austral (southern hemispheric) spring. The phenomenon called the "ozone hole" and other reported downward trends in column O_3 on a global scale have led many atmospheric scientists to conclude that emissions of CFCs and other halogenated compounds into the atmosphere, their transport into the stratosphere, and their subsequent photolytic destruction and production of Cl atoms is a problem of enormous gravity requiring regulatory initiatives to phase out and ban the use of long-lived halocarbons containing Cl and bromine (Br).

Ozone Layer Dynamics

The O_3 layer represents an atmospheric environment in which O_3 is continually formed and destroyed. The natural formation and destruction of O_3 in the stratosphere is initiated by high-intensity, short-wave UV radiation (<242 nm) received from the sun.

$$O_2 + h\nu \rightarrow O(^1D) + O(^1D) \qquad (4.9)$$

$$O(^1D) + O_2 + M \rightarrow O_3 + M \qquad (4.10)$$

$$O_3 + h\nu \rightarrow O_2 + O(^1D) \qquad (4.11)$$

$$O_3 + O(^1D) \rightarrow 2O_2 \qquad (4.12)$$

Ozone formation and destruction in the stratosphere involves reactions involving O_2 with an odd number of atoms, $O(^1D)$ and O_3, called odd oxygen. Ozone is initially formed by the UV photodestruction of free molecular O_2 into two separate atoms that react with molecular O_2 to produce O_3 (Equations 4.9 and 4.10). The subsequent absorption of UV light by O_3 results in its breakdown to diatomic and singlet atomic O (Equation 4.11). Reaction of O_3 with atomic oxygen would then produce two molecules of O_2 (Equation 4.12). This series of reactions converts solar energy into the energy of chemical bonds and ultimately to heat. This heat causes the rise in temperature with height that characterizes the stratosphere.

Ozone concentrations in the stratosphere are dynamic. They undergo significant changes over time in their vertical and geographic distribution. The solar sunspot cycle significantly affects stratospheric O_3 levels. Approximately 1.8% more O_3 is present in the stratosphere at the solar maximum in the sunspot cycle, as compared to the solar minimum. These changes are associated with increases in UV radiation at 200 nm that the atmosphere receives during maximum sunspot activity. Sunspot-induced variations may also be responsible for changes in the vertical distribution of O_3. In the mid-1980s, higher O_3 concentrations at lower altitudes were observed. This phenomenon caused a problem for flight crews in trans-Pacific commercial aircraft travel. Ozone levels in aircraft cabins were sufficiently high (~ 0.20 ppmv) to cause respiratory irritation in flight attendants who, because of the nature of their work (their physical exertion and higher respiratory demands), had a greater degree of exposure to O_3 than other cabin occupants.

Ozone variation in the stratosphere is also associated with the quasibiennial oscillation, a cycle of varying wind direction in the equatorial stratosphere that can influence the flow of O_3 to the north or south in different phases of the cycle. Atmospheric motions play a major role in determining the global distribution of O_3, and significant seasonal differences are observed.

Nitrogen Oxides

As previously described, nitrogen oxides (primarily NO) associated with human activities were perceived to be a threat to the O_3 layer because of emissions from SSTs and from nuclear weapons testing. Nitric oxide can react with O_3 to produce NO_2 and O_2. Nitrogen dioxide can subsequently react with atomic oxygen to regenerate NO. Thus, NO catalytically destroys O_3 because it is repeatedly regenerated:

$$NO + O_3 \rightarrow NO_2 + O_2 \qquad (4.13)$$

$$NO_2 + O(^1D) \rightarrow NO + O_2 \qquad (4.14)$$

Nitrogen oxides such as NO_2 and NO are emitted into the lower atmosphere in large quantities. Because of a variety of chemical reactions and sink processes, these emissions pose little threat to the O_3 layer. Ozone depletion associated with NO_x must therefore involve emissions directly into the stratosphere from extensive high-altitude aircraft traffic, nuclear explosions, or the destruction of stable nitrogen compounds such as N_2O. Nitrous oxide can migrate from the troposphere, where it can be oxidized by singlet atomic oxygen, $O(^1D)$, yielding two molecules of NO.

$$O(^1D) + N_2O \rightarrow 2NO \tag{4.15}$$

It can also be formed by reactions initiated by the absorption of cosmic rays and solar protons.

A major sink mechanism for NO in the stratosphere is reaction with HO_2 to produce NO_2, which is further oxidized to nitric acid (HNO_3):

$$HO_2 + NO \rightarrow OH + NO_2 \tag{4.16}$$

$$NO_2 + OH + M \rightarrow HNO_3 + M \tag{4.17}$$

Nitric acid may serve as a mechanism by which the NO_x is removed from the stratosphere. It may also serve as a reservoir for subsequent NO_x regeneration.

The reaction of NO_x with chlorine monoxide (ClO) scavenges Cl in the stratosphere. The resultant product can serve as a reservoir for Cl atoms (see following section).

Chlorofluorocarbons and Other Halogenated Hydrocarbons

Chlorofluorocarbons were suggested to be a potential threat to the O_3 layer in the early 1970s when scientists working independently at the National Center for Atmospheric Research, the University of Michigan, and Harvard University reported that free Cl atoms could catalytically destroy O_3. These research results were soon followed by the discovery by University of California, Irvine scientists Rowland and Molina that CFCs (which had no known tropospheric sink) could be photolytically destroyed by UV light in the wavelength spectrum <230 nm to yield free Cl and F atoms. Because such wavelengths do not penetrate to the lower atmosphere, CFCs must be transported to the middle and upper stratosphere (30 to 50 km) by atmospheric motions and molecular diffusion. This altitudinal range is above most of the atmospheric O_3 and O_2.

Chlorine atoms released in photolytic destruction of CFCs and other chlorinated HCs diffuse downward to the O_3 layer where they react with O_3 in a series of chain reactions:

$$Cl + O_3 \rightarrow ClO + O_2 \tag{4.18}$$

$$ClO + O(^1D) \rightarrow Cl + O_2 \tag{4.19}$$

On reaction with a Cl atom, O_3 is destroyed, producing O_2 and chlorine monoxide (ClO), which on reaction with atomic oxygen regenerates a Cl atom. This series of reactions can theoretically result in a destruction of 100,000 O_3 molecules before the removal of Cl from the stratosphere as hydrochloric acid (HCl). The lifetime of Cl from formation to atmospheric removal is estimated to be 1 to 2 years. Sink processes involving Cl include reactions with methane (CH_4) to yield HCl and ClO with NO_2 to yield chlorine nitrate ($ClONO_2$):

$$Cl + CH_4 \rightarrow HCl + CH_3 \qquad\qquad (4.20)$$

$$ClO + NO_2 + M \rightarrow ClONO_2 + M \qquad\qquad (4.21)$$

The CFCs are not the only halogenated compounds which have the potential to produce O_3-destroying halogens in the stratosphere. Compounds determined to pose a potential threat to the O_3 layer include carbon tetrachloride (CCl_4), methyl chloroform (CH_3CCl_3), methyl chloride (CH_3Cl), and methyl bromide (CH_3Br).

Carbon tetrachloride and CH_3CCl_3 are anthropogenic in origin. Both substances are eventually transported into the stratosphere where their photolytic destruction contributes to its Cl burden. Carbon tetrachloride has no tropospheric sink and thus has a relatively long atmospheric lifetime of 40 years. Methyl chloroform which does have a tropospheric sink (oxidation by OH) has an estimated atmospheric lifetime of less than 5 years. Methyl chloride, which is of natural origin, is reported to be second only to CFC-12 (CCl_2F_2) as a source of stratospheric Cl. It fortunately is relatively short-lived with an estimated atmospheric lifetime of 1.5 years.

Methyl bromide is produced naturally in the world's oceans and in biomass burning. It is also widely used as a soil fumigant, a major atmospheric source. Methyl bromide and other longer-lived bromocarbons such as $CBrClF_2$ and $CBrF_3$ used in fire extinguishing may also be major sources of stratospheric Br.

Bromine compounds can be photolytically destroyed in the stratosphere yielding Br atoms that react with O_3 in the same series of reactions reported for Cl:

$$Br + O_3 \rightarrow BrO + O_2 \qquad\qquad (4.22)$$

$$BrO + O(^1D) \rightarrow Br + O_2 \qquad\qquad (4.23)$$

Bromine is much more efficient on a per atom basis in destroying O_3 (by a factor of about 40) and may be responsible for up to 20% of Antarctic O_3 depletion and 5 to 10% of current global O_3 loss discussed in the following sections.

The Antarctic Ozone Hole

In 1985 the Rowland–Molina hypothesis of stratospheric O_3 depletion associated with CFCs went from the realm of an unverified theory to the stark reality of alarming fact. British scientists conducting ground-based measurements of total column O_3 above Halley Bay (76 degrees S latitude) were surprised by unusually large declines in O_3 levels associated with the beginning of the southern hemispheric spring. Because their observations had not been previously reported (measurements were also being made by satellites), they initially were skeptical of their results. After careful reevaluation of the accuracy of their instruments, they concluded that they, in fact, observed significant declines in O_3 levels. Examination of original NASA satellite data (which excluded extreme results in

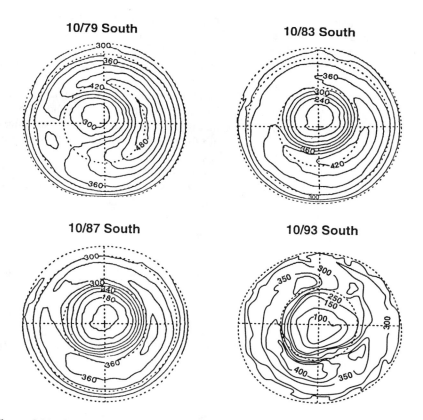

Figure 4.11 Changes in column O_3 levels (Dobson units) during the Antarctic spring from 1979 to 1992. (From NRC. 1989. National Academy Press, Washington, DC. With permission.)

data summaries) revealed the same phenomenon. NASA data indicated that the phenomenon began occurring in the mid-1970s. Satellite data revealed that the area affected by the Antarctic O_3 hole extended over several million square kilometers. Since its discovery, the Antarctic O_3 hole has been characterized by increasing geographic area affected, increasing duration, and the continuing decline in total column O_3 levels (reported in Dobson units; Figure 4.11). The region affected by the O_3 hole has increased from a few million square kilometers in 1979 to over 24 million km² in 1994, with a low value of 90 Dobson units with depletions of nearly 100% at altitudes of 15 to 20 km. The O_3 hole is a seasonal phenomenon, usually lasting 5 to 6 weeks or so at the beginning of the austral spring. Recent observations indicate that the O_3 hole has persisted into early December. Debris from the Mt. Pinatubo volcanic eruption in the Philippines in 1991 are believed to have contributed to its expansion and to historically low values in 1992 and 1993.

The Antarctic O_3 hole was a complete surprise to atmospheric scientists. Their models were based on gas-phase chemistry and focused on the stratosphere

over and near source regions where temperatures were more favorable for stratospheric chemistry. A number of theories were initially proposed to explain its
formation. Some scientists suggested that the unique polar circulation patterns
were responsible, whereas others theorized that it was being produced by chemical
reactions involving anthropogenic contaminants.

Studies conducted in 1986, 1987, and years thereafter have demonstrated the
presence of elevated concentrations of Cl and ClO in the region of the O_3 hole.
These studies provide strong evidence that the O_3 hole was caused by chemical
reactions involving Cl in conjunction with the unique meteorological conditions
in the Antarctic. Ozone destruction occurs as a consequence of complex heterogeneous chemical reactions involving polar stratospheric clouds (PSCs), HCl,
$ClONO_2$, and chlorine (Cl_2). Both HCl and $ClONO_2$ are produced by sink processes which remove Cl from O_3 destruction processes. Polar stratospheric clouds
at the beginning of the austral spring promote the conversion of stable inorganic
chlorine compounds such as HCl and $ClONO_2$ into photolytically active Cl_2.

$$HCl + ClONO_2 \left(\xrightarrow{\text{ice}} \right) HNO_{3(s)} + Cl_2 \qquad (4.24)$$

$$Cl_2 + hv \rightarrow 2Cl \qquad (4.25)$$

$$2[Cl + O_3 \rightarrow ClO + O_2] \qquad (4.26)$$

$$ClO + NO_2 \rightarrow ClONO_2 \qquad (4.27)$$

$$\text{Net}: \ HCl + 2O_3 + NO_2 \left(\xrightarrow{\text{ice}} \right) HNO_{3(s)} + 2O_2 + ClO \qquad (4.28)$$

The reaction of ClO with NO_2 to form $ClONO_2$ (Equation 4.27) slows down
O_3 destruction. Polar stratospheric clouds facilitate the conversion of NO_2 to
HNO_3 which remains mostly in the condensed phase. This reduction in NO_2
greatly increases the efficiency of O_3 destruction by Cl radicals. Reactions
described in Equations 4.24 to 4.27 are mediated by surface chemistry on both
Type 1 and Type II PSCs. The former are believed to consist primarily of nitric
acid trihydrate which condense on H_2SO_4/H_2O aerosols at temperatures 2 to 4°K
above the frost point of water ice (about 196°K); the latter consist of ice particles
which condense at temperatures just below the ice frost point (about 188°K).
Type II PSCs mediate the conversion of $ClONO_2$ to hypochlorous acid (HOCl)
which reacts with HCl to yield Cl_2, which is then photolytically converted to Cl
and ClO.

$$ClONO_2 + H_2O_{(s)} \rightarrow HOCl + HNO_3 \qquad (4.29)$$

$$HOCl + HCl \rightarrow Cl_2 + H_2O \qquad (4.30)$$

In Equations 4.24 and 4.29, nitrogen oxides are sequestered, thus significantly diminishing their role as Cl scavengers and enhancing O_3 depletion.

The actual O_3 loss is much greater than would be predicted from the above reactions. It appears that a large percentage of O_3 depletion takes place as a result of ClO dimer (ClOClO) formation in an alternate ClO reaction pathway:

$$ClO + ClO + M \rightarrow ClOClO + M \qquad (4.31)$$

$$ClOClO + hv \rightarrow Cl + ClOO \qquad (4.32)$$

$$ClOO + M \rightarrow Cl + O_2 + M \qquad (4.33)$$

$$2[Cl + O_3 \rightarrow ClO + O_2] \qquad (4.34)$$

As previously indicated, PSCs play a crucial role in the polar stratosphere in the winter and spring months. Surface-catalyzed reactions on PSC particles generate Cl compounds which photolyze to yield chlorine radicals (Cl and ClO). Polar stratospheric clouds can form only under very low temperature conditions. Both types play significant roles in O_3 destruction chemistry. Recent studies indicate that Cl radical precursors form readily not only on Type I and Type II PSCs but on liquid H_2SO_4 solutions and on solid H_2SO_4 hydrates.

Ozone Trends

Ozone concentrations in the vertical dimension are determined by measurements of total column O_3 using ground-based Dobson spectrophotometers and satellites using Total O_3 Mapping Spectrometers (TOMS). Data are reported in Dobson units. Ozone levels can also be measured as a function of altitude using a variety of instruments. Trends in O_3 levels over the Antarctic continent during austral spring can be seen in Figure 4.11.

There is increasing evidence that O_3 depletion may be occurring on a global scale. The Ozone Trends Panel of the National Academy of Science has reported significant decreases in total column O_3 over the northern middle latitudes (30 to 64° N latitude) with the greatest decrease occurring during the winter months. Significant decreases in total column O_3 levels have also been reported in southern hemisphere mid-latitudes.

Analysis of updated O_3 records through early 1991 indicate continued losses of total column O_3 in mid-latitudes. The wintertime maximum rate of O_3 decrease near 40° N latitude is slightly greater than 6% per decade with an average 12 month decrease per decade of 5%. Normalized trends for North America and Europe can be seen in Figure 4.12. Most losses have occurred in the lower stratosphere in the altitude range of 17 to 24 km. Altitudinal trends from 1970 to 1990 are shown in Figure 4.13 for measurements made over Payerne, Switzerland. These data as well as those from other stations show a significant increase in O_3 concentrations in the troposphere and a significant decrease in the lower

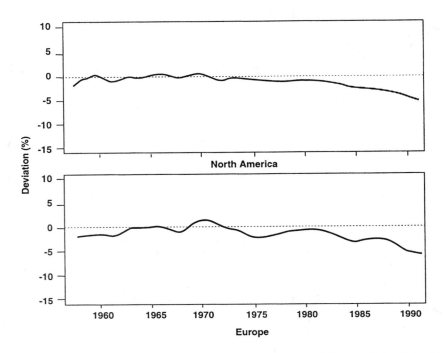

Figure 4.12 Trends in column O_3 levels over North America and Europe. (From Stolarski, R. et al. 1992. *Science* 256:342–349. With permission.)

stratosphere where most O_3 is located. Tropospheric O_3 increases as a function of altitude range from 5 to 20% per decade, with stratospheric decreases of 5 to 15% per decade.

In addition to O_3 decreases, stratospheric temperatures have been observed to decrease globally by $-1.7°C$ between altitudes 25 to 55 km from 1979 to 1987. These are consistent with those expected from decreases of stratospheric O_3.

Changes in Surface Ultraviolet Irradiance

The effect of O_3 depletion would be to increase the amount of UV reaching the earth's surface. A 1% decrease in column O_3 would be expected to increase the surface flux of ultraviolet-B (UV-B) by 2%. In an initial study of eight geographical locations in the U.S., UV-B irradiance appeared to decrease between 1974 and 1985. This paradoxical result has been suggested to be due to the shielding effect of tropospheric O_3 and aerosols.

Subsequent studies conducted in Canada have indicated that UV-B irradiance has been increasing most notably during the winter months (Figure 4.14). The intensity of UV (at 300 nm) measured near Toronto has increased by 35% per year in winter and 7% per year in summer from 1989 to 1993. The wavelength dependence of these levels indicates that they are consistent with a downward trend in column O_3 over Toronto, Canada.

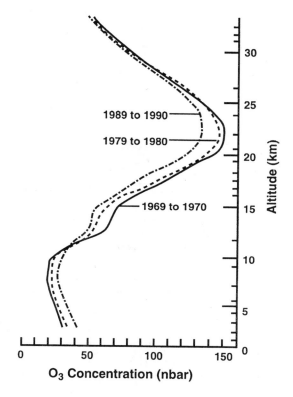

Figure 4.13 Altitudinal trends in O_3 levels over Payerne, Switzerland. (From Stolarski, R. et al. 1992. *Science* 256:342–349. With permission.)

Ozone Depletion Effects

The O_3 layer absorbs 99% of UV wavelengths less than 320 nm. Ozone absorption in the UV-B region (290 to 320 nm) is, however, less efficient than at shorter wavelengths, and relatively significant fluxes of both UV-B and UV-A (320 to 400 nm) reach the earth's surface. UV-B, on absorption by organisms, can cause significant adverse biological effects. As a result, it is often described as damaging UV radiation (DUV).

Significant short- and long-term environmental exposures to DUV may cause sunburn, cataracts, or skin cancer. There are three basic forms of skin cancer: basal cell carcinoma, squamous cell carcinoma, and melanoma. Carcinoma-type skin cancers have clearly been demonstrated to be associated with sunlight, with the greatest prevalence observed among lightly complected individuals living in so-called sunbelt latitudes in North America. Skin carcinomas, fortunately, are the least life-threatening of all forms of cancer, with a mortality of less than 5%. Because tumors must be surgically removed, skin cancers are disfiguring.

Melanoma tumors develop from the darkly pigmented melanin structures or moles that many Caucasians develop on their skin, particularly during midlife.

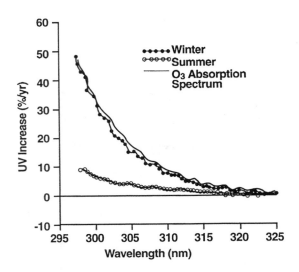

Figure 4.14 Summer and winter trends in UV-B radiation over Toronto, Canada from 1989 to 1993. (From Kerr, J.B. and C.T. McElroy. 1993. *Science* 262:1032–1034. With permission.)

Melanoma is a much more deadly form of skin cancer because the tumors metastasize rapidly, spreading malignant cells to other parts of the body through the bloodstream and lymphatic system. The mortality rate is approximately 20%. The incidence of melanoma in the U.S. has increased about sixfold since 1960.

No dose–response relationship between melanoma and exposure to sunlight and UV radiation appears to exist. Rather, melanoma appears to be associated with severe sunburn experienced early in life. Melanoma has also been linked to building environments with fluorescent lighting. Paradoxically, melanoma tumors apparently associated with UV radiation exposures from fluorescent lighting occur on nonexposed parts of the body.

Increases in UV-B radiation fluxes at the earth's surface as a result of stratospheric O_3 depletion by even a few percent would be expected to cause a significant increase in all forms of skin cancer among Caucasians and other lightly pigmented individuals. This would mean thousands of new cases of skin cancer in the U.S. each year.

One immediate concern associated with stratospheric O_3 depletion is the potential for significant adverse biological changes in phytoplankton and other parts of the food chain in the upper zones of ocean waters surrounding the Antarctic continent. A number of laboratory studies have shown that exposures to UV-B radiation can decrease the productivity of algae and adversely affect various forms of aquatic larvae and other organisms. There is also evidence to suggest that UV-B levels incident on ocean surfaces may adversely affect populations of marine organisms. In a recent study in the seas around Antarctica, the thinning of the O_3 layer was observed to be associated with an increase in the ratio of depth-dependent UV-B as a function of total irradiance and a UV-B related decrease in photosynthesis in several species of marine algae. Severe inhibition of photosynthesis and DNA damage in algal species has been observed in short-term studies, with no apparent reduction in algal growth over extended periods of time. Increased UV-B levels have the potential for adversely affecting phytoplankton, the base of the Antarctic marine food chain that includes the very abundant krill (the small, shrimplike animal that feeds on phytoplankton) and other higher-order species such as fish, birds, and sea mammals (e.g., seals and whales). How severely Antarctic life is or may be affected depends on the ability of individual organisms to tolerate increased UV irradiances and how quickly species can adapt (genetically) to increased UV levels.

Increased UV irradiance which may be occurring on a global scale has the potential for adversely affecting a large variety of life forms. Of note is the affect of UV on plants. Ultraviolet light is a well-known plant growth regulator. Under ambient conditions stem elongation is controlled directly by the level of UV exposure. Plants grown in full sunlight tend to be relatively short. Similar plants grown in partial to full shade tend to have elongated very "leggy" stems. Increases in environmental exposures to UV-B may result in reduced stem elongation in some crop species resulting in decreased productivity. The degree of this effect is species and, in many cases, variety dependent. For some crops yield reductions may be significant. The potential effect of increased UV on plants, particularly crops and forest systems, may be a major future concern as we must continue to provide for the needs of an ever-expanding world population.

A worldwide decline in many different amphibian species has been observed in the past several decades. Though there are likely to be many contributory causes, limited experimental studies conducted in the American Cascade Mountains indicate that a number of amphibian species are very sensitive to near-surface UV levels. A potential causal relationship among stratospheric O_3 depletion, increased near-surface UV radiation levels, and amphibian mortality has been suggested.

GLOBAL CLIMATE

Natural Variations

The atmosphere driven by the energy of the sun is in continual motion and change. This change or flux viewed in the context of hours, days, or weeks is called weather. Weather includes precipitation events, fluctuations in temperature, humidity and wind speed, and the degree of cloud cover or incoming solar radiation. These phenomena viewed within the context of geographical areas and extended periods of time (years, decades, centuries, millennia) describe climate. Climate is characterized by average temperatures, temperature extremes, precipitation, the quantity and periodicity of solar radiation, and other atmospheric phenomenon. Though it is a much broader concept, climate is often described as being average weather.

Weather, which is based on events in the short term, can be and often is quite variable. Because climate is based on averaged values, its variability is, of course, considerably less. Climate is subject to change, but the timeframe of such change may be relatively long. Evidence of small to large climatic changes may be found in historical accounts, various aspects of the geological record, and analytical results of measurements made on deep sea sediments, polar ice, and tree rings.

Climatic change occurs naturally. It may also occur as a result of human activity. Natural regional or global climatic changes may occur by a variety of geophysical mechanisms. These include (1) quasiperiodic ocean temperature changes, (2) increased sea ice formation associated with decreased Arctic Ocean salinity, (3) long-term self-fluctuation of the climatic system, (4) changes in the earth's orbital geometry, (5) variations in solar radiation, (6) volcanic dust, (7) changes in greenhouse gases such as CO_2, (8) changes in surface albedo, and (9) latitudinal changes associated with continental drift. Orbital changes and variations in solar radiation are discussed below to elucidate the effects of two major natural phenomena on the earth's climate. This will be followed by an extended discussion of the potential effect of anthropogenic activities on the earth's climatic system.

Glaciation and Orbital Variation

The formation of glaciers represents one of the most extreme cases of climatic change over what could be described as relatively short time periods. Based on geological evidence, it appears that the northern hemisphere has experienced six major glaciations in the past 1 to 1.5 million years. Glacial periods lasting thousands of years were characterized by the formation and movement of massive ice sheets several kilometers thick. Glaciations were followed by climatic warming and glacial retreat. The last North American glacier receded over 12,000 years ago.

A number of theories have been proposed to explain the development and recession of large-scale glaciers and subsequent warming. One of the most prominent of these was proposed by Milankovitch in the 19th century. According to Milankovitch, predictable changes in the earth's geometry can affect the distribution of solar radiation over the earth's surface. These solar radiation changes could then affect the development of major glaciations.

Cyclic changes in the earth's orbital geometry include changes in obliquity and precession. Obliquity or tilting of the earth's axis slowly varies from 22.1 to 24.5°, with a complete cycle every 41,000 years. Present obliquity is about 23.5°. In addition to changes in obliquity, the earth's axis precesses or changes in direction. This precession describes a circle among the stars, with the center of the circle being due north (Figure 4.15). This precession, or wobbling of the earth's axis, has two cycles of 19,000 and 23,000 years.

The obliquity of the earth's axis is responsible for seasonal changes in climate over much of the earth. As this obliquity changes from 22.1 to 24.5°, the contrasts between seasons may become greater. Precession also affects seasonal differences because it determines the point on the earth's elliptical orbit when summer and winter occur. Winters that occur when the earth is closest to the sun (perihelion) would be warmer than those that occur when the orbital distance is the greatest

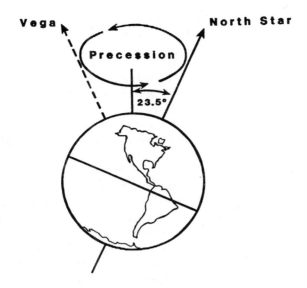

Figure 4.15 Precession of the earth among the stars.

(aphelion). Although these orbital changes do not affect the amount of solar radiation reaching the surface, they do influence the amount received at a given latitude or geographical location.

A third change in orbital geometry is called eccentricity, the elliptical nature of the earth's orbit around the sun. The 100,000-year orbital cycle changes the amount of sunlight reaching the earth by 0.1%.

Recent geological evidence (oxygen isotope and deep sea sediment studies) indicates global climatic cycles of approximately 23,000, 41,200, and 100,000 years corresponding to changes in the earth's orbital geometry. The geological record demonstrates that the earth's eccentricity and glacial cycles have been in step for the last 800,000 years. Because of the correspondence between the geological record, orbitally induced variations in solar radiation received at the earth's surface have been described as being the "pacemaker" of the ice ages.

Solar Radiation

Average global temperature and therefore global climate are in part determined by the amount of solar radiation reaching the earth's atmosphere and surface. Astronomers have theorized that, at least for long time periods, solar output and, hence, the quantity of solar radiation received by the earth, is essentially constant. The solar constant is the total solar radiation received on a flat surface oriented perpendicular to the sun external to the earth's atmosphere. Its value is 1.94 cal/cm²/sec. Satellite measurements have recently shown that solar radiation does vary and that the solar constant is not constant when viewed in the short term. Short-term changes in solar output are directly related to sunspot cycles. The peak value of solar irradiances occurs near the maximum of solar activity. The peak-to-peak variation in the sun's total irradiance during the sunspot cycle is approximately 0.08%.

Sunspots are dark, relatively cool regions on the sun's surface. Heat flow is inhibited from the region of a sunspot. Interestingly, the sun's total irradiance is greater rather than lesser during peak periods of sunspot activity. Apparently, there is a concurrently active source of solar emission that more than compensates for the decreased output from sunspot regions. Increased solar emissions are associated with faculae and plages. Both faculae and plages are bright, active regions where emissions of solar radiation are enhanced above those for the surrounding quieter regions of the sun. Plages make up an order of magnitude or more of the sun's surface and are longer-lived than sunspots.

Changes in the sun's luminosity associated with sunspots and other sun surface activity have been proposed as a cause of significant climatic change. The Little Ice Age, a period of global cooling observed worldwide from the 17th to the 19th century, has been associated with a drastic decline of sunspot activity from 1645 to 1715 (the Maunder Minimum). The Little Ice Age was accompanied by (1) the advance of European mountain glaciers, (2) lower average winter temperatures in central Britain, (3) the abandonment of cereal cultivation in Iceland and vineyards in England, and (4) the winter freezing of the North Sea and the canals of Venice. It is probable that global surface temperatures were >1°C less during that period, and that precipitation patterns were also changed.

Changes in the sun's luminosity associated with sunspot cycles have from time to time captured the interest of some climatologists in attempting to explain climatic change. Recent ice core and ocean temperature analysis appear to show significant correlations between temperature changes and the 11- and 22-year sunspot cycles. These correlations imply that changes in solar luminosity could have been responsible for as much as half of global warming over the past century.

Anthropogenic Effects

A variety of human activities have the potential for significantly affecting climate on a regional scale. Deforestation and overgrazing have contributed to increased desertification in the Middle East and North Africa. On a global scale there is increasing scientific and public policy concern that human activities such as anthropogenic emissions of greenhouse gases, aerosols of anthropogenic origin, and massive deforestation may result in major perturbations of the earth's climatic system. In the 1960s and 1970s, atmospheric scientists theorized that emissions of CO_2 to the atmosphere from the combustion of fossil fuels as well as the clearing of tropical forests would result in global warming by enhancing the greenhouse effect. Other scientists focusing on the potential effect of atmospheric aerosols produced from human activity theorized that climate would be changed in the direction of cooling. With an increased understanding of earth's climatic system and its history, most atmospheric scientists became convinced that anthropogenic emissions of greenhouse gases was a major cause for concern and that significant global warming was possible and even probable.

The Greenhouse Effect and Global Warming

The atmosphere is relatively transparent in the visible portion of the solar spectrum and is much less so in the infrared wavelengths emitted by the surface of the earth on absorption of sunlight. On a global average basis the earth's surface emits 390 W/m^2 and the atmosphere 237 W/m^2. The difference, 153 W/m^2, is retained in the lower atmosphere somewhat like the panes of a greenhouse retain heat. A heat balance, however, is maintained with emissions through and by the atmosphere balancing out solar energy absorbed by the earth's surface.

The lower atmosphere is warmed by the sorption of infrared energy and by reradiation which warms the earth's surface as well. Without this extra warmth, the surface would often be below freezing. The greenhouse effect increases temperatures of the lower atmosphere by 32 to 34°C, maintaining a life-sustaining average temperature of 15°C.

The greenhouse effect described above results from the absorption and reradiation of infrared energy by trace gases such as CO_2, H_2O, CH_4, N_2O, O_3, and more recently, man-made substances such as CFCs. The emissivity of the atmosphere varies across the infrared spectrum. As can be seen in Figure 1.3, much of infrared absorption is due to CO_2 and H_2O vapor. These gases, as well as O_3, absorb very little of the infrared energy emitted by the surface of the earth in the spectral region of 7 to 13 μm, the so-called atmospheric window. Eighty percent of the infrared energy emitted by the earth's surface in the 7 to 13 μm range is readily lost to space.

Concentrations of greenhouse gases (excluding H_2O vapor) which have been affected by human activities (for 1990 and the preindustrial period prior to 1800) are summarized in Table 4.1. The table also includes annual increases (based on 1990 levels), the radiative effectiveness of greenhouse gases relative to CO_2, and estimated contribution to global warming.

Table 4.1   Greenhouse Gases and Their Relationship to Global Warming

	CO_2	CH_4	CFC-11	CFC-12	N_2O	O_3[a]
Concentration						
Preindustrial	280 ppmv	0.8 ppmv	0	0	288 ppbv	11 ppbv
1990	353 ppmv	1.72 ppmv	280 pptv	489 pptv	310 ppbv	21 ppbv
Annual increase						
1990	1.8 ppmv (0.5%)	0.015 ppmv (0.9%)	9.5 pptv (4%)	17 pptv (4%)	0.8 ppbv (0.25%)	0.02 ppbv (1%)
Radiative effectiveness (relative to CO_2)	1	21	13,400	15,800	206	2000
Relative contribution to global warming (1990)[b]	55%	11%	(24%)		6%	

[a] Tropospheric O_3.
[b] Stratospheric water vapor 4%.

From Elsom, D.M. 1992. Blackwell Publishers, Oxford, UK. With permission.

Carbon dioxide concentrations have increased by more than 20% since 1800 and are expected to double sometime after the year 2040. The annual average increase is approximately 1.8 ppmv or 0.5% per year. Emissions of CO_2 are increasing by approximately 2.5% per year. Much of this increase is due to the combustion of fossil fuels. Land-use changes (where high carbon density ecosystems such as forests are replaced by lower carbon density agricultural or grazing land) have also had significant effects on CO_2 release to the atmosphere. Such land-use changes in the future are expected to contribute proportionally less to atmospheric CO_2 levels as compared to fossil fuels.

Most of the CO_2 removed from the atmosphere is absorbed by the upper 70 to 100 m of the earth's oceans. Forests appear to be an important sink for CO_2 as well. As a result, only about 40% of CO_2 emitted to the atmosphere remains there. The atmospheric lifetime for CO_2 is estimated to be 50 to 100 years.

As can be seen in Table 4.1, CO_2 is the greenhouse gas believed by atmospheric scientists to be primarily responsible for global warming trends in the past century and for projected future increases in surface air temperatures. The contribution of CO_2 to global warming trends is estimated to be 55%.

Although not shown in Figure 1.3, both CH_4 and N_2O are strong infrared absorbers. Though atmospheric concentrations are relatively low compared to CO_2 and H_2O vapor, their potential for enhancing the greenhouse effect is large because of significantly higher infrared absorption potentials compared to CO_2. Methane and N_2O absorb strongly in the atmospheric window with a relative (to CO_2) radiative effectiveness of 21 and 206, respectively. Their contributions to global warming trends are estimated to be approximately 11 and 6%.

Methane concentrations in the atmosphere have more than doubled from preindustrial times with an annual increase (1990) of 15 ppbv or 0.9% per year. Major emission sources of CH_4 appear to include anaerobic bacterial activity in swamps and rice paddies, the digestive systems of ruminant animals (particularly livestock) and termites, and landfill sites. Other sources include coal mines and leaking natural gas pipe lines. The latter was a particularly notable source of emissions in the countries of the former Soviet Union. Repairs to such gas lines are believed by some scientists to have been responsible for the anomalous absence of an increase in atmospheric CH_4 levels in 1994.

Nitrous oxide is produced naturally by biological processes in the soil. It is also emitted to the atmosphere in fossil fuel and biomass burning, soil disturbance, and from animal and human wastes. The increased use of nitrogen fertilizers may also be a significant source of atmospheric N_2O. Atmospheric N_2O levels appear to have increased by about 20 ppbv since preindustrial times with an annual average increase of 0.8 ppbv or 0.25%.

Chlorofluorocarbons are man-made and therefore new to the atmosphere. Though most environmental concern associated with increasing atmospheric CFC levels has focused on stratospheric O_3 depletion, CFCs are also strong thermal absorbers (not surprising given their widespread use as refrigerants). Though concentrations of the two most widely used CFCs (CFC-11 and CFC-12) are in the pptv range, the potential contribution of CFCs to global warming appears to be large. CFCs absorb strongly in the atmospheric window. On a molecule-to-molecule basis, they absorb infrared energy 12,000 to 16,000 times more effectively than CO_2. The estimated contribution of CFCs to global warming is approximately 25%. With the phaseout of CFCs that is currently occurring, annual increases in atmospheric concentrations should reach zero with subsequent declines beginning in the early part of the 21st century.

As can be seen in Table 4.1, tropospheric O_3 has a significant potential for absorbing thermal radiation with a radiative effectiveness relative to CO_2 of 2,000. Though O_3 levels in the lower atmosphere have approximately doubled, the effects of this increase are uncertain since they may have been mitigated by decreases in lower stratospheric O_3 levels in the past two decades.

Global Warming Trends and Projections

Potential global warming associated with the enhancement of the greenhouse effect by CO_2 and other trace gases associated with human activities has been a subject of intense scientific activity and public policy debate. Atmospheric models of enormous complexity have been developed by scientists to predict the impact of greenhouse gas enhancement on global and regional climate. These models are typically described as general circulation or global climatic models (GCMs). All GCMs developed and used to date are in general agreement in that they predict increases in global surface temperatures in the same range.

By grouping CO_2 and all other greenhouse gases, concentrations equivalent to a doubling of the CO_2 level are expected to occur in the year 2030. Such a doubling is predicted to cause surface air temperatures to rise by 2 to 5°C with about half of this (1 to 3°C) occurring by year 2030. Because of positive feedback mechanisms at high latitudes such as decreased albedo associated with the melting of snow and ice, thinner sea ice allowing greater flow of heat from the ocean, and increased evaporation enhancing the water vapor greenhouse effect, temperature increases in these regions are expected to be 2 to 3 times higher than in the tropics where enhanced convection may act as a negative feedback mechanism. Various GCMs have predicted mean winter surface air temperatures to increase by 1 to 3°C at low latitudes, 4 to 8°C in middle latitudes, and 6 to 10°C at polar and subpolar latitudes.

Most climatic models predict that global warming of approximately 1°C should have occurred in the past century. This is double the increase in surface air temperatures (0.5°C) actually observed. Changes in global surface air temperatures from 1880 to 1990 can be seen in Figure 4.16. In the 1980s and early 1990s, global surface air temperatures were higher than any time in the past century with eight of the warmest years occurring from 1980 to 1992. In addition, sea surface temperatures have increased by 0.8°C since 1900. Borehole studies also indicate a rise in ground-surface temperatures in the past century.

Global climatic models appear to overpredict the contribution of greenhouse gases relative to observed increases in global surface temperatures in the last century. Recent evidence suggests that these differences are due to atmospheric aerosols over major industrialized areas of the world. Aerosols appear to play an important climatic role in that they reflect solar radiation back to space. It has

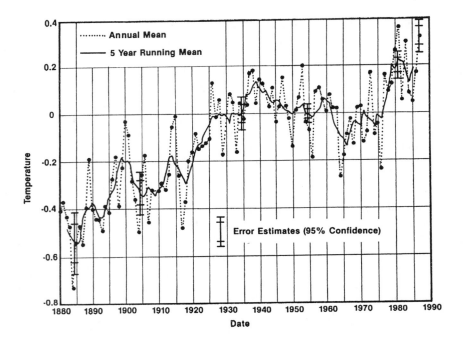

Figure 4.16 Changes in average global surface temperatures. (From Hansen, J. and Leb-edoff, S. 1988. *Geophys. Res. Letter* 15: 323–326. With permission.)

been estimated that natural and anthropogenic aerosols over the eastern U.S. have the potential for reducing the amount of sunlight reaching the ground by as much as 7%. The reflection of solar radiation back to space by sulfate aerosols is estimated to have reduced by one-half the warming effect associated with increases in greenhouse gases. In addition, aerosols may have an indirect effect by increasing the albedo of clouds. Combined direct and indirect effects of sulfate aerosols could cancel the potential warming effects associated with atmospheric increases in greenhouse gases.

Evidence to support a significant role of sulfate aerosols in counteracting global warming include the observed short-term cooling associated with the eruption of Mt. Pinatubo in 1991. Mt. Pinatubo emitted massive quantities of SO_2 into the atmosphere producing a long-lasting stratospheric haze of sulfuric acid and nocturnal warming patterns. In the U.S., the former Soviet Union, and China, annual maximum average temperatures remained unchanged or rose only slightly in the past 10 years. Minimum temperatures, however, have increased significantly. This increase in nighttime temperatures is consistent with the theory that aerosols are moderating global warming by reflecting sunlight. Even though atmospheric aerosols appear to be slowing the effects of greenhouse gases, in the long run they are expected to lose out to enhanced greenhouse warming because aerosol concentrations are not expected to increase significantly in the future whereas greenhouse gases are. Additionally, sulfate aerosols show significant geographic differences. As such, greenhouse warming would be expected to be more pronounced in the Southern Hemisphere and in some of the "cleaner" areas of the Northern Hemisphere. Such a phenomenon could alter weather patterns.

Global Warming Uncertainties

The recently confirmed role of sulfate aerosol associated with regional haze over developed areas of the world in producing a cooling effect points out how difficult it is to predict what the effects of anthropogenic activities may be on something as complex as the earth's climatic system. As a result, a fair amount of uncertainty exists in predicted climatic changes at global and regional levels. One major uncertainty in GCMs is cloud behavior. Global warming would be expected to increase cloud formation as well as cloud distribution and characteristics. By increasing cloud cover, more sunlight would be reflected back to space. An increase of the earth's albedo of 0.5% would be expected to halve the greenhouse warming associated with a doubling of CO_2. If cirrus clouds were to increase in abundance, they would be expected to absorb heat, thus warming the atmosphere. If low-lying clouds produced over the ocean were to increase, more sunlight would be reflected back to space causing a cooling effect. Differences in the nature of cloud cover would affect the predicted increase in global warming associated with a doubling of CO_2 (or its equivalent) by as much as 3°C in some GCM projections.

Other uncertainties are associated with the various feedback mechanisms among land, oceans, the atmosphere, and the earth's climatic system. Global warming would, for example, increase evaporation resulting in increased H_2O vapor concentrations. This would be expected to enhance the greenhouse effect. Global warming would be expected to melt sea ice and snow cover allowing underlying surfaces to absorb more solar radiation, thus decreasing the earth's albedo and increasing evaporation. By increasing the temperature of the world's oceans, the solubility of CO_2 would be decreased. As the upper layers of the world's oceans become saturated with CO_2, less CO_2 would be removed from the atmosphere. In addition, as global temperatures rise there is a potential for large amounts of CH_4 to be released from frozen tundra in the Northern Hemisphere.

A variety of both positive and negative feedback mechanisms exist which can either enhance or mitigate global warming. Most feedback mechanisms that have been proposed would enhance warming. Most of these feedback mechanisms are not included in the GCMs being used today.

Effects of Global Warming

The warming of the earth's atmosphere resulting from increased atmospheric CO_2 and other trace gases is expected to cause profound regional and global climatic changes. An increase of global temperatures of several degrees Celsius would be expected to cause a sea level rise of 0.5 to 1.5 m in the next 50 to 100 years. This rise would endanger human settlements in low-lying coastal areas in many parts of the world. It would also result in the destruction of estuaries and coastal wetlands. Other effects would include changes in regional temperature and rainfall patterns. Substantial changes in tropical rainfall have been predicted. Some GCMs also predict that temperatures would increase and rainfall would decrease simultaneously in the central plains of the U.S., with significant soil moisture decreases. Global warming could result in an expansion of the earth's arid zones and increase desertification. Taken together, projected climatic changes would be expected to have major impacts on agriculture, forests, land use, and biological systems, as well as on humans themselves.

Climate Changes and Stratospheric Gases

Global climatic changes may also occur from changes associated with O_3 depletion and other sources of stratospheric contamination. The effect of O_3 decreases in the stratosphere is to increase the amount of UV radiation that reaches the ground, resulting in stratospheric cooling. Increased stratospheric CO_2 levels would also contribute to a strong cooling trend in the stratosphere. Several studies have shown that the lower stratosphere has cooled since 1965. It is unknown as to what the effect of stratospheric cooling would be on the earth's climate. However, coupling between climate phenomena in the stratosphere and troposphere has been reported and widely accepted by atmospheric scientists.

As previously discussed, CH_4 is a major source of stratospheric H_2O. Stratospheric H_2O can enhance the greenhouse effect. A doubling of stratospheric H_2O vapor would have the same global warming effect as the 70-ppmv rise in CO_2 levels that has occurred since 1880. Stratospheric H_2O levels are increasing as a result of increased CH_4 levels in the atmosphere.

READINGS

1. Adams, D.D. and W.P. Page (Eds.). 1985. *Acidic Deposition — Environmental, Economic and Policy Issues*. Plenum Publishers, New York.
2. Bach, W. 1971. "Atmospheric Turbidity and Air Pollution in Greater Cincinnati." *Geogr. Rev.* 61:573–594.
3. Beaty, C.B. 1978. "The Causes of Glaciation." *Amer. Sci.* 66:452–459.
4. Bridgman, H.A. 1994. *Global Air Pollution — Problems in the 1990s*. John Wiley & Sons, New York.
5. Brown, C.A., P. Hamill, and J.C. Wilson. 1995. "Particle Formation in the Upper Tropical Troposphere: A Source of Nuclei for the Stratosphere." *Science* 270:1650–1653.
6. Chandler, T.J. 1962. "London's Urban Climate." *Geophys. J.* 128:279–302.
7. Charlson, R.J., S.E. Schwartz, and S.E. Hales. 1992. "Climate Forcing by Anthropogenic Aerosols." *Science* 255:423–430.
8. Deming, D. 1995. "Climatic Warming in North America: Analysis of Borehole Temperatures." *Science* 268:1576–1577.
9. Elsom, D.M. 1992. *Atmospheric Pollution — A Global Problem*. 2nd ed. Blackwell Publishers, Oxford, UK.
10. Gates, D.M. 1993. *Climate Change and Its Biological Consequences*. Sinuer Associates, Sunderland, MA.

11. Hansen, J. and S. Lebedeff. 1988. "Global Surface Air Temperatures: Update Through 1987." *Geophys. Res. Letter* 15:323–326.

12. Hidy, G.M., D.A. Hansen, R.C. Henry, K. Ganesan, and J. Collins. 1984. "Trends in Historical Acid Precursor Emissions and Their Airborne and Precipitation Products." *JAPCA* 34:333–354.

13. Hidy, G.M. 1984. "Source-Receptor Relationships for Acid Deposition: Pure and Simple? A Critical Review." *JAPCA* 34:518–531.

14. Husar, R.B. and D.E. Patterson. 1987. *Haze Climate of the United States.* NTIS PB 87- 141057/AS.

15. Jennings, S.G. (Ed.). 1993. *Aerosol Effects on Climate.* University of Arizona Press, Tucson, AZ.

16. Kahn, J.R. 1988. "The Surface Temperature of the Sun and Changes in the Solar Constant." *Science* 242:908–910.

17. Kerr, R.A. 1979. "Global Pollution: Is the Arctic Haze Actually Industrial Smog?" *Science* 205:290–293.

18. Kerr, J.B. and C.T. McElroy. 1993. "Evidence of Large Upward Trends of Ultraviolet-B Radiation Linked to Ozone Depletion." *Science* 262:1032–1034.

19. Kiehl, J.T. and B.B. Briegleb. 1993. "The Relative Roles of Sulfate Aerosols and Greenhouse Gases in Climate Forcing." *Science* 260:311–314.

20. Kraljic, M.A. 1992. *The Greenhouse Effect.* H.W. Wilson Co., New York.

21. Logan, Z. 1983. "Nitrogen Oxides in the Troposphere — Global and Regional Budgets." *J. Geophys. Res.* 88:10785–10802.

22. Lucier, A.A. and S.G. Haines (Eds.) 1990. *Mechanisms of Forest Response to Acidic Deposition.* Springer-Verlag, New York.

23. McElroy, M.B. and R.J. Salawitch. 1989. "Changing Composition of the Global Stratosphere." *Science* 243:763–770.

24. National Acid Precipitation Assessment Program. 1987. *Interim Assessment: The Causes and Effects of Acidic Deposition.* U.S. Government Printing Office, Washington, DC.

25. National Research Council. 1984. *Global Tropospheric Chemistry — A Plan for Action.* National Academy Press, Washington, DC.

26. National Research Council. 1988. *Acidic Deposition: Long-Term Trends.* National Academy Press, Washington, DC.

27. National Research Council. 1989. *Ozone Depletion, Greenhouse Gases and Climate Change.* Proceedings of a joint symposium by the Board of Atmospheric Sciences and Climate and the Committee on Global Change. National Academy Press, Washington, DC.

28. National Research Council. 1991. *Rethinking the Ozone Problem in Urban and Regional Air Pollution.* National Academy Press, Washington, DC.

29. Office of Technology Assessment. 1984. *Acid Rain and Transported Air Pollutants: Implications for Public Policy.* OTA-0-205.

30. Peters, R.L. and T.E. Lovejoy. 1992. *Global Warming and Biological Diversity.* Yale University Press, New Haven, CT.

31. Ramanthan, V. 1988. "The Greenhouse Theory of Climatic Change: A Test by an Inadvertent Global Experiment." *Science* 240:293–299.

32. Rodriguez, J.M. 1993. "Probing Stratospheric Ozone." *Science* 261:1128–1129.

33. Rowland, F.S. 1989. "Chlorofluorocarbons and the Depletion of Stratospheric Ozone." *Amer. Sci.* 77:36–45.

34. Saitoh, T.S., T. Shimade, and H. Hoshi. 1996. "Modeling and Simulation of the Tokyo Urban Heat Island." *Atmos. Environ.* 30:3431–3442.

35. Schneider, S.H. 1989. "The Greenhouse Effect: Science and Policy." *Science* 243:771–780.

36. Schneider, S.H. 1994. "Detecting Climate Change Signals: Are There Any "Fingerprints"?" *Science* 263:341–347.

37. Schwartz, S.E. 1989. "Acidic Deposition: Unraveling a Regional Phenomenon." *Science* 243:753–762.

38. Silver, C.S. 1990. *One Earth, One Future: Our Changing Global Environment.* National Academy Press, Washington, DC.

39. Smith, R.C., B.B. Prezelin, K.S. Baker, R.R. Bidgare, N.P. Boucher, T. Coley, D. Karentz, S. MacIntire, H.A. Matlick, D. Menzies, M. Ondiusek, Z. Wan, and K.J. Waters. 1992. "Ozone Depletion: Ultraviolet Radiation and Phytoplankton Biology in Antarctic Waters." *Science* 255:952–958.

40. Stolarski, R., R. Bojkov, L. Bishop, C. Zerefos, J. Staehlin, and J. Zawodny. 1992. "Measured Trends in Stratospheric Ozone." *Science* 256:342–346.

41. Thompson, D.J. 1995. "The Seasons, Global Temperature, and Precession." *Science* 268:59–68.

42. Tolbert, M.A. 1994. "Sulfate Aerosols and Polar Stratospheric Cloud Formation." *Science* 264:527–528.

43. USEPA. 1979. *Research Summary — Acid Rain.* EPA/600/8-79-028.

44. USEPA. 1979. *Protecting Visibility. An EPA Report to Congress.* EPA/450/5-79-008.

45. USEPA. 1982. *Air Quality Criteria for Particulate Matter and Sulfur Oxides.* EPA/600/8- 82-029a-c.

46. USEPA 1995. *Air Quality Criteria for Particulate Matter. Vol. 1.* EPA/600/AP-95/001a.

47. Van de Kamp. 1992. *Health Effects of Global Warming.* Dept. of Health and Human Services, Bethesda, MD.

48. Warneck, P. 1988. *Chemistry of the Natural Atmosphere.* Academic Press, New York.

49. Wuebbles, D.J., K.E. Grant, P.S. Cornell, and J.E. Penner. 1989. "The Role of Atmospheric Chemistry in Climate Change." *JAPCA* 39:22–28.

QUESTIONS

1. How do the concepts of acid rain, acid precipitation, and acidic deposition differ?

2. What is the effect of the following factors on light scattering by particles: (1) particle size, (2) relative humidity, and (3) chemical composition?

3. How do particle size and number affect the probability of precipitation?

4. How does acidic deposition affect the ecology of freshwater lakes?

5. Assuming that the acidity of precipitation has increased in the past two decades, how can one explain this in light of the fact that emission levels of SO_2 have decreased or remained static since 1970?

6. What is the potential effect of the following on global surface temperatures: (1) increased atmospheric turbidity and (2) decreased atmospheric emissivity?

7. Describe four ways in which human activities could result in stratospheric ozone depletion.

8. What evidence, if any, is there to support a hypothesis that anthropogenic emissions of CO_2 have, or will have, a significant effect on global climate?

9. What was the most likely cause of large-scale glaciation in the Northern Hemisphere?

10. A belt of sulfuric acid extends around the earth with minimum concentrations at 18 km altitude. What is its origin? What climatic significance, if any, does it have?

11. How are visibility reduction and atmospheric turbidity related? How do these two concepts differ?

12. Specifically, how do atmospheric particles reduce our ability to see an object clearly?

13. What are the unique factors responsible for the Antarctic O_3 hole?

14. What are the relative relationships among CO_2, CFCs, CH_4, and N_2O and their potential effects on global warming?

15. What evidence is there to indicate that stratospheric O_3 depletion is occurring outside of the Antarctic?

16. What is heterogeneous-phase chemistry? What is its role in stratospheric O_3 depletion?

17. What are some of the environmental and public health consequences of strato-spheric O_3 depletion?

18. What is the relationship between greenhouse gas increases and haze aerosol over industrialized areas of the world?

19. How does acidic deposition affect soil chemistry?

20. What effect, if any, does acidic deposition have on forests and crops?

21. What is the significance of HCl and $ClONO_2$ in the stratosphere?

22. What are potential environmental effects associated with global warming?

5

AIR POLLUTION DISASTERS

Deaths and severe acute illness reported for disastrous episodes of high pollutant concentrations in the Meuse Valley, Belgium in 1931, Donora, PA, in 1948, and London, England in 1952 and 1956, provide stark evidence that ambient air pollution can seriously endanger public health. In each case persistent (3 to 6 days) thermal inversions combined with significant industrial, and in the case of London, domestic emissions were responsible for causing very high pollutant concentrations and human exposures.

There were 60 deaths reported in the Meuse Valley disaster, 20 in Donora, and approximately 4000 in London in 1952. In each case, deaths occurred among individuals who had existing respiratory and/or cardiovascular disease. In London, many of the 4000 deaths were caused by pneumonia. Retrospective studies of other London fogs have indicated that excess mortality was associated with London fogs as early as 1873. In each of these disasters, many individuals reported becoming ill. Acute illness symptoms were characterized by cough, shortness of breath, chest pain, and eye and nose irritation. People of all ages were affected. In the Donora area, more than half of the 14,000 residents reported being ill as a result of exposure to atmospheric pollutants.

From these disasters and other episodes, it is evident that exposures to elevated levels of ambient pollutants can cause acute illness and even death. Fortunately, such exposure conditions are unlikely to occur in the U.S. and other countries of the developed world where significant pollution control efforts have been in place over the past three decades. They are more likely to occur in rapidly industrializing countries such as China, Mexico, and countries of eastern Europe.

HEALTH CONCERNS ASSOCIATED WITH "NORMAL" EXPOSURES

Ambient pollution exposures which result in severe community-wide illness and death are relatively rare. Pollutant exposures in our cities and towns vary considerably, both in types of pollutants present and prevailing concentrations. These exposures may reflect the nature and extent of industrial activities, dispersion characteristics of the surrounding environment, motor vehicle emissions, atmospheric chemistry and, in some mountain valleys, the use of wood heating.

Exposures to the normal range of pollutants or pollutant concentrations in our cities and towns may result in a variety of health effects. These may include acute but transient symptoms such as eye, nose, and throat irritation and, in some sensitive individuals, asthmatic attacks. Acute effects such as asthmatic attacks are of regulatory concern. However, the primary focus of most regulatory activity in North America and other developed countries is to protect exposed or potentially exposed populations from the chronic effects of relatively low level exposures. These may include respiratory and cardiovascular disease, neurotoxic effects, and cancer.

CAUSE–EFFECT RELATIONSHIPS

A good understanding of the health effects or potential health effects of pollutants at levels typical of our cities and towns is essential to all health-based regulatory activities. Ideally, all pollution control programs should have scientific information sufficient to establish a cause–effect relationship between pollutant exposures and illness symptoms among those exposed, and to determine an adequate margin of safety. Because humans experience a variety of exposures and other stresses, establishing definitive cause–effect relationships is very difficult.

A strong relationship between individual or a combination of atmospheric pollutants and health effects is suggested when there is a convergence of evidence from epidemiological and toxicological studies, and studies of those who have been occupationally exposed. Scientific information from such studies is of particular importance in addressing chronic exposures of humans to ambient pollutants.

Epidemiological Studies

Epidemiological studies are conducted to determine potential relationships between a variety of environmental factors and human disease. They are characterized by the application of statistical techniques to data collected on the health status of a population or populations of individuals and a variety of exposures and potential confounding variables. Such studies have the potential for providing evidence of causal relationships between pollutant exposures and reported health effects. In general, epidemiological studies become more important as the risk attributable to atmospheric pollutants becomes smaller and the duration of exposure required to produce effects becomes longer. They are useful in determining potential associations between pollutant exposures and the development of chronic disease and in identifying potential relationships between acute effects and pollutant exposures. These may include pulmonary function changes, asthmatic attacks, and increased mortality, particularly under episode conditions of elevated pollutant levels.

Epidemiological studies vary in design. They may be cross-sectional (assess the relationship between pollutant exposures and health effects over a cross-section of a population), longitudinal (assess exposures and health effects over a period of time), or case-control (exposed and non- or less-exposed populations are compared). They may be prospective (data to be collected) or retrospective (evaluation of existing data). Study designs differ in their ability to identify potential causal relationships and statistical confidence in results obtained.

The power of an epidemiological study to show a causal relationship when one indeed exists depends on the strength of the study design, the availability of good exposure and health effects data, and an adequately sized study population. It also requires careful evaluation of a number of potential confounding variables or factors.

Confounding Factors. Epidemiological assessment of the potential relationship between exposures to air pollutants and observed parameters of human health may be confounded by a variety of coexisting factors. These may include the presence of existing disease, age, individual sensitivity, and meteorological conditions. They may also include sex, race, socioeconomic status, tobacco smoking, lifestyle, and occupation. Assessment of potential effects may also be confounded by interaction between (1) two or more pollutants, (2) pollutants and meteorological variables such as temperature and relative humidity, (3) pollutants and other exposures (occupational, smoking, and/or indoor air pollution), and (4) pollutants and infectious disease. Such interactions may explain differences observed between epidemiological and toxicological studies.

Interaction Effects

Interactions Between Pollutants. Because community air is a complex mixture of gases and particulate matter (PM), it is likely that some of these pollutants may interact to modify the physiological effects of others. This may occur in several ways. For example, one pollutant may affect the site of deposition of another. This is apparently the case for gases such as sulfur dioxide (SO_2). Because of its solubility in watery fluids, SO_2 would normally be removed in the upper respiratory system. The absorption/adsorption of SO_2 on particles, however, allows the transport of SO_2 deep into the pulmonary system, where SO_2 toxicity may occur. Thus, particles may potentiate the effect of SO_2. Also, harmful aerosols such as sulfuric acid can be produced by the interaction of gaseous pollutants in the warm moist environment of the lung.

Interaction among pollutants may result in effects that may be additive, synergistic, or antagonistic. Additive effects occur when the exposure to several pollutants produces an effect equal to the sum of the effects of pollutants acting alone. Synergistic effects are those in which the total toxic response to two or more pollutants is greater than the sum of the effects of each acting alone. The occurrence of synergistic effects poses a significant dilemma for regulators because most control efforts to date have been designed to control individual pollutants rather than pollutant combinations. Epidemiological and toxicological information on synergistic effects is limited. Antagonistic effects occur when one pollutant apparently weakens the effect of another, reducing its harmful consequences. Antagonistic effects have not heretofore been demonstrated for community air pollutants.

Interaction Between Pollutants and Meteorological Factors. Most major air pollution episodes have been associated with meteorological extremes which have confounded the assessment of the causal role of ambient air pollution in the genesis of mortality and morbidity. During the many London fogs and disasters in the Meuse Valley and Donora, the weather was cold and damp. The independent effect of cold temperatures on mortality has been confirmed many times. Indeed, studies in the U.K. have indicated that mortality is more closely related to cold and humidity than to fog frequency. Although it is difficult to distinguish the respective contribution of air pollution and low temperatures to mortality, epidemiological evidence indicates that air pollution health effects are exaggerated in cold, damp weather.

Meteorological variables may contribute to air pollution-induced health effects in two distinct ways. Specifically, they may weaken natural defense mechanisms against air pollutants. Cold, damp days, for example, are known to decrease pulmonary resistance. Upper respiratory infections are also associated with cold weather. Meteorological variables such as temperature and relative humidity may also interact with pollutants to increase their toxic effect. As an example, studies have shown that high relative humidities promote the formation of acid aerosols and influence droplet size.

Although most major air pollution episodes have been associated with cold, damp weather, photochemical oxidant episodes in Los Angeles occur during very hot weather. Studies in Los Angeles have noted a marked increase in mortality during periods of elevated photochemical oxidant levels. Most investigators have, however, concluded that observed excess mortality was due to heat rather than to air pollution. Heat waves, in general, are one of the greatest stresses on the elderly and infirm, and usually result in excess mortality. There is a possibility, however, that a synergism exists between heat and high photochemical oxidant levels. Laboratory studies have shown, for example, that elevated temperatures enhance the toxicity of some gases. For O_3 and peroxyacyl nitrate (PAN), increasing temperature results in increased mortality in laboratory-exposed animals.

Interaction Between Ambient Air Pollutants and Other Pollutant Exposures. In addition to exposures received in the ambient environment, many individuals may be exposed to high pollutant levels from cigarette smoking in their occupational and/or domestic environments. Such exposures may increase the health risk of exposure to ambient air pollutants. Indeed, cigarette smokers and those occupationally exposed have been identified as groups at special risk.

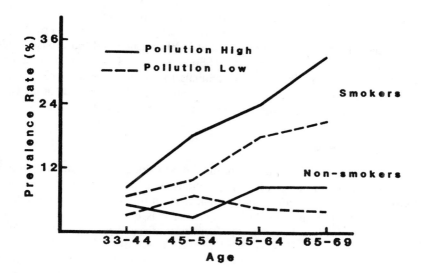

Figure 5.1 Prevalence of chronic bronchitis associated with tobacco smoking and exposure to ambient air pollution. (From Lambert, P.M. and D.D. Reid. 1970. *Lancet* 5:853–857. With permission.)

The vulnerability of cigarette smokers to community air pollution may be due to several factors. First, cigarette smoking may initiate respiratory and cardiovascular disease. Individuals with such disease, no matter what the cause, may be more sensitive to community air pollution exposures. Second, in the absence of such diseases, smoking may significantly augment the effects of community air pollution exposures (Figure 5.1). Studies indicate that when both factors are present, effects are more than additive. These effects may be due to the impairment of respiratory clearance mechanisms. For example, studies have shown that dust clearance in smokers is approximately five times slower than in nonsmokers.

Many individuals are exposed to relatively high levels of gaseous and/or particulate pollutants in their place of employment. These exposures are usually much greater than those experienced in community air. Because of occupational exposures, some individuals may be at higher risk to ambient exposures. This increased risk may be due to occupationally induced respiratory and/or cardiovascular disease, and/or impaired defense mechanisms.

The assessment of the health effects of community air pollutants may be confounded by the fact that individuals spend more time indoors than they do in the ambient environment. On the average, Americans only spend 2 hours/day outdoors. Therefore, their ambient pollution exposure may be quite different than, let's say, a policeman's or construction worker's. An expanded discussion of indoor air pollution and its health implications can be found in Chapter 11.

Interaction Between Pollutants and Infectious Disease. Some epidemiological studies have suggested that air pollutants increase the incidence and/or severity of infectious diseases, such as influenza and pneumonia. For example, in the London fog of 1952, many of the reported excess deaths were due to pneumonia. Apparently, the high pollutant levels predisposed those individuals and others to pneumonia infections. Laboratory studies indicate that combined influenza infection and exposures of animals to O_3 increases influenza-related mortality. The increased mortality is dose-dependent. Exposures of animals to NO_2 and simultaneous pneumonia infection have shown similar results.

Exposure Assessment

Most epidemiological studies are limited by inadequate information on pollutants present and levels to which populations are exposed. In general, ambient exposures are heterogeneous, poorly characterized mixtures of gases and particles. In quantifying an individual's exposure, it is desirable to have as much information as possible on past and present exposures to specific pollutants. Even in the best designed studies, such information is not available. Because of the paucity of exposure data, many epidemiological studies have used the place of residence as an index of exposure severity. The place of residence of an exposed population may, however, introduce a bias because it may be associated with peculiar ethnic or cultural traits, living standards, occupations, and exposure to infectious agents. In addition, the use of residence as an indicator of exposure may be complicated by the problem of mobility; that is, the duration of time that an individual has lived in the area.

Population Susceptibility

Epidemiological assessment of air pollution-related health effects may also be complicated by variations in population susceptibility. Such assessments can, however, be simplified by the identification of populations at risk. These may include the aged, the very young, those with existing respiratory and/or cardiovascular disease, those occupationally exposed, and cigarette smokers. These populations may be more sensitive to normally encountered pollutant levels than the average healthy individual. Epidemiological studies of sensitive populations are more likely to identify an air pollution–health effects association when one indeed exists. Our present air-quality standards (Chapter 8) are designed to protect populations at special risk.

Despite confounding factors and limitations described, many epidemiological studies have been able to provide valuable insight into defining the probability that air pollution exposures contribute to the incidence or prevalence of specific community health problems. A strong association or a high probability of disease causation can be inferred from such studies when: (1) there are a number of different populations in which a similar association is observed, including different kinds of people, locations, climate, and times of year; (2) the incidence and/or severity of the health effect increases with increasing exposure and, conversely, decreases with decreasing exposure; and (3) a plausible biological mechanism can be hypothesized for the observed association.

The value of epidemiological studies is increased when they are used in conjunction with controlled biological (toxicological) studies on humans or animals. Epidemiological studies are important for identifying possible associations which can be tested under controlled laboratory conditions. In addition, they can be used to evaluate human health risks suggested by laboratory exposures.

Toxicological Studies

Toxicological or controlled biological studies may be conducted on humans or animals to determine the functional, structural, and biochemical effects of toxic substances. Historically, toxicological studies have been conducted to determine the toxicity of a substance when administered in varying doses, with the dose being a function of the concentration of the substance and the duration of exposure. Significant differences in toxic response occur when the same quantity of toxic material is administered over different exposure periods. Acute responses result when an organism is subjected to a high dose, that is, a high concentration of a substance over a relatively short period of time. The response to the exposure is sudden, severe, and usually lasts for a brief period. If the dose is sufficiently high, death may result. Indeed, the term toxic implies that a substance has the potential for causing death. Lower doses, that is, lower concentrations over longer periods, generally do not directly cause mortality. As the dose decreases, the response is generally less severe, may take longer to develop, and may be more persistent. In chronic exposures, adverse effects may take years, if not decades, to develop.

In traditional toxicology, animals (and sometimes humans) are exposed to varied doses of toxic substances to determine their ability to cause physiological and pathological changes and to develop dose–response relationships. In addition to the evaluation of such physiological and pathological changes, the scope of toxicology in recent years has been expanded to include studies of carcinogenesis (cancer induction), teratogenesis (induction of birth defects), mutagenesis (production of mutations), and gametotoxicity (damage to sex cells).

One major advantage of toxicological studies is that the investigator can control many variables that often confound epidemiological studies. Controllable variables in animal studies include age, sex, environmental conditions such as temperature and relative humidity, genetic constitution, nutrition, and pollutant dose. By controlling dose (i.e., concentration of the pollutant as a function of exposure duration), dose–response relationships can be evaluated. Human exposures are also controllable, but to a more limited degree. In addition to those variables described for animals, factors peculiar to humans such as smoking, occupation, and health status are to some degree under the experimenter's control.

Human Studies

Ideally, toxicological studies of humans should provide the strongest, best scientific evidence available in establishing a cause–effect relationship between an air pollutant or pollutant combination and adverse health effects. In such studies it is possible to expose humans under controlled exposure conditions and to monitor them for signs indicating the onset of disease. Most human studies are, however, limited to short-term acute exposures. Because of the long exposure periods required to induce chronic responses and the irreversibility of some of these, human studies are not usually conducted at realistic concentrations of pollutants that occur in ambient atmospheres. In addition, ethical considerations which are imperative in designing human exposure studies limit the kind of information that can be acquired.

In theory, a true cause–effect relationship can only be developed by human experimentation. Ethical and other limitations in such studies make it, however, virtually impossible to prove such a relationship. Because of these limitations, most controlled biological studies are conducted on animals.

Animal Studies

Animal exposures can provide valuable information on the effects of pollutants when there are no other acceptable means of obtaining such information. Because animals can be sacrificed for pathological and/or biochemical analyses, they are ideally suited for toxicological studies. Additionally, such studies lend themselves well to chronic exposures.

The use of animals, of course, has limitations. Animals are not humans. Although they may have the same general organ systems, these may be structurally and physiologically different. In addition, species differ in their sensitivity to pollutants. For example, dogs are 10 times more resistant to NO_2 than rodents. Also, animals are not as long-lived as humans. Consequently, results obtained from animal studies can only be extrapolated to humans with some element of uncertainty. Uncertainty is reduced when the results can be demonstrated in more than one mammalian species, especially primates, and the systems under study are closely analogous to those of humans.

Exposure concentrations in controlled studies often have to be in excess of ambient levels to produce toxicological responses. This may be due to a variety of factors as well as to the artificial nature of controlled exposures. In general, controlled exposures are limited to one or two carefully defined toxic substances. In urban environments, exposure is usually to a heterogeneous mixture of pollutants whose concentrations are quite variable.

Studies of Occupational Exposures

Although ethical and legal constraints limit the use of human volunteers for the experimental evaluation of the effects of toxic substances, such constraints have not limited the voluntary or involuntary exposure of industrial workers. For many air pollutants of interest, exposures in the work environment occur at much higher levels than in ambient air. Because of potential health risks and legal liability, some industries maintain detailed exposure and health data on employees. As a consequence, occupational exposures may be quite well-defined and may provide an important source of information on symptom expression, job-related disease, and dose–response relationships. In the absence of human toxicological studies, occupational exposures provide the only available human dose–response data.

Studies of occupationally exposed workers have limitations. Although one or more pollutants may be common to both occupational and ambient environments, the overall pollutant mix may be considerably different. In addition, occupational exposure is limited to workday/workweek cycles. Community exposures are far more variable, being further complicated by differences in indoor and ambient air quality and the amount of time individuals are in such environments.

A major limitation of occupational studies is that the population of workers does not well reflect the general population. In many industries, workers are primarily healthy males between the ages of 18 and 65. Studies of such workers may provide important information on disease syndromes induced by specific pollutants or pollutant combinations. They cannot, however, provide adequate information on those who are not industrially exposed and who may be at special risk. In such instances, information on the health effects on healthy individuals may not be as important as responses of individuals who are most likely to be sensitive. Such individuals may be excluded from the work force, or, because of existing health problems, may be excluded from specific occupations.

IMPACT OF POLLUTANTS ON THE HUMAN BODY

Pollutant effects are normally manifested in specific target organs. These may be direct; that is, pollutants come in intimate contact with the organ affected. Such is the case for eye and respiratory irritation. Effects may also be indirect. For example, pollutants may enter the bloodstream from the lungs or gastrointestinal system by respiratory clearance. Effects may then be distant to the immediate organ of contact. Indeed, a target organ may not have immediate intimate contact with air contaminants. The principal target organs or organ systems are the eyes and the respiratory and cardiovascular systems (Figure 5.2).

Eye Irritation

Eye irritation is one of the most prevalent manifestations of pollutant effects on the human body. It is most often associated with exposure to aldehydes and photochemical oxidants. The threshold for eye irritation by oxidants is approximately 0.10 to 0.15 ppmv (reported as O_3). Eye irritation increases with increasing oxidant concentrations although O_3 and NO_2, the two principal oxidants, do not cause eye irritation. Apparently oxidant levels are indicators of the eye irritation potential of photochemical air pollutants such as PAN, acrolein, formaldehyde (HCHO), and other photochemically derived compounds. Some questions exist as to whether eye irritation should be categorized as a significant health effect,

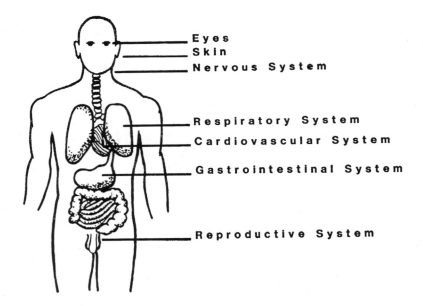

Figure 5.2 Potential target organs of atmospheric pollutant exposures.

because no other physiological changes are detectable and eye irritation disappears quickly after the exposure has ceased.

Cardiovascular System Effects

Pollutants such as CO and lead are absorbed into the bloodstream and may have both direct and indirect effects on the cardiovascular system. These effects will be discussed in subsequent sections. Cardiovascular disease may also result from the indirect effects of other air pollution-incited disease. For example, some individuals die of *cor pulmonale*, a heart failure resulting from the stress of severe chronic respiratory disease.

Respiratory System Effects

The respiratory system is the principal mechanism of gas exchange and, therefore, it receives direct exposure to airborne contaminants. Before pollutant effects on the respiratory system are discussed, it would be useful to put this problem into perspective by first discussing the structural and functional anatomy of the respiratory system as well as its defense mechanisms.

The principal function of the respiratory system is to supply O_2 for body metabolism. In addition, it removes waste CO_2 from the bloodstream and tissues. These respiratory functions are facilitated by the three major units of the respiratory tract: nasopharyngeal, tracheobronchial, and pulmonary systems (Figure 5.3).

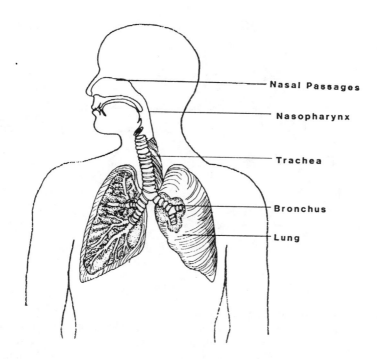

Figure 5.3 General anatomy of the human respiratory system.

The Nasopharyngeal System. The nasopharyngeal system or upper airway consists of the nasal passages, nasopharynx, oropharynx, and glottis. The nasal region is subdivided into two cavities by a septum. These cavities, which are just beyond the nares or nostrils, are ringed with coarse hairs. Projecting from the walls of the cavities are bony turbinates which force inspired air to flow along a winding course where, on contact with the large surface area of the nasal passages, it is warmed and moistened. The nasal surfaces are covered by a mucus layer and hairlike projections called cilia which remove relatively large particles.

The nasopharynx begins at the terminus of the bony turbinates and extends to the base of the soft palate, which is located in the back of the mouth. The nasopharynx is covered by a layer of mucus and ciliated cells.

The oropharynx extends from the soft palate to the glottis, which is located at the junction of the trachea and the esophagus, where eating and breathing functions are separated. The glottis consists of the vocal cords and surrounding tissue. The larynx, or voice box, is that portion of the glottis that makes sound.

The Tracheobronchial System. The tracheobronchial system is a series of tubes or ducts which transport inspired air to lung tissue, where O_2 and CO_2 can be exchanged with blood. This system consists of a large tube — the trachea — which subdivides to form smaller-diameter tubes, the bronchi. The bronchi, in turn, branch to form smaller and smaller airways. From the trachea to the lung

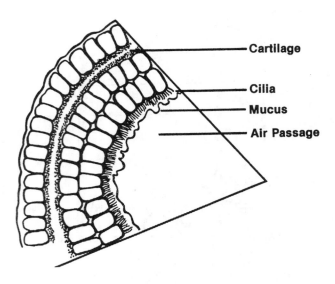

Figure 5.4 Partial cross-section of a respiratory airway.

tissue there may be as many as 23 to 32 generations of branching, with each generation being smaller in diameter and length. The bronchi, or bronchial tubes, compose the first 12 to 22 generations of branching. They have cartilaginous plates and are surrounded by a muscular layer which constricts the bronchi when they are entered by foreign substances. The bronchi are followed by the bronchioles, which do not have cartilaginous plates and are usually less than 1 mm in diameter. The interior surfaces of the trachea, bronchi, and bronchioles are lined by ciliated cells interspersed with mucus-secreting cells (Figure 5.4). The mucus layer produced by these cells is moved toward the oropharynx by the rhythmically beating cilia. The terminal bronchiole is the last bronchiole to have cilia before entering the pulmonary system.

The Pulmonary System

The pulmonary system consists of respiratory bronchioles, alveolar ducts, and alveoli (Figure 5.5). The alveoli are membranous sacs which terminate the bronchioles. A fully developed lung is estimated to have approximately 300 million alveoli. The large surface area of alveolar sacs facilitates the efficient exchange of CO_2 and O_2 between the lung and the blood capillaries which cover the alveoli.

The movement of air into (inspiration) and out of (expiration) the lungs is aided by the muscles of the chest and diaphragm which alternately compress and expand the lungs. During inspiration, air pressure in the alveoli decreases, becoming slightly negative. Because of this negative pressure, air flows into the lung. During expiration, alveolar air pressure rises, causing air to flow out of the lungs.

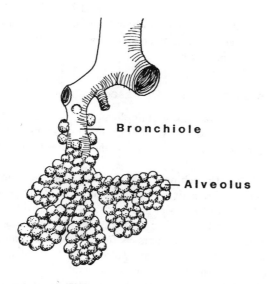

Figure 5.5 External structure of terminal bronchioles and lung alveoli.

Defense Mechanisms. The respiratory system is protected from foreign substances by a variety of defense mechanisms. In the nasal region, for example, large particles may be removed by the stiff nasal hairs or by impingement on the mucus layer of the winding passages of the turbinates. Cilia sweep the mucus layer and entrapped particles toward the back of the mouth where they are swallowed or expectorated. Foreign substances may also be removed from the upper airway by the sneeze reflex. The irritant effect of these substances initiates a response in which the respiratory muscles contract, closing the vocal cords, building up a high pressure. The vocal cords suddenly open, expelling the pressurized air and any foreign particles in its path.

Other defense mechanisms serve to remove and/or prevent foreign substances (mainly particles) from entering the tracheobronchial and pulmonary systems. When, for example, particles enter the upper portion of the bronchial tree, muscle layers constrict the bronchi, narrowing bronchial diameter, thus reducing the amount of particles that can enter the lungs. The cough reflex, which is similar to the sneeze reflex, assists this process by eliminating mucus in which particles have been trapped or removed from deeper lung tissue. The cilia which line the tracheobronchial system also clean airways by propelling mucus and particles upward, where they are removed by coughing and expectoration or by swallowing. Air pollutants may in this fashion enter the gastrointestinal system.

Despite these defense mechanisms some very fine particles and/or gases enter pulmonary tissue. Within the alveoli and bronchioles of the pulmonary region, specialized cells called phagocytes ingest deposited matter. These phagocytes and the matter they contain are normally transported out of the lungs in mucus by the respiratory cilia. However, they may also penetrate the alveolar membrane and enter either the lymphatic or circulatory systems. Additionally, soluble components deposited in the alveoli may be absorbed by adjacent tissue and enter the lymphatic or circulatory systems without being phagocytized. Although these mechanisms effect clearance from the respiratory system, they may expose other body systems to toxic materials.

Air Pollution and Respiratory Disease

Community air pollution has been implicated as a causal or aggravating agent in diseases of the respiratory system such as chronic bronchitis, pulmonary emphysema, lung cancer, bronchial asthma, and respiratory infections.

Chronic Bronchitis. Bronchitis is a respiratory disease characterized by an inflammation of the membrane lining bronchial airways. Bronchitis may be caused by pathogenic infections or by respiratory irritants such as those which occur in cigarette smoke, industrial exposures, and ambient and indoor air pollution.

When bronchial inflammation persists for 3 months or longer it is classified as chronic bronchitis. The primary cause of chronic bronchitis is extended irritation of the bronchial membrane. Chronic bronchitis is characterized by a persistent cough and excessive mucus or sputum production. It is often accompanied by the destruction of cilia and thickening of bronchial epithelia. Chronic bronchitis has a significant effect on respiratory function, as airway resistance is increased by the occlusion of the bronchi, by the swelling of the inflamed membrane, and concomitant mucus production (Figure 5.6). Consequently, persons afflicted with the disease have difficulty in breathing.

Severe cases of chronic bronchitis are often followed by the development of pulmonary emphysema. Combined chronic bronchitis and pulmonary emphysema are usually diagnosed and described as chronic obstructive lung disease (COLD).

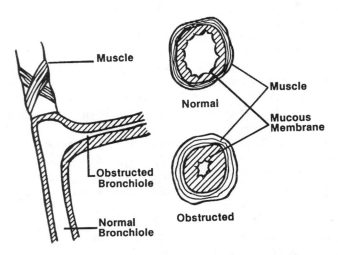

Figure 5.6 Changes in respiratory airways associated with bronchial inflammation.

Some epidemiological studies suggest that ambient pollutants have a role in initiating or aggravating chronic bronchitis. The problem of identifying ambient air pollutants as causal agents in the etiology of chronic bronchitis is confounded to a considerable degree by the significant causal role of cigarette smoking in initiating this disease. It is further obscured by the effects of occupational exposures. However, available epidemiological and toxicological evidence suggests that exposure to a combination of ambient pollutants such as PM and SO_2 may contribute to the initiation of chronic bronchitis as a result of long-term community air pollution exposures.

Pulmonary Emphysema. Whereas chronic bronchitis is a disease which affects the upper portion of the lung system, pulmonary emphysema is a disease of the deep lung, that is, the lung tissue which normally facilitates gas exchange with the blood.

Emphysema is often a disease of old age. It is characterized by a destruction or degeneration of the walls of the alveolar sacs (Figure 5.7). As a result, the total surface area of pulmonary tissue is reduced, diminishing aeration of blood. Individuals afflicted with emphysema usually develop pulmonary hypertension, as the destruction of the alveolar walls also results in the destruction of blood vessels. In addition to elevation of arterial pressure, pulmonary resistance is increased. Pulmonary emphysema is consequently accompanied by shortness of breath and difficulty in breathing.

Emphysema patients usually have difficulty in exhaling, because air remains trapped in the lungs and overinflates alveoli beyond the terminal bronchioles. This overinflation contributes to the "barrel chest" characteristic of most emphysema patients. Exhalation difficulties are due to the compression and collapse of some of the smaller airways and the overall decrease in lung elasticity common to emphysema victims.

Little epidemiological evidence exists to implicate ambient air pollutants as a contributing factor to the initiation and development of pulmonary emphysema.

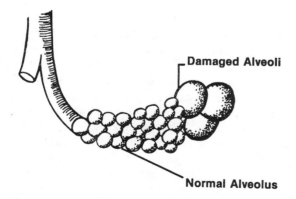

Figure 5.7 Changes in lung alveoli associated with emphysema.

Animal studies suggest, however, that chronic exposures to NO_2 can initiate pre-emphysematous lung changes.

Lung Cancer. Cancer is one of the leading causes of death in the U.S., accounting for about 370,000, or one out of five deaths each year. Of these, approximately 130,000 are due to lung cancer, which is the leading cause of cancer death in males. Common to all cancers is the characteristic unrestrained cell growth that produces malignant tumors, which have a higher rate of growth than the normal tissue which surrounds them. As a malignant tumor develops, it compresses, invades, and/or destroys normal tissue. Most cancers metastasize and spread to other parts of the body. The tendency for lung cancer cells to metastasize is particularly great; as a result, the prognosis for lung cancer patients is bleak, with an approximate 92% mortality.

Cancer is usually a disease of the elderly and individuals in late middle age. This pattern is due in part to the fact that cancers are latent diseases. That is, there is a delay of several decades or more between initial exposure to a carcinogenic agent and the development of disease. For lung cancer the latent period may be as long as 30 years or more, with the peak incidence at age 50. The long interval between the initial exposure to the carcinogenic (cancer-producing) agent and disease onset makes it especially difficult to conclusively identify a specific causal agent. During this latency period, an individual may undergo changes in occupation, socioeconomic status, place of residency, nutrition, and personal habits such as smoking. These changes tend to confound the epidemiological evaluation of urban air pollution in the etiology of lung cancer. The latency period not only confounds the identification of causal agents in diseases such as lung cancer, but also tends to give individuals a false sense of security about the safety of a variety of chemical or physical exposures.

The most common form of lung cancer is bronchiogenic (i.e., it originates in the bronchial membrane and invades the tissue of the bronchial tree). The malignancy may spread to the rest of the lung and eventually to other parts of the body. Available epidemiological evidence indicates that a large portion of the approximately 130,000 lung cancer deaths each year are caused by chronic exposure to tobacco smoke. Lung cancer has also been linked to occupational exposures to asbestos, arsenic and radioactive gases (radon), and dusts. Evidence for a causal role of urban air pollution in the etiology of lung cancer is far less conclusive. If indeed exposures to urban air pollution result in lung cancer, their contribution to the overall lung cancer rate is small.

A causal role of ambient air pollution in the development of some lung cancers is suggested from several types of epidemiological investigations, including: (1) comparisons of urban and rural populations, (2) comparisons of lung cancer rates of migrants, and (3) regression statistical analyses of the relationships between lung cancer deaths and various indices of air pollution.

Comparisons of rural and urban nonsmoking populations indicate that urban lung cancer death rates are approximately twice those found in rural areas. These studies suggest that there is an apparent urban factor which contributes to the overall incidence of lung cancer. However, these studies do not identify the characteristics of the urban environment that are primarily responsible.

Studies of individuals migrating from countries which have had high ambient air pollution levels to countries where air pollution levels were much lower indicate that lung cancer rates in such migrants tend to be intermediate between the country of origin and their new homeland. Even when differences in smoking habits are considered, lung cancer rates of migrants from high air pollution countries such as Great Britain were greater than countries such as New Zealand. These data suggest that ambient air pollution exposures may produce effects which persist long after exposures have ceased.

Evidence for a causal role of ambient air pollution in the induction of lung cancer is mixed. Arguments supporting its causal role include: (1) the presence of potent carcinogens such as benzo(α)pyrene (BaP) and other polycyclic aromatic hydrocarbons (PAHs) in urban air, and (2) epidemiological studies of migrants and urban and rural populations which indicate a statistical association between lung cancer and urban residence and, by implication, ambient air pollution. Arguments which tend to question the role of ambient air pollution as a causal factor include: (1) an increase of lung cancer incidence with increasing air pollution cannot be detected; (2) in England, where lung cancer rates and air pollution rates have been high, a strong positive correlation exists with population density, not air pollution; (3) the urban factor affects women less than men; and (4) decreased urban levels of BaP have not been observed to result in a decrease in urban lung cancer. Because the available evidence is inconsistent, it is not possible to conclude that ambient air pollution per se is a contributory factor to excess lung cancers in urban areas.

Bronchial Asthma. Asthma is an acute respiratory reaction characterized by constriction of muscles and swelling of the lining of the respiratory airways, excessive mucus production, and increased resistance to air flow. Such reactions are episodic, being described as asthmatic attacks. They may occur suddenly and without warning. Overt symptoms include severe shortness of breath, wheeze, chest tightness, and cough. Attacks may occur for a few minutes to several hours.

Asthma is a relatively common respiratory ailment with a prevalence rate of approximately 4% in the U.S. or an estimated 10 to 15 million people. The prevalence of asthma has increased dramatically in the U.S. since 1980, particularly among children where the increase has been greater than 50%. In addition, the death rate from severe asthmatic attacks has also increased.

Asthma, in most cases, appears to be induced by exposures to one or more allergenic substances. Asthmatic attacks may result from repeated exposures to asthmagenic substances or to a variety of nonspecific irritants. These may include SO_2 and other gas- or particulate-phase irritants.

HEALTH EFFECTS OF REGULATED AIR POLLUTANTS

Though polluted atmospheres may be contaminated with hundreds of different substances, only a relatively small number have been identified as having the potential to cause adverse health effects under exposure conditions normally encountered in the ambient environment. Most notable are those for which National Ambient Air Quality Standards (NAAQS) have been promulgated. These include carbon monoxide (CO), sulfur oxides (SO_x), ozone (O_3), PM, nonmethane hydrocarbons (NMHCs), nitrogen oxides (NO_x), and lead. A variety of other substances are regulated under National Emissions Standards for Hazardous Pollutants (NESHAP), commonly referred to as "air toxics." Because of their ubiquitous presence, population exposure potential, and potential health effects, pollutants regulated under NAAQS are discussed in detail below.

Carbon Monoxide

Potentially harmful exposures to CO occur from a variety of sources and environments. At very high concentrations (>1,000 ppmv), CO exposures may be lethal with death resulting from asphyxiation. At lower concentrations (several hundred ppmv), CO may cause a variety of neurological symptoms including headache, fatigue, nausea, and, in some cases, vomiting. Asphyxiation and sublethal symptoms are usually caused by poorly vented combustion appliances, idling motor vehicles in closed environments, excessive CO production and inadequate ventilation associated with a variety of industrial occupational activities, and smoke inhalation from structural fires. Carbon monoxide poisonings, though not uncommon, represent exposures which are considerably higher than those which occur in the ambient air of our cities and towns, the kind of exposure conditions addressed by the NAAQS for CO. Such exposures rarely exceed the 35 ppmv 1-hour standard.

The principal mechanism of action of CO toxicity is tissue hypoxia associated with the preferential bonding of CO with hemoglobin [to form carboxyhemoglobin (COHb)]. Carbon monoxide has a hemoglobin affinity which is 240 times that of O_2. In addition, CO binds more tightly with hemoglobin than O_2. As CO binds with hemoglobin, the blood's capacity to carry O_2 is decreased. The body must compensate by increasing blood flow to maintain the O_2 needs of vital organs such as the brain. Carbon monoxide also binds to intracellular hemoproteins such as myoglobin, cytochrome oxidase, tryptophan oxidase, and dopamine hydroxylase. Such binding may result in extravascular effects.

The quantity of COHb formed in the blood of a human depends on a number of factors, including the concentration and duration of CO exposure, exercise, temperature, health status, and the metabolism of the individual exposed. At low exposure concentrations, the quantity (% COHb) of hemoglobin tied up by CO is a linear function of exposure concentration and duration (Figure 5.8). Note the effect of exercise on lung ventilation rate and COHb formation in the blood.

The formation of COHb is reversible. Depending on initial COHb levels, the elimination halftime is approximately 2 to 6.5 hours. Because elimination is slow,

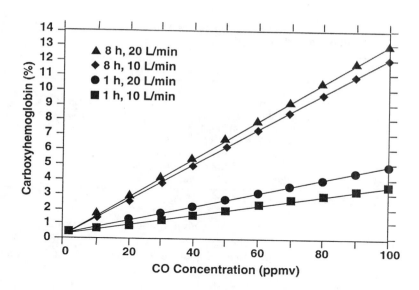

Figure 5.8 Carboxyhemoglobin levels in the blood as a function of exposure concentration and duration. (From USEPA 1991. EPA/600/8-90/045f.)

COHb may accumulate on continuous exposure. Approximately 10 to 50% of the body store of CO is external to the vascular system. Extravascular CO is commonly found in skeletal muscle. The primary health concerns associated with low-level ambient (<50 ppmv) exposures to CO are effects on the cardiovascular system (particularly in sensitive individuals) and on neurobehavior.

There are more than 500,000 deaths in the U.S. each year from heart attacks and 5 to 7 million people are estimated to have a history of heart attack, angina (heart-related chest pain), or both. Scientific support for the current CO NAAQS comes from studies of patients with stable angina pectoris associated with cardiovascular heart disease. The lowest observed physiological effect level in patients with exercise-induced ischemia (tissue blood deficiency) is somewhere between 3 and 4% COHb. Baseline values in nonexposed nonsmokers is approximately 0.5 to 0.7%. Because nonsmokers may be exposed from a variety of sources, average COHb levels are approximately 1%; COHb levels in smokers vary from 3 to 8% with an average of approximately 4%.

Exposure to CO levels sufficient to produce 6% COHb appears to be sufficient to cause arrhythmia in exercising patients with coronary artery disease. There is suggestive evidence that CO exposure may increase the risk of sudden death from arrhythmia in such patients.

In addition to the results of controlled exposure studies described above, a variety of epidemiological studies have indicated that a significant relationship may exist between increased ambient CO levels and increased mortality from heart attacks or increased cardiovascular complaints. However, because of inadequate exposure characterization, such epidemiological studies are not conclusive.

Table 5.1   Human Responses to CO Exposures Expressed as Percent
             Carboxyhemoglobin Saturation

Blood COHb levels	Effect
0–1%	NONE
2.5%	Impairment of time interval discrimination in nonsmokers.
2.8%	Onset of angina pectoris pain shortened in exercising patients; duration of pain lengthened.
3.0%	Changes in relative brightness thresholds.
4.5%	Increased reaction time to visual stimuli.
10%	Changes in performance in driving simulation.
10–20%	Headache, fatigue, dizziness, loss of coordination.

From NACPA. USDHEW. 1970. Publication No. AP-62.

There is conflicting evidence from animal studies that exposures to CO contribute to the development of atherosclerosis (a form of arteriosclerosis). There is very little data to indicate that an atherogenic effect would occur at commonly encountered ambient CO levels.

The brain is the most sensitive body organ to reduced tissue O_2 levels associated with CO exposures. Though it can, under normal circumstances, compensate for CO-induced hypoxia by increasing blood flow and tissue O_2 extraction, some neurological or neurobehavioral effects can be anticipated. Behaviors that require sustained attention or performance appear to be very sensitive to COHb-induced changes. Human studies of hand–eye coordination, vigilance, and continuous performance have reported consistent effects of CO exposures at COHb as low as 5%. Effects at lower COHb values appear to be small and are a subject of controversy.

Key health effects that have been clearly demonstrated to be associated with different COHb blood concentrations are summarized in Table 5.1. These effects occur above the target level of 2.1% COHb which the current NAAQS for CO is intended to achieve.

Levels of 3 to 6% COHb required to produce the cardiovascular or neurobehavioral changes described are likely to occur only in worst case atmospheric conditions in the U.S., in downtown areas with heavy traffic, and very poor atmospheric mixing conditions. They are more likely to occur in unregulated environments of large cities in developing countries. High-population, high-altitude Mexico City would be particularly at risk.

The NAAQS for CO, like other NAAQS, are designed to protect the most sensitive population, individuals with existing cardiovascular disease. Based on both theoretical and experimental research, there may be other subpopulations at potential risk. The most probable subpopulations include the fetus, infants, the elderly, individuals with preexisting disease that decrease the availability of O_2, and individuals using certain medications and recreational drugs. Exposure to other pollutants and residence at high altitude may also increase risks associated with CO exposure.

Based on an evaluation of air-quality trends for CO measurements made at fixed sites, >99% of the most sensitive nonsmoking adult population would be expected to have COHb levels below 2.1%, the NAAQS target levels. However, results of both exposure monitoring and field modeling studies indicate that individual personal exposures do not correlate with CO concentrations measured at fixed site monitors. This poor correlation appears to be due to the mobility of people, spatial and temporal variability of CO, and exposures to CO associated with other-than-ambient air. Expired air measurements of a sample of individuals in Washington, DC would suggest that at least 6% of the population is exposed to CO levels which result in COHb concentrations that exceed 2%. Highest CO exposures appear to occur during motor vehicle travel.

Sulfur Oxides and Particulate Matter

Although individual NAAQS have been promulgated for SO_x and PM, the individual health effects of these pollutants are not easily separated. Because they are usually produced by a common source, such as the combustion of coal, high SO_x levels are often associated with high PM levels. There is also some scientific evidence to suggest that PM potentiates the toxic response to SO_x.

Epidemiological studies of acute air pollution episodes in the U.S. and Great Britain have indicated an association between increased mortality/morbidity and elevated levels of SO_x and PM. Epidemiological studies of longer-term, lower-level exposures to SO_x and PM have shown an association with certain adverse health effects. For example, British studies have shown significantly higher mortality from bronchitis and lung cancer in areas that experience relatively high SO_x and PM exposures. British schoolchildren have also been shown to have a higher frequency and severity of respiratory disease in such areas. Italian investigators have observed an increased frequency of cough, sputum, labored breathing, and bronchitis in people living in cities moderately polluted by SO_x and PM. The frequency of bronchitis showed a significant correlation with average SO_x concentrations. Other investigations also implicate SO_x and PM as being causal agents in inciting community health problems. The epidemiological studies mentioned here are far from definitive, as they are subject to all the limitations and possible misinterpretations which are characteristic of assessing health effects of complex atmospheric mixtures, and a variety of confounding variables.

Sulfur Dioxide

Much of the available information on the acute and chronic effects of SO_2 is based on laboratory animal exposures considerably in excess of exposure concentrations found in polluted atmospheres. Because of its high solubility in watery fluids, SO_2 is almost entirely removed in the nasal passages, throat, and mouth. Less than 1% of inspired SO_2 reaches lung tissue (alveoli). Exposure of the lower airway and alveoli increases considerably during exercise and mouth breathing. The major physiological effects of SO_2 exposures are changes in the mechanical function of the upper airways. These changes include an increase in nasal flow resistance and a decrease in nasal mucus flow rate. Laboratory animal exposures at high concentrations indicate that SO_2 exposures can cause chronic bronchitis.

When the primary air quality standard for SO_2 [0.03 ppmv (80 µg/m³), annual mean; 0.14 ppmv (365 µg/m³), maximum 24-hour concentration] was promulgated in 1971, no health effects had been reported for short-term exposures (<1 hour) at levels observed in the ambient environment (≤1 ppmv, 2.67 mg/m³). In the years following 1980, numerous challenge studies were conducted on asthmatic individuals in the concentration range of 0.25 to 1.0 ppmv (0.67 to 2.67 mg/m³). These studies have shown that moderately exercising asthmatics exposed to 0.5 to 1.0 ppmv (1.34 to 2.67 mg/m³) SO_2 experienced significant bronchoconstriction (airway narrowing) which was not observed in normal subjects.

Bronchoconstriction was evidenced by increased airway resistance, decreased forced expiratory volume (FEV), and symptoms such as chest tightness, shortness of breath, and wheezing. Bronchoconstriction occurred within 5 to 10 minutes of exposure and was brief in duration with lung function returning to normal for most subjects within 1 hour of exposure. This duration is similar to that observed for exercise. In some severe responses, bronchoconstriction required the cessation of activity, medical intervention, or both.

Mild-to-moderate asthmatic children and adults are at the greatest risk for short-term SO_2-induced respiratory effects. Individuals with more severe asthma, ironically, would be at lower risk because their low exercise tolerance would deter them from engaging in exercise sufficient in intensity to cause effects.

A substantial percentage (20 to 25%) of mild-to-moderately asthmatic individuals exposed for 5 to 10 minutes at concentrations of 0.6 to 1.0 ppmv (1.60 to 2.67 mg/m³) SO_2 during moderate exercise or activity would be expected to have changes in respiratory function and symptom induction that clearly would exceed those associated with daily variation in lung function or in response to other stimuli. The severity of effects may be sufficient to cause disruption of ongoing activities, use of bronchodilator medication, and possibly seeking medical attention at SO_2 exposures below 0.5 ppmv (1.34 mg/m³). Less than 20% of mild-to-moderately asthmatic individuals exposed to 0.2 to 0.5 ppmv (0.53 to 1.34 mg/m³) SO_2 during moderate exercise are likely to experience lung function changes larger than they daily experience. Such responses are less likely to be perceptible and of immediate public health concern.

Available epidemiological data do not show any evidence of excess asthma mortality associated with SO_2 exposures in urban areas. From an evaluation of exposure data, it appears that the probability of asthmatic individuals exposed to peak levels of SO_2 >0.5 ppmv (1.34 mg/m³) for 5 to 10 minutes is low viewed from a national perspective. Exposure data indicate that peak SO_2 concentrations >0.5 ppmv (1.34 mg/m³) do occur, suggesting that mild-to-moderately asthmatic individuals residing in the vicinity of certain sources or source types may be at increased risk.

Evidence of a health risk for a relatively small portion of the asthmatic population has posed a significant regulatory dilemma for the U.S. Environmental Protection Agency (USEPA). In the late 1980s, the Clean Air Scientific Advisory Committee (CASAC) recommended consideration of a 1-hour standard in the range of 0.2 to 0.5 ppmv (0.53 to 1.34 mg/m^3) to protect against 5-minute peak exposures of 0.4 to 1.0 ppmv (1.07 to 1.34 mg/m^3) SO_2. Because of the uncertainties associated with the need and appropriateness of a new 1-hour standard, USEPA decided not to revise the primary standard for SO_2 at that time.

Particulate Matter

Toxic effects of PM may be due to the direct irritant effects of particles such as sulfuric acid or to substances which are readily adsorbed to the large surface area of small particles. The concentration of adsorbed substances may be considerably greater than in the ambient atmosphere. Adsorbed substances of particular concern include SO_2, PAHs, and heavy metals which tend to predominate in respirable submicrometer particles. The PAHs such as BaP are of concern because of their known carcinogenicity.

Respiratory Deposition and Retention. The health consequences of atmospheric PM depend on its ability to penetrate respiratory defense mechanisms. In general, defense mechanisms are adequate to remove inhaled particles larger than 10 μm from the inhaled air stream. Particles smaller than this are described as "inhalable"; that is, they can enter and be deposited in the respiratory system. Particles less than 2.5 μm are called "respirable"; they can enter and be deposited in pulmonary tissue. Particles larger than 2.5 μm are mostly removed in the upper respiratory system. Deposition of particles as a function of particle size is illustrated in Figure 5.9.

Because of changes in flow patterns in the tracheobronchial zone, particles tend to be deposited at or near airway bifurcations (branches). As nerve endings are concentrated at these sites, the mechanical stimuli of deposited particles often lead to reflex coughing and bronchoconstriction. The sensitivity of nerve endings to chemical stimuli results in increased breathing rates and reduced pulmonary compliance (ability of the lung to yield to increases in pressure without disruption).

Deposition of particles is not only influenced by particle size but also by mass concentration, molecular composition, pH, and solubility. Deposition also varies among nonsmokers, smokers, and individuals with lung disease. Tracheobronchial deposition is slightly higher in smokers, and greatly increased in individuals with lung disease.

After particles have been deposited, their retention may be a function of the rate of clearance, which varies greatly among the different regions of the respiratory tract. In the ciliated airways of the nose and upper tracheobronchial zone, clearance in healthy individuals is achieved in less than a day. In general, as the site of deposition becomes more distal the time required for clearance increases. Clearance of particles in the alveolar region may take weeks to months. Slow

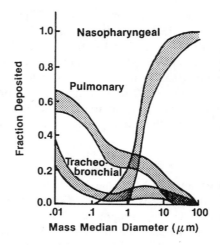

Figure 5.9 Deposition of particles of different sizes in the nasopharyngeal, tracheobronchial, and pulmonary regions of the human respiratory system. (From NAPCA. USDHEW. 1969. Publication No. AP-44.)

clearance of particles from the respiratory system of humans is generally considered to be detrimental because toxic substances are in contact with sensitive tissue for longer periods of time.

Alveolar deposition of particles is most efficient in the range of 0.1 to 2.5 μm. The effective toxicity of these small particles may be greater than that of larger particles, because concentrations of toxic substances such as lead, zinc, chromium, mercury, sulfates, nitrates, etc., increase with decreasing particle size. In addition, the enormous surface area of small particles allows for high reaction and dissolution rates for toxic chemical species. Their relatively long retention in the alveolar region permits substances such as lead to be extracted and transported to other parts of the body.

The air-quality criteria document for PM and SO_x issued by USEPA in 1982 concluded that the most clearly defined effects of ambient exposures to PM were sudden increases in the number of deaths occurring on a day-to-day basis during episodes of very high pollution such as those which occurred in the Meuse Valley in 1931, Donora in 1948, and London in 1952. The 1982 criteria document also concluded that mortality was substantially increased when 24-hour PM concentrations exceeded 1000 μg/m³ with SO_2 levels in excess of 0.38 ppmv (1000 μg/m³); the elderly and others with preexisting respiratory or cardiovascular disease were most affected. After reviewing the historical London data and additional studies, USEPA in 1986 concluded that there was an increased risk of mortality among at-risk populations in the PM range of 500 to 1000 μg/m³ with SO_2 concentrations in excess of 0.27 ppmv (700 μg/m³). Additionally, there appeared to be convincing evidence that relatively small but significant increases in mortality occurred at PM concentrations below 500 μg/m³, with no indication of a threshold.

Since 1986, numerous epidemiological studies have been published which have evaluated potential associations between human mortality and morbidity and acute exposures at or below the 24-hour PM_{10} standard of 150 $\mu g/m^3$. Several investigators have reported very small but statistically significant associations between increased relative risk (RR) for mortality and various measures of PM for many different U.S. cities as well as other countries. The 24-hour PM_{10} total mortality effect estimate generally falls in the RR range of 1.025 to 1.05, which represents a 2.5 to 5% increase in the risk of death over daily background rates for which a 50 $\mu g/m^3$ increase in ambient PM_{10} concentration could be a contributing factor.

A consistent trend has been observed between PM exposure and respiratory mortality. In studies in Santa Clara, CA, the PM-respiratory mortality was 4.3 times as large as deaths taken as a whole; for Philadelphia, PA, 2.7 times; for Utah Valley, 2.5 times; for Birmingham, AL, 1.5 times; and for Santiago, Chile, 1.8 times. The PM RR in these studies ranged from 50 to 400% higher for respiratory disease-related deaths than for all other causes of death, indicating that increases in respiratory deaths were the major contributor to increased mortality associated with PM exposure. For a 24-hour average 50 $\mu g/m^3$ PM_{10} increase, a 30 to 40% higher RR of mortality is suggested for the elderly subpopulation and those with preexisting respiratory disease.

A variety of epidemiological studies have attempted to assess potential associations between ambient PM exposures and morbidity (illness). Major health endpoints considered included hospital admissions, asthmatic attacks, respiratory symptoms, and lung function. A review of these studies indicates a coherence of effects across these health endpoints (including mortality) with most showing a 1 to 3% change per 10 $\mu g/m^3$ PM_{10}. Hospital admission studies for chronic obstructive pulmonary disease and pneumonia have shown a moderate but statistically significant risk in the range of 6 to 25% with an increase of 50 $\mu g/m^3$ or its equivalent. Acute pulmonary function studies suggest a short-term effect with peak flow decreases in the range of 30 to 40 ml/sec, with an increase of 50 $\mu g/m^3$ PM_{10} averaged over 24 hours.

Epidemiological studies which have examined the association between mortality or morbidity and acid aerosol exposures have been limited due to availability of exposure data. Some studies do suggest that health effects may be associated with exposure to ambient acid aerosols. Evidence from London smog episodes suggests that elevated daily aerosol concentrations equivalent to 400 $\mu g/m^3$ H_2SO_4 may be associated with acute excess mortality when present with elevated levels of PM and SO_2. Significant association between acid aerosols and mortality in London during nonepisode conditions ($\leq$30 $\mu g/m^3$ H_2SO_4) has also been observed.

Acid aerosol exposures have been associated with increased hospital admissions for respiratory health problems. This association has been observed for the London episode of 1952 and under present-day conditions as well. Summertime respiratory hospital admissions in Toronto, Canada and New York state have been observed to be associated with acid aerosol episodes, with the greatest effect in acid aerosol episodes $\geq$10 $\mu g/m^3$ H_2SO_4, which occur two to three times a year in eastern North America. Other studies indicate significant increases in upper respiratory symptoms in healthy children associated with daily concentrations of H_2SO_4. Significant associations have been observed between H^+ concentrations and the prevalence of bronchitis and related symptoms, as well as pulmonary function. Chronic exposures to acid aerosol may have significant effects on a variety of measures of children's respiratory health.

Potential mechanisms which may help explain the phenomenon of PM-related mortality include: (1) hastening of death of individuals near death, (2) increased susceptibility to infectious disease, and (3) exacerbation of chronic underlying pulmonary or cardiovascular disease.

The first and second hypotheses indicated above are not supported by evidence available. As such, the exacerbation of existing disease appears to be the most likely cause of mortality associated with acute 24-hour average exposures.

As indicated, certain population groups appear to be at higher risk of mortality and/or morbidity associated with elevated PM_{10} levels within the range of the current 24-hour standard of 150 µg/m³. These include persons with preexisting respiratory disease, children, and the elderly. These populations may be affected by lower levels of PM than other subpopulations, and the impact of effects may be greater in magnitude. Subpopulations with diminished respiratory function such as the elderly and individuals with asthma, emphysema, and chronic bronchitis would be expected to be more adversely affected by a given decrease in pulmonary function than more healthy and normal groups.

Hydrocarbons

Air-quality standards for nonmethane HCs have been promulgated to achieve standards for O_3, because HCs are key reactants in photochemical processes. The NMHC standard is therefore not based on the health effects of HC chemicals. Although controlled biological studies have been conducted on several HC species, little evidence is available to indicate that specific HC chemicals found in ambient urban atmospheres occur at sufficient levels to cause adverse health effects.

Hydrocarbon chemicals or their derivatives of health concern include carcinogenic PAHs, such as BaP, and eye irritants including HCHO, acrolein, and PAN. The PAHs are produced as a result of the incomplete combustion of high molecular weight HC species. BaP is the most abundant PAH in urban air. It is normally found adsorbed on aerosol particles. Controlled animal studies indicate that BaP exposures far in excess of ambient levels are required to induce cancer tumors.

The PAH levels of urban areas in the U.S. have been reduced by a factor of three since the mid-1950s. This decrease has been due to a change of fuel usage from coal to natural gas and oil for home space heating. If the PAHs in urban areas were contributing to the urban excess of lung cancer, then there should have been a decrease in lung cancer attributed to the urban factor. There is no evidence that this has occurred.

Hydrocarbon derivatives such as HCHO, acrolein, and PAN are of health interest because they contribute to eye and mucous membrane irritation experienced in photochemical smog. The threshold for HCHO-induced eye irritation ranges from of 0.1 to 1.0 ppmv; for acrolein, moderate eye irritation can be detected at 0.25 ppmv; for PAN, 20 ppbv. In general, eye irritation experienced in urban areas is likely produced by the combined effects of these substances and other HC derivatives.

Nitrogen Oxides

Nitric oxide is a relatively nonirritating gas and is therefore believed to pose no health threat at ambient levels. Its importance lies in the fact that it is rapidly oxidized to NO_2, which has a much higher toxicity.

Unlike SO_2, which is rapidly absorbed in fluids of the upper tracheobronchial zone, NO_2 is less soluble and thus can penetrate deep into the lungs where tissue damage may occur. In those occupationally exposed to high NO_2 levels, adverse effects, such as pulmonary edema, only manifest themselves several hours after exposure has ended. Elevated NO_2 exposures have been known to occur from exposures in the manufacture of nitric acid, electric arc welding, use of explosives in mining, and in farm silos.

In toxicological studies of animals, abnormal pathological and/or physiological changes have only been observed at concentrations in excess of those found under ambient conditions. At the lowest NO_2 exposure levels (0.5 ppmv) at which adverse effects have been detected, pathological changes have included the destruction of cilia, alveolar tissue disruption, and obstruction of respiratory bronchioles. Exposures at higher levels have caused more severe tissue damage, which in rats and rabbits resembles emphysema.

Animal toxicological studies also indicate that NO_2 may be a causal or aggravating agent in respiratory infections. There is evidence to indicate that NO_2 can damage respiratory defense mechanisms, allowing bacteria to multiply and invade lung tissues. In mice, hamsters, and monkeys infected with pneumonia-causing bacteria, exposure to elevated NO_2 levels has resulted in increased mortality, reduced life span, and reduced ability to clear bacteria from the lungs.

A variety of epidemiological studies have been conducted to determine whether NO_2 exposures at ambient levels are sufficient to cause respiratory symptoms or disease. Studies of community exposures to ambient levels of NO_2 have not demonstrated any significant relationships. A causal relationship between low-level NO_2 exposures and respiratory symptoms or disease is suggested from epidemiological studies that have attempted to evaluate the potential health effects associated with the use of gas cooking stoves and kerosene heaters. Study results associated with gas cooking stoves have been mixed, with some studies indicating a significantly increased risk of children (under the age of seven) experiencing one or more respiratory symptoms in homes with gas stoves compared to homes with electric ones. Exposure to NO_2 emitted from gas cooking stoves has been suggested as the cause. In kerosene heater studies conducted in Connecticut, children under the age of seven exposed to ≥ 0.016 ppmv (30 $\mu g/m^3$) NO_2 were found to have a more than twofold higher risk of lower respiratory symptoms (fever, chest pain, productive cough, wheeze, chest cold, bronchitis, pneumonia, or asthma) than children who were not exposed. They also had an increased risk of upper respiratory symptoms (fever, sore throat, nasal congestion, dry cough, croup, or head cold).

Ozone

Ozone is one of the most ubiquitous and toxic pollutants regulated under ambient air quality standards in the U.S. and other countries. Because of seasonal weather conditions, O_3 levels vary from year to year. Twenty percent or more of the U.S. population lives in areas in which people are exposed to concentrations above 0.12 ppmv, 1-hour average, the NAAQS for O_3 in the U.S.

The effects of O_3 exposures in the concentration ranges observed in polluted atmospheres have been evaluated in experimental studies on animals and humans. Additionally, several epidemiologic studies have attempted to relate ambient O_3 levels to a variety of health endpoints.

Because of its low solubility, inhaled O_3 is readily delivered to terminal respiratory airways and alveolar tissue, the major target sites for its effects. It injures tissue membranes by oxidizing amino acids, sulfhydryl (SH) groups in enzymes and other proteins and polyunsaturated fatty acids. The effect of lipid peroxidation and resultant production of free radicals is to increase membrane permeability, thereby causing the leakage of essential electrolytes and enzymes. It also results in the swelling and disintegration of cellular organelles such as lysosomes and mitochondria and an inhibition of metabolic pathways. Peroxidation products and cell injury may result in the production of inflammatory cells which contribute to more tissue injury.

These toxic effects may result in a variety of histopathological and physical changes, including structural changes in cells lining alveoli, thickening of the alveoli themselves, the production of lesions, and decreased elasticity of lung tissue.

At the physiological level, O_3 exposures result in significant changes in lung function. These include increased respiratory rates, increased pulmonary resistance, decreased tidal volume, and changes in respiratory mechanics. Changes appear to be transient, returning to normal after exposure is discontinued or after several days of adaptation on continuous exposure. Pulmonary function changes are dose-dependent and increase with increased lung ventilation associated with exercise.

Because of the irritancy of O_3, it is logical to assume that more sensitive populations such as asthmatics, the young, the elderly, and those with chronic obstructive lung disease would be more responsive to O_3 effects than healthy adolescents and young adults. For the most part, increased pulmonary responsiveness in challenge studies have not been observed in these subpopulations.

Of increased recent concern is the potential health effects of low O_3 concentrations (<0.12 ppmv) over exposure periods longer than 1 hour. Several experimental studies with humans have shown significant effects on pulmonary function at concentrations <0.12 ppmv (240 $\mu g/m^3$) in exercising adults. The lowest level that significant effects have been observed is 0.08 ppmv (160 $\mu g/m^3$) for a multi-hour exposure. At present, the NAAQS for O_3 does not reflect more than a 1-hour exposure. At present, USEPA is considering an 8-hour standard of 0.07 to 0.09 ppmv (140 to 180 $\mu g/m^3$) in addition to the current standard or by itself. It has been estimated that between 1991 and 1993, 8-hour concentrations exceeded 0.085 ppmv (170 $\mu g/m^3$) four times in regions where an estimated 27 million children live.

The potential for O_3 to aggravate asthma has been a major public health concern since early studies in Los Angeles indicated a significant association between emergency hospital visits by asthmatic patients and ambient oxidant (presumably primarily O_3) levels. Since then, a number of studies have shown strong correlations between emergency hospital admissions for asthmatic symptoms and ambient O_3 levels. One most notable study was conducted in central New Jersey over 5 consecutive years. Hospital admissions were 26% higher on days when O_3 levels were >0.06 ppmv (120 $\mu g/m^3$) compared to days when levels were <0.06 ppmv.

The epidemiological studies, which demonstrate a strong link between ambient O_3 levels and worsening of asthma symptoms, are for the most part not supported by the majority of challenge studies where enhanced bronchial responsiveness was not observed among asthmatic subjects. Based on animal studies, it has been suggested that O_3 may have an indirect effect on the severity or occurrence of asthma by increasing the potential for sensitization to inhaled allergens because of increased airway mucosal permeability. It is also possible that ambient O_3 levels are an indicator of photochemical pollutant levels, which may be responsible for the observed worsening asthma symptoms, or other pollutants which combine with O_3.

Several studies have been conducted to evaluate the potential effects of O_3 exposures in contributing to respiratory infections. Based on laboratory animal studies, O_3 has a variety of effects which can impair the body's ability to defend itself against infection. These include: (1) inhibition of mucociliary transport as well as cilia morphology and function; (2) impaired killing of bacteria; and (3) impaired function of macrophages. Macrophages engulf bacteria and particles and, as a result, serve as a pulmonary defense mechanism. Ozone, on the other hand, can inactivate certain viruses, including those that cause influenza. Elevated O_3 levels may, therefore, reduce the transmission of viral infections. Epidemiological studies have not, however, confirmed a relationship between elevated O_3 exposures and the prevalence or severity of respiratory infections.

There appears to be an adaptation to or attenuation of the effects of O_3 exposures on repeated exposures to health-affecting O_3 levels. In animal studies, rats have been observed to develop a tolerance to O_3 in long-term (8 hour/day) exposures. In human studies, the effect of O_3 is greatest on the second day of exposure, returning to normal by the fifth day. In southern California, studies have shown that pulmonary function changes caused by a relatively high episode (0.5 ppmv, 1000 $\mu g/m^3$) during a low O_3 season are more severe than when these same levels occur in a high pollution season.

There is limited information on O_3's potential to cause cancer. Studies on microorganisms, plant root tips, and tissue culture indicate that O_3 is genotoxic or mutagenic. Substances which are mutagenic appear to have a high potential for being carcinogenic as well. There is no conclusive evidence linking O_3 exposure and lung cancer in either epidemiologic or animal studies. In the latter, the data are conflicting. In rat studies, no significant increases in lung tumors have been observed. Ozone exposures appeared to cause a marginal increase in lung tumors in male mice with a more pronounced effect in females.

The present air-quality standard is based on regulatory concerns associated with the effects of O_3 on healthy people who exercise regularly outdoors and on asthmatics. The latter appear in epidemiologic studies to have increased symptomatic responses and visits to emergency rooms for treatment at relatively low ambient O_3 levels (>0.06 ppmv).

The air-quality standard concept, in addition to protecting the most sensitive subpopulations, is supposed to provide an adequate margin of safety to account for uncertainties associated with threshold levels. Many of the studies on which the above discussion is based show apparent O_3-related health effects at concentrations below 0.12 ppmv and/or exposure durations greater than 1 hour. It appears, therefore, that the current NAAQS for O_3 is not adequate to protect public health. Both a lower acceptable level and a longer exposure duration for an O_3 standard appears to be warranted. In contrast to the U.S., Canada has a 1-hour air-quality standard for O_3 of 0.082 ppmv (164 $\mu g/m^3$); and Japan, 0.06 ppmv (120 $\mu g/m^3$). The World Health Organization recommends 1-hour guideline values of 0.075 to 0.10 ppmv (150 to 200 $\mu g/m^3$) and an 8-hour guideline of 0.05 to 0.06 ppmv (100 to 120 $\mu g/m^3$).

Recently the Clean Air Scientific Advisory Committee compared responses of children playing outdoors at the present O_3 1-hour exposure standard with children playing outdoors at a more stringent 0.07 ppmv (140 $\mu g/m^3$) 8-hour exposure standard. Differences were observed to be small and overlapped all health endpoints evaluated. The panel concluded that there was no threshold concentration above background concentrations for biological effects. As such, selecting a standard at the lowest observable-effects level and providing a margin of safety is not possible. The panel recommended, however, that the 1-hour exposure standard be replaced with an 8-hour standard because health effects studies indicated that 6 to 8-hour exposures were more relevant to health than a 1-hour exposure.

Lead

USEPA regulated lead as an additive to gasoline in 1974 because it was deemed necessary to protect catalytic converters from being "poisoned" (coating catalytic materials, rendering them inactive) by lead. When USEPA began its phaseout of lead in automobile fuels, it used the relatively well-known health effects of lead to justify its actions. Environmental groups reasoned that because the health effects of lead were so significant, USEPA should promulgate an air-quality standard for it. Because of the onerous task of achieving air-quality standards which it had already promulgated, USEPA was initially reluctant to regulate lead in a similar way. Federal courts, in response to lawsuits brought by public interest groups, required USEPA to promulgate a NAAQS for lead.

Lead is a heavy metal which occurs naturally in soil, water, and even in air. Because of its usefulness and its relative ease in being smelted from mineral ores, lead has been used for thousands of years. As a consequence, humans have received and continue to receive lead exposures significantly above background levels. Notable historical exposures have included ingestion from liquids stored in lead containers (particularly during Roman times), ingestion of food in lead-soldered cans, ingestion of lead paint and dusts by children, and inhalation of airborne lead dusts from motor vehicle emissions and emissions from primary and secondary lead, as well as other metal smelters. Significant lead exposures have also occurred (and continue to occur) among lead workers. Much of our understanding of the adverse effects of lead on humans is based on worker exposures as well as exposures of children to ingested lead paint chips and dust.

Unlike a variety of other trace elements, lead is not required by the body to conduct any metabolic or physiological functions. Because of the widespread use of lead, total body burdens of humans today are approximately 300 to 500 times greater than in our preindustrial ancestors.

The lead burden of the body results from food and water ingestion and inhalation of lead-containing particles. On entering the body, lead is absorbed from its source where it is transported by the bloodstream to all parts of the body. Though commonly found in the blood and soft tissues, because of its similarities to calcium, lead tends to be deposited in bone, which serves as a depot for it.

The actual amount of lead extracted from food, water, ingested dust, and airborne particles depends on the form of lead, a person's nutritional status (poorly nourished individuals absorb more lead), metabolic activity, and prior exposure history. In the case of inhalation exposures, about 20 to 40% of airborne lead particles inhaled will be deposited in lung tissue where 50% or more of lead present will be absorbed and will enter the bloodstream. Some of the inhaled particles will be removed by pulmonary clearance mechanisms which, on swallowing, will enter the bloodstream through the gastrointestinal tract. Lead absorption from particles entering the gastrointestinal system is about 40% in children and about 10% in adults.

On exposure to significant levels of lead, an individual may develop symptoms of acute or chronic lead poisoning. Principal target organs include the blood, the brain and nervous system, and the renal and reproductive systems. Acute exposures (blood levels > 60 $\mu g/dL$) may produce colic, shock, severe anemia, nervousness, kidney damage, irreversible brain damage, and even death. Chronic poisoning may result in a variety of symptoms depending on the exposure. Acute and chronic symptom responses based on blood levels are summarized in Table

Table 5.2   Blood Lead Levels and Associated Health and Physiologic Effects in Children and Adults

Blood lead levels (μg/dL)	Children	Adults
<10		Early signs of hypertension ALA-D inhibition
10–15	Crosses placenta Neurodevelopmental effects ALA-D inhibition Impairment of IQ Increased erythrocyte protoporphyrin Reduced gestational age and birth weight	
15–20		Increased erythrocyte protoporphyrin
20–30	Altered CNS electrophysical response Interference with Vitamin D metabolism	
30–40	Reduced hemoglobin synthesis Peripheral nerve dysfunction	Systolic hypertension Altered testicular function Peripheral nerve dysfunction
40–50		Reduced hemoglobin synthesis Overt subencephalopathic neurologic symptoms
60	Peripheral neuropathies	Reproductive effects in females
70	Anemia	
80		Anemia Encephalopathy symptoms
80–100	Encephalopathy symptoms Chronic nephropathy Colic and other gastrointestinal symptoms	Chronic nephropathy Encephalopathy symptoms

From CDC. 1991. *Preventing Lead Poisoning in Children*. DDHS. October.

5.2. Chronic exposures on the more extreme end may cause severe brain damage and damage to the kidneys and blood-forming systems. At the low end of blood lead levels where chronic effects have been observed, health effects may include neurodevelopmental effects in children, increased blood pressure and related cardiovascular effects in adults, and possibly cancer.

Because lead accumulates in the body and is only slowly removed, repeated exposures over months to years will result in elevated blood lead levels. Because lead is deposited in bone, blood lead levels reflect relatively recent exposures (past 1 to 3 months) and/or lead mobilized from bone and other depots. Though not indicative of the total body burden of lead, blood lead is indicative of levels to which soft tissues are exposed. Blood lead is reproducible and can be used to indicate relative levels of exposure.

Hematological changes (i.e., effects on blood chemistry and associated physiological changes), are one of the earliest manifestations of chronic lead exposure. Lead interferes with the synthesis of the heme component of hemoglobin, the

O_2-carrying substance in red blood cells. Heme is also a component of cytochrome P-450 and other electron-transferring cytochromes. Heme is essential to the proper functioning of cells in vital organ systems such as the brain, kidney, liver, and blood-forming tissues. Lead's effects on hemoglobin synthesis may result in anemia, a common symptom associated with chronic lead poisoning. It can also inhibit the functioning of heme-dependent liver enzymes which can increase an individual's vulnerability to the harmful effects of other toxic substances. Lead's effects on heme production can also interfere with the production of Vitamin D and may play a role in causing neurological effects. Inhibition of Vitamin D production has been observed at blood lead levels of 30 µg/dL in adults and as low as 12 µg/dL in children. Inhibition of enzymes involved in heme biosynthesis have been observed at blood lead levels as low as 10 µg/dL. A threshold blood lead level for these effects has not been determined.

Lead exposures have also been associated with adverse sexual and reproductive effects. Sperm abnormalities, reduced fertility, and altered testicular function have been observed in male industrial workers at blood lead levels of 40 to 50 µg/dL. Lead may also pose significant risks to the fetus in exposed pregnant females. Because lead readily crosses the placenta, the fetus can be exposed to elevated maternal blood lead. Lead appears to be mobilized from bone during pregnancy. Several studies have shown that blood lead levels in pregnant females may be higher than population averages. Several epidemiological studies have reported an association between maternal blood lead levels with preterm delivery and low birth weight. In the latter case, an inverse relationship between blood levels and birth weight was observed in one study, with blood lead levels down to 15 µg/dL. Other studies failed to observe such a relationship.

The nervous system appears to be a major target of lead exposure. At high blood levels (>80 µg/dL), lead causes encephalopathy (brain damage). Children with blood levels in this range may experience permanent neurological damage such as severe mental retardation and recurrent convulsions. There is evidence to suggest that lead may impair peripheral nerve conduction in children at blood lead levels as low as 20 to 30 µg/dL. Brain wave changes have been observed at levels as low as 15 µg/dL, with no apparent threshold.

Many prospective epidemiological studies have been conducted to evaluate the relationship between blood lead levels and neurodevelopmental effects in children. These studies, taken together, indicate an association between general measures of intelligence (IQ) and pre- and postnatal blood levels as low as 10 to 15 µg/dL. These effects appear to be associated with both pre- and postnatal exposures. They suggest that, up to at least the age of seven, exposure to relatively low lead levels (10 to 40 µg/dL) may be associated with neurodevelopmental effects (decreased intelligence, short-term memory loss, reading and spelling underachievement, impairment of visual motor function, poor perception integration, disruptive classroom behavior, and impaired reaction time). A threshold level for neurodevelopmental effects based on blood lead levels has not been identified. Based on these studies, the U.S. Centers for Disease Control (CDC) has proposed 10 µg/dL as the blood lead level of concern for children. CDC also recommends that when many children in a community have blood lead levels between 10 and 14 µg/dL, a community-wide lead poisoning prevention program should be initiated.

In addition to these effects, there is evidence from occupational exposures that low-level lead exposures are related to increased blood pressure in adults. A relationship appears to exist between diastolic blood pressure and a range of blood lead levels down to 7 µg/dL for middle-aged white males. There is also some evidence for similar effects in women and other age groups. Elevated blood pressure is a significant risk factor for cardiovascular disease such as heart attack and stroke.

Lead has been shown to be genotoxic (i.e., it can cause gene mutations). It can also cause cell transformation and interference with DNA synthesis in mammalian tissue culture. Rodent studies have shown that lead compounds can induce kidney tumors. Though not conclusive, studies are suggestive of a causal relationship between lead exposures and cancer. Based on available evidence, USEPA has designated lead as a Group B2 human carcinogen, which means there is sufficient evidence of carcinogenicity from animals studies, but epidemiologic studies are not conclusive.

As can be seen from the above discussion, young children (<7 years old) appear to be the population at the greatest risk from lead exposure. They are at greater risk for the following reasons: (1) a less developed blood-brain barrier and therefore greater neurologic sensitivity; (2) a faster resting inhalation rate; (3) a tendency to breathe through the mouth when at play; (4) a faster metabolic rate, resulting in a greater daily intake of lead through food; (5) absorption of a greater amount of lead through intestines, especially in children younger than age two; (6) hand-to-mouth behaviors that result in the ingestion of lead from soil and dust; and (7) likelihood of pica (abnormal ingestion of nonfood items). Children from low-income areas are especially vulnerable because they have diets low in calcium and iron — elements that suppress lead absorption.

Lead exposure from inhalation has declined considerably since the mid-1970s. In survey studies conducted by CDC from 1976 to 1980 and 1989 to 1991, blood lead levels were observed to have decreased by approximately 78% in all age groups. Decreases in blood lead levels from 1976 to 1980 can be seen in Figure 5.10. From 1976 to 1980, 88% of children ages one to five were estimated to have had blood lead levels greater than or equal to 10 µg/dL. From 1988 to 1991, less than 10% of children in this age group were determined to have blood lead levels ≥ 10 µg/dL. Decreases in blood levels were also observed across various subgroups stratified by race and ethnicity, gender, urban status, and income levels. However, lead levels were disproportionately higher for Hispanic and African-American children.

The observed decline in blood levels coincided in time with the phaseout of lead in gasoline, reduction in use of lead-soldered cans for food and soft drinks (beginning in 1980), and limits on lead in paint (1978). The CDC has concluded that the major contributor to the observed decline in blood lead levels has been due to reduced use of lead in gasoline.

The major remaining sources of lead which have the potential for causing adverse health effects include (1) deteriorating lead-based painted surfaces in

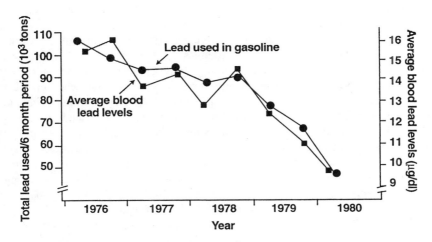

Figure 5.10   Changes in average blood lead levels in the U.S., 1976–1980. (From CDC. 1991. *Preventing Lead Poisoning in Children*. DDHS, October.)

older buildings, (2) accumulated lead in dust and soil, and (3) stationary sources (e.g., primary and secondary lead smelters).

A strong and consistent association exists between ambient concentrations of lead and blood lead levels in children and adults. It has been estimated that a blood lead to air lead relationship for adults is 1.8 µg/dL per µg/m³; for children, 4.2 µg/dL. Even exposures of 1 µg/m³ lead may be expected to have some adverse effects on children and adults, assuming that no threshold exists for neurodevelopmental effects in children and hypertension in adults. Fortunately, ambient lead concentrations in urban areas unaffected by emissions from stationary sources are now relatively low in the U.S. In California, a state with a significant motor vehicle population, average ambient lead levels of 0.06 µg/m³ are at their lowest values in 50 years.

Though both ambient and blood lead levels have decreased significantly in the U.S. over the past two decades, lead exposures from ambient air pollution is likely to be a significant public health threat in both developed and developing countries where lead is still widely used as a gasoline additive and where major stationary lead sources are unregulated.

HAZARDOUS AIR POLLUTANTS

Under Section 112 of the 1970 Clean Air Act (CAA) Amendments, USEPA was authorized to promulgate national emission standards for substances designated as hazardous air pollutants. Such substances were given special regulatory status because they were deemed to produce exposure hazards greater than pollutants regulated under air-quality standards. These substances were believed to cause unique exposure risks because of their innate toxicity and because there was no apparent threshold for their effects. Additionally, exposures from such substances were usually limited to the near vicinity of a source and were, in general, not considered to be community-wide exposure problems.

By the time the CAA Amendments were passed in 1990, only seven hazardous pollutants had been designated and emission standards applied to sources. These included mercury, beryllium, asbestos, vinyl chloride, benzene, inorganic arsenic, and radionuclides. Of these, only mercury and beryllium were uniquely toxic in the classical sense. The other five pollutants had been identified as human carcinogens from animal exposure studies, tests of genotoxicity, and epidemiological studies. Asbestos exposures, for example, have been definitively linked to increased risk of lung cancer and mesothelioma (a cancer of the chest cavity) in asbestos workers; vinyl chloride with angiosarcoma (a form of liver cancer) and other cancers among rubber/plastics workers; arsenic with lung cancer among smelter workers; benzene with leukemia among a variety of workers exposed to benzene; and radionuclides with a number of different cancers.

In the 1990 amendments to the CAA, Congress listed another 182 substances or groups of substances for regulation as hazardous pollutants. Such a large number of substances, of course, can be expected to have a variety of toxic effects. They include many substances regulated in the workplace. The major health concerns associated with these newly listed hazardous air pollutants include acute and chronic toxicity, neurotoxicity, reproductive toxicity, carcinogenicity, mutagenicity, and teratogenicity.

Though the toxic effects of many of the substances listed as hazardous air pollutants are well known from animal studies and workplace human exposures, little is known about ambient concentrations and health risks associated with such exposures.

ROLE OF HEALTH EFFECTS IN REGULATING AIR POLLUTANTS

The major goal of air pollution control during the last 25 years has been the protection of public health. To achieve this goal, USEPA has promulgated air-quality standards for seven pollutant categories. The regulation of these pollutants is based on the assessment of the health risk from the available scientific evidence. Because health effects data are less than definitive, the NAAQS represent both scientific judgment and political and economic realities.

In the U.S., health-based air-quality standards assume a threshold value below which no health damage will occur. Air-quality standards not only reflect this presumed threshold but also a margin of safety. The safety factor may be one half, one third, or one tenth of the concentration which the best evidence indicates is the threshold, or critical health effects, level. In many cases, air-quality standards are designed to protect the health of exceptionally sensitive population groups, that is, those at special risk. This emphasis on protecting exceptionally sensitive, relatively small population groups has received considerable criticism from industrial opponents of stringent air pollution control measures. They contend that it is not practical or economically desirable to protect the most sensitive individuals. On the other hand, the assumption that a threshold value exists may not be valid. A growing number of health scientists believe that for toxic materials there is no threshold value. If no threshold exists, then any permissible level above zero would be potentially harmful to someone, and a higher margin of safety would be desirable. Based on one's perception of risk, efforts to protect public health are either excessive or inadequate. Regulators are faced with a dilemma. If they favor the more high-risk groups, their decisions will impose real economic costs on some members of society who may fail to appreciate these as valid expenditures. On the other hand, regulators may not wish to make decisions as to who is or is not protected from the adverse consequences of toxic air pollutants.

PERSONAL AIR POLLUTION

It is paradoxical that as a nation we have spent hundreds of billions of dollars over the past 25 years in an effort to control ambient air pollution in order to safeguard public health, while at the same time nearly 40 million Americans voluntarily expose themselves to a form of air pollution whose health consequences are orders of magnitude greater. This is more of a paradox when we consider that the scientific evidence to support the establishment of air-quality standards for some pollutants is relatively limited, whereas the scientific case for a causal role of cigarette smoking in a number of health problems is overwhelming. Why such a contradiction? Apparently it is the nature of the risk. Smoking is a voluntary risk; exposure to ambient air pollution is not.

The health consequences of cigarette smoking are considerable, resulting in approximately 500,000 deaths each year. Of these more than 130,000 can be attributed to lung cancer, 25,000 to other cancers, 300,000 to cardiovascular disease, and about 20,000 to chronic pulmonary disease. According to reports issued by the U.S. Surgeon General, cigarette smoking is the single most important environmental factor contributing to premature mortality in the U.S.

The excess overall mortality of smokers (compared to nonsmokers) is 70%. Mortality is a function of the quantity smoked and smoking duration. Life expectancy significantly decreases with increased cigarette consumption and years of smoking. For a male smoking two packs per day, life expectancy is decreased by 8 to 9 years.

In addition to the risk of premature death that smoking poses, smokers are also likely to suffer more illness, including chronic bronchitis, emphysema, and cardiovascular disease. Before clinical disease is manifested, smokers usually suffer the discomfort of excessive coughing, phlegm production, and hoarseness.

One major effect of cigarette smoking is the induction of chronic respiratory diseases such as chronic bronchitis and emphysema. It is probable that a high percentage of diagnosed cases of such chronic obstructive lung diseases have been caused by cigarette smoking, as smokers have a higher incidence of these respiratory disabilities compared to nonsmokers.

Cigarette smoking is a major causal factor in human cancer. Strong epidemiological evidence exists to implicate smoking with lung cancer and cancers of the larynx, oral cavity, esophagus, urinary bladder, kidney, and pancreas. Smoking has made lung cancer the leading cause of cancer deaths in males. Although the incidence of lung cancer is less in females, it is rapidly increasing as more females smoke for longer periods of time. Overall, the lung cancer risk to smokers is about 10 to 12 times greater than that for nonsmokers. For smokers who are occupationally exposed to asbestos, the risk becomes significantly greater on the order of about 55 times. Such a synergistic effect in lung cancer induction is also observed in smokers who are exposed to rubber fumes and radon in uranium mines.

In the public mind, lung cancer is the most important health consequence of smoking. It is not generally realized by the public or smokers that smoking-related mortality from cardiovascular disease is approximately twice that of all forms of smoking-related cancer. Smoking is one of the three major independent risk factors for heart attacks in adults. It also appears to contribute to more severe and extensive atherosclerosis of the aorta and coronary arteries.

The voluntary risk accepted by smokers becomes less than voluntary to the fetus of a pregnant female. Significant scientific evidence exists that maternal smoking retards the rate of fetal growth and increases the risk of spontaneous abortion and fetal and neonatal death in otherwise normal infants. There is also increasing evidence that children of smoking mothers may have measurable deficiencies in physical growth and intellectual and emotional development.

Of the thousands of identified chemical constituents in cigarette smoke, CO, nicotine, and tar are the most likely contributors to the health hazards associated with smoking. Carbon monoxide levels are especially high, with levels in the mainstream smoke in the range of 1.5 to 5.5% (15,000 to 55,000 ppmv) and sidestream levels up to three times higher. The levels of CO to which smokers are exposed are exceptionally high. The effect of these high levels would be greater if it were not for the intermittent nature of the exposure. Inspiration of air between puffs greatly diminishes the effective dose. The major effect of CO in cigarette smoking appears to be to increase the risk of angina pectoris patients to myocardial infarction and sudden death. Because cigarette smoke contains numerous harmful substances, it is difficult to specifically assess the harmful effects of CO and its exact role in cardiovascular disease. Indeed, the effects of cigarette smoking on the incidence of clinical cardiovascular disease may not be due to CO alone. Other factors such as nicotine, cyanide, and trace elements may also be important. Nicotine is considered to be a major contributor to coronary atherosclerotic disease. Tar and nicotine make up the particulate phase of cigarette smoke. Tar consists of a variety of PAHs considered to be the likely causes of smoking-related cancers.

READINGS

1. Boushey, H.A. 1989. "Ozone and Asthma". 214–217. In: Utill, M.Z., and R. Frank (Eds.). *Susceptibility to Inhaled Pollutants*. ASTM STP 1024. American Society for Testing Materials, Philadelphia.

2. Brunekreef, B., D.W. Dockery, and M. Krzyzanowski. 1995. "Epidemiological Studies on Short-Term Effects of Low Levels of Major Ambient Air Pollution Components." *Environmental Health Perspectives* 103:3–14 (Suppl.).

3. California Air Resources Board/California Environmental Protection Agency. 1996. *Proposed Identification of Inorganic Lead as a Toxic Air Contaminant.* Part B Health Assessment.

4. Carnow, B.W. and P. Meir. 1973. "Air Pollution and Pulmonary Cancer." *Arch. Environ. Health* 27:207–218.

5. Centers for Disease Control. 1991. *Preventing Lead Poisoning in Children.* DHHS, October.

6. Coffin, D.L. and H.E. Stokinger. 1977. "Biological Effects of Air Pollutants." 232–345. In: A.C. Stern (Ed.). *Air Pollution. Vol. II. The Effects of Air Pollution.* 3rd ed. Academic Press, New York.

7. Ferris, B.G. 1978. "The Health Effects of Exposure to Low Levels of Regulated Pollutants. A Critical Review." *JAPCA* 28:483–497.

8. Frampton, M.W. and N.J. Roberts. 1989. "Respiratory Infection and Oxidants." 182–191. In: Utill, M.J. and R. Frank (Eds.). *Susceptibility to Inhaled Pollutants.* ASTM STP 1024. American Society for Testing and Materials, Philadelphia, PA.

9. Goldsmith, J.R. and L.T. Friberg. 1977. "Effects of Air Pollution on Human Health." 458–553. In: A.C. Stern (Ed.). *Air Pollution. Vol. II. The Effects of Air Pollution.* 3rd ed. Academic Press, New York.

10. Goldstein, B. 1983. "Toxic Substances in the Atmospheric Environment. A Critical Review." *JAPCA* 33:454–466.

11. Goldstein, B.D. 1986. "Critical Review of Toxic Air Pollutants." *JAPCA* 36:367–370.

12. Higgins, I. 1983. "What is an Adverse Health Effect?" *JAPCA* 33:661–663.

13. Horstmann, D.H. and L.J. Folinsbee. 1989. "Sulfur Dioxide-Induced Bronchoconstriction in Asthmatics Exposed for Short Durations Under Controlled Conditions: A Selected Review". 195–206. In: Utill, M.Z. and R. Frank (Eds.). *Susceptibility to Inhaled Pollutants.* ASTM STP 1024. American Society for Testing and Materials, Philadelphia, PA.

14. Koenig, Z.Q., D.S. Covert, and W.E. Pierson. 1989. "Acid Aerosols and Asthma: A Review." 207–213. In: Utill, M.Z. and R. Frank (Eds.). *Susceptibility to Inhaled Pollutants.* ASTM STP 1024. American Society for Testing and Materials, Philadelphia, PA.

15. Lambert, P.M. and D.D. Reid. 1970. "Smoking, Air Pollution and Bronchitis in Britain." *Lancet* 5:853–857.

16. Lerkauf, G.D., L.G. Simpson, J. Santrock, Q. Zhao, J. Abbinate-Nissen, S. Zhou, and K.E. Driscoll. 1995. "Airway Epithelial Cell Responses to O_3 Injury." *Environ. Health Perspect.* 103:91–96 (Suppl.).

17. Lipmann, M. 1989. "Health Effects of Ozone: A Critical Review." *JAPCA* 39:672–694.

18. NAPCA. USDHEW. 1969. *Air Quality Criteria for Particulate Matter.* Publication No. AP-44.

19. NAPCA. USDHEW. 1970. *Air Quality Criteria for Carbon Monoxide.* Publication No. AP-62.

20. National Research Council, Committee on Medical and Biological Effects of Environmental Pollutants. 1977. *Nitrogen Oxides.* National Academy of Sciences, Washington, DC.

21. National Research Council, Committee on Medical and Biological Effects on Environmental Pollutants. 1977. *Ozone and Photochemical Oxidants*. National Academy of Sciences, Washington, DC.

22. National Research Council, Committee on Medical and Biological Effects of Environmental Pollutants. 1977. *Carbon Monoxide*. National Academy of Sciences, Washington, DC.

23. National Research Council, Committee on Medical and Biological Effects of Environmental Pollutants. 1979. *Airborne Particles*. National Academy of Sciences, Washington, DC.

24. Selikoff, I.J., and D.H.K. Lee. 1978. *Asbestos and Disease*. Academic Press, New York.

25. USEPA. 1982. *Air Quality Criteria for Particulate Matter and Sulfur Oxides*. EPA/600/8- 82-029a-c.

26. USEPA. 1986. *Air Quality Criteria for Lead*. EPA/600/8-83-028a-f.

27. USEPA. 1990. *Air Quality Criteria for Lead*. Supplement to the 1986 addendum. EPA/600/8-90/049f.

28. USEPA. 1991. *Air Quality Criteria for Carbon Monoxide*. EPA/600/8-90/045f.

29. USEPA. 1994. *Review of the National Ambient Air Quality Standards for Sulfur Oxides: Assessment of Scientific and Technical Information*. Supplement to the 1986 OAQPs Staff Paper Addendum. EPA/452/R-94-013.

30. USEPA. 1995. *Air Quality Criteria for Particulate Matter. Vol. III*. EPA/600/AP-95-001c.

31. Watson, A.Y., R.R. Bates, and D. Kennedy (Eds.). 1988. *Air Pollution, The Automobile and Public Health*. Health Effects Institute. National Academy Press, Washington, DC.

QUESTIONS

1. In conducting epidemiological studies of pollutant exposures and health effects, a variety of factors confound statistical analysis and make it more difficult to establish a causal relationship. Most notable are interaction effects between ambient air pollutants and weather, between ambient air pollutants, and between ambient air pollutants and other pollutant exposures. Describe in specific terms each type of potential interaction.

2. What evidence is there to support a hypothesis that ambient pollutant exposures increase one's risk to lung cancer?

3. The respiratory system has evolved mechanisms to remove particles which are in inspired air. Describe each of these, how they function, and general particle size ranges involved.

4. Ambient exposures to SO_2, PM_{10}, and O_3 may individually cause what type of health effects?

5. Compare the known health risks of tobacco smoking to those associated with major ambient pollutants. How do they compare?

6. Asthma is a very prevalent respiratory disease. What is the relationship between exposures to ambient air pollutants and the induction of the asthmatic condition? of asthmatic attacks?

7. Exposure to ambient air pollutants has been associated with premature mortality. Describe circumstances under which such mortality may occur.

8. Respiratory responses to the inhalation of SO_2 and NO_2 differ significantly. Why?

9. Average blood lead levels in the U.S. declined dramatically in the late 1970s. What is the most probable reason?

10. Air-quality standards have been promulgated for seven pollutants and pollutant categories. There is considerable scientific evidence that two of these standards may not be adequate to protect public health. What are they?

11. How do pollutants regulated as hazardous pollutants differ from those regulated under air-quality standards?

12. Exposure to which ambient pollutants is likely to cause or contribute to the following health effects?

    1. asthmatic attacks   2. learning disabilities
    3. heart disease   4. premature death
    5. emphysema

13. What factors determine penetration and deposition of particles in the lungs?

14. What health concerns are associated with exposures to polycyclic aromatic hydrocarbons?

15. Epidemiological and toxicological studies are both important in determining causation. What is the specific relevance of toxicological studies?

6

WELFARE EFFECTS

In developing control programs for ambient air pollution, we as a nation have placed most of our emphasis on the protection of public health. Effects other than health have been designated as secondary in importance. Nonhealth-related effects are usually referred to as welfare effects. These include damage to vegetation, damage to materials, injury to livestock, odor pollution, and reduced visibility. Visibility impairment has been discussed in Chapter 4. This chapter discusses those welfare effects not previously discussed in this volume.

PLANTS

Although our present control efforts are almost entirely concerned with protecting human health, it has been plants which have served as sentinels of the biological damage that ambient air pollutants are capable of producing. Effects of phytotoxic pollutants such as sulfur dioxide (SO_2), hydrogen chloride (HCl), and hydrogen fluoride (HF) were reported in Europe as early as the mid-19th century. Particularly severe damage to vegetation has been associated with emissions of SO_2 and heavy metals from primary metal smelters. In the U. S., the area surrounding Copper Hill Smelter in Ducktown, TN has often been cited as one of the most extreme examples of air pollution-induced damage to vegetation. Copper Hill emissions devastated the surrounding countryside, producing a landscape resembling spoil banks of an unreclaimed strip mine. Much of this devastation was produced before the turn of the 20th century.

During the 1920s, emissions from a smelter in Trial, British Columbia were carried across the border into the U. S., causing significant damage to agricultural crops in the Columbia River Valley. This was one of the first major international air pollution incidents and helped to bring about an awareness of the problem of transboundary transport. Significant destruction of vegetation has also been reported around the giant nickel smelter near Sudbury, Ontario and other primary zinc, lead, copper, and aluminum smelters throughout the world. Significant plant injury has also been reported from a variety of other industrial point sources, including coal-fired power stations, phosphate fertilizer plants, and glass manufacturing.

Until the 1940s and 1950s, plant damage was generally recognized to be an isolated problem associated with single-point sources. As a result of intensive scientific investigations, it became apparent that widespread injury to agricultural crops reported as early as the mid-1940s in the Los Angeles Basin was due to phytotoxic air pollutants such as ozone (O_3) and peroxyacyl nitrate (PAN). Indeed, it was a plant biochemist, Dr. Arie Hagen-Smit, often referred to as the father of air pollution studies, who first began to unravel the chemical complexities of smog formation. His pioneering work on the chemistry of photochemical smog was an outgrowth of his observations of air pollution-induced injury on greenhouse plants which were to be used in plant biochemistry studies. This is a classic example of how events affect the future course of scientific discovery and, of course, the career of the scientist.

After Hagen-Smit elucidated the chemical basis of O_3 formation in smog, other plant scientists confirmed that O_3 was the causal agent of observed injury on a variety of agricultural crops and other plant species. The first such confirmation came in 1958 when it was shown that O_3 caused what had been referred to as "stipple" on grapes. Investigations by chemists at the University of California, Riverside led to the identification of PAN in 1961 and the subsequent confirmation that this phytotoxic substance was responsible for injury previously referred to as smog injury.

Since these early California discoveries, O_3 injury on sensitive vegetation has been observed in many parts of the U. S. Indeed, elevated O_3 levels resulting from photochemical conversion of anthropogenic emissions are widespread. Levels sufficient to cause injury on very sensitive vegetation are reported in most areas east of the Mississippi River. Because of its ubiquitous distribution and high phytotoxicity, O_3 is the most important of the identified phytotoxic air contaminants.

Control efforts and changes in operating practices have significantly reduced the localized plant damage that has been associated with many point sources. Paradoxically, one of these changes, the use of tall stacks for the more effective dilution of emissions from coal-fired power stations, has not only resulted in decreased injury to vegetation in the vicinity of these sources but has inadvertently contributed to the problem of long-range transport and acidic deposition distant from sources of acid precursors. It is ironic that in our attempts to solve a localized direct plant effects problem we have contributed to a large-scale acid problem whose ecological consequences may be much greater.

Pollutants which have a long history of being known to cause significant plant injury under ambient conditions of exposure include SO_2, fluorides, O_3, PAN, and ethylene. Although it has been suspected to be a major plant-injuring pollution problem for some time, acidic deposition has only recently been suggested by scientists to be causally related to plant injury under ambient conditions (e.g., the decline of red spruce in high-altitude forests).

Other pollutants are also known to cause injury to plants. These are the so-called minor pollutants and include NO_2, Cl_2, HCl, NH_3, and particulate matter. They are classified as minor pollutants because their emission from continuous point sources at levels sufficient to cause injury is relatively limited. Plant injury associated with exposures to gases described as minor pollutants is often associated with accidental releases in industrial operations or in transportation. Injury induced by particulate dusts, on the other hand, is usually associated with continuous emissions from point sources.

Plant Structure

The induction of air pollution injury and subsequent symptom expression is dependent on several physical and biological factors. A plant basically has four organs: roots, stems, leaves, and reproductive structures. Each can be affected by phytotoxic pollutant exposures. However, the leaf is the principal target organ, because it is the organ involved in gas exchange. Consequently, most air pollution-induced visible injury on plants is foliar. Subtle effects such as growth reduction may be generalized, affecting all organs.

Although leaf function is consistent from species to species, significant differences in leaf structure and anatomy exist. These differences are particularly striking among different plant groups. Historically, the effects of phytotoxic pollutants on two groups of higher plants have been of major interest. These include the angiosperms and the gymnosperms. The angiosperms are those plants that produce seeds in enclosed reproductive structures and include all of the flowering plants. Within this group are two subgroups: broad-leaved species called dicots and narrow-leaved species called monocots. These subgroups differ not only in external but also internal leaf structure. The gymnosperms are members of a plant group that produce naked seeds on the top of scales which collectively are referred to as cones. The gymnosperms are therefore coniferous plants. Their external leaf structure is characteristically needle-like and thus considerably different from those of the angiosperms. The internal anatomy of gymnosperm leaves or needles is different as well. The anatomical differences among the major plant groups described are major factors in the distinct symptom patterns observed when they are exposed to phytotoxic pollutants.

Anatomy of a Dicot Leaf

For illustrative purposes the internal anatomy of a dicot leaf is described. Dicots are the dominant plant group and are therefore most likely to be affected by phytotoxic pollutant exposures. As can be seen in Figure 6.1, the upper surface is overlain by a waxy layer referred to as the cutin. Below the cutin is a layer of colorless cells, the upper epidermis. Both the cutin and the upper epidermis protect the leaf from desiccation and mechanical injury. Located beneath the upper epidermis is a layer or two of photosynthetically active cells whose axes are at right angles to the leaf surface. This tissue is the palisade mesophyll. Contiguous with the palisade cells is the spongy mesophyll. Cells

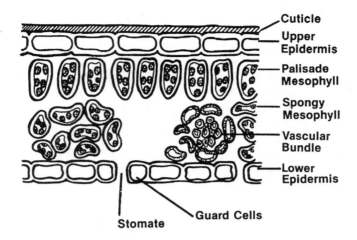

Figure 6.1   Cross-sectional anatomy of a dicot leaf.

of this tissue are irregular in shape and are loosely arranged. The loose arrangement provides for large intercellular spaces where gas exchange occurs. The lower surface of the leaf is bounded by the lower epidermis, which also functions to protect the leaf.

In most species, large numbers of specialized units called stomates are located in the lower epidermis. In some species stomates are located on both the lower and upper surfaces. The stomates consist of a variable-sized pore and two crescent-shaped guard cells (Figure 6.2). The movement of the guard cells determines pore diameter and regulates the exchange of metabolic gases such as CO_2 and H_2O vapor. The pore of the stomate is contiguous with large intercellular spaces in the spongy mesophyll. It is through these stomate pores that phytotoxic gases enter and react with the metabolically functioning tissues of the leaf. The entry of the phytotoxic gas depends on physiological and environmental conditions which determine the diameter of the stomate pore. In addition to these tissue

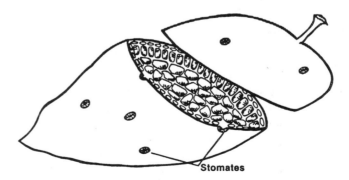

Figure 6.2   Stomates on lower surface of a dicot leaf.

layers, the leaf is subtended by a network of conducting tissues called veins. These serve to transport water and minerals upward into the leaf and metabolic products such as sugars downward into stems and roots.

Plant Injury

Injury to plants can be manifested as visible and/or subtle effects. Visible effects are identifiable changes in leaf structure, which may include chlorophyll destruction (chlorosis), tissue death (necrosis), and pigment formation. Visible symptom patterns may result from acute or chronic exposures. Acute injury often results from brief exposures (several hours) to elevated levels of a phytotoxic pollutant. Tissue necrosis is generally the dominant symptom pattern elicited as a result of acute exposures. The necrotic foliar patterns are usually specific for a given pollutant.

Chronic injury usually results from intermittent or long-term exposures to relatively low pollutant concentrations, with chlorophyll destruction and chlorosis as the principal symptoms of injury. However, chlorosis is a nonspecific symptom associated with several pollutants, natural senescence (aging), and injury caused by other factors such as nutritional deficiencies.

Some pollutants may produce symptom patterns other than, or in addition to, the classic patterns of necrosis and chlorosis associated with acute and chronic injury. These are usually referred to as growth abnormalities. They may be subtle, such as growth reduction, or visible, such as accelerated senescence of flowers, bolting of flower buds, abscission of plant parts, and a curvature of the leaf petiole called epinasty.

Symptom Patterns of Phytotoxic Pollutants

Sulfur Dioxide. Sulfur dioxide enters stomates and immediately comes into contact with spongy mesophyll cells in the near vicinity of the pore, where the toxic response is initially manifested. As the exposure continues, there is a progressive expansion of injury and tissue collapse (Figure 6.3). As injury develops, affected tissue may have a grayish-green water-soaked appearance which on drying becomes ivory-to-white, red, brown or even black, depending on the species. The injury extends from the bottom to the top of the leaf and is visible on both surfaces.

The severity of the injury, that is, the amount of leaf tissue affected, is dependent on the dose to which the plant has been exposed. The greater the dose or exposure, the more severely injured are individual leaves as well as the whole plant. The entire leaf, or in some extreme cases the entire plant, may be killed.

The severity of observed injury on individual leaves is also dependent on leaf maturity. Young, fully expanded leaves are most sensitive to SO_2. Older leaves and those not fully expanded are less sensitive.

In dicots, acute SO_2 injury is usually manifested as an interveinal necrosis, although necrosis in some species may appear on the margins. In some instances,

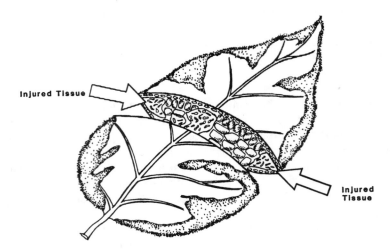

Injured Tissue

Injured
Tissue

Figure 6.3 Typical tissue injury associated with SO₂ exposure.

chlorosis may be associated with the necrotic areas. In monocots, such as grains, lilies, and gladiolus, injury may appear as irregular bifacial necrotic streaks between the larger veins. At low-to-moderate concentrations, SO_2 may cause leaf tip injury on monocots. Sulfur dioxide injury on conifers is observed as reddish-brown to brown tip necrosis which may have a banded appearance from repeated exposures. Older needles may become chlorotic under moderate exposures and may shed prematurely. Sulfur dioxide injury on plants in each of these three plant groups is illustrated in Figures 6.4 to 6.6.

Different species and varieties vary in their sensitivity to SO_2 exposures. The relative sensitivity of plant species to SO_2 is summarized in Table 6.1. Of those listed, alfalfa is the most sensitive species, with acute injury observed at a dose of 1 ppmv for 1 hour. Because of its hypersensitivity, alfalfa has been used as a bioindicator of phytotoxic SO_2 levels in the ambient atmosphere.

Ozone. Ozone enters the leaf through the stomates on the lower epidermis. In dicots it bypasses the spongy cells and injures the palisade mesophyll cells preferentially (Figure 6.7). As a consequence, symptoms of acute injury are usually visible only on the upper leaf surface. In monocots, conifers, and dicots without a palisade mesophyll, cells near the stomates are injured, with symptoms appearing on either or both surfaces. Developing leaves ranging from 65 to 95% of their full size are most sensitive to O_3 injury. The maturity of the plant also determines sensitivity, with younger plants being more sensitive and older plants more resistant.

The most common O_3-induced symptom patterns observed on dicots are upper surface flecks. These flecks are produced when groups of palisade cells are killed or injured, resulting in chlorotic or necrotic lesions. The adjacent epidermal cells may in some cases collapse. The flecks may be white, tan, or yellow. If flecking is extensive, the upper leaf surface may appear bronzed or

Figure 6.4   Sulfur dioxide injury on Shummard oak.

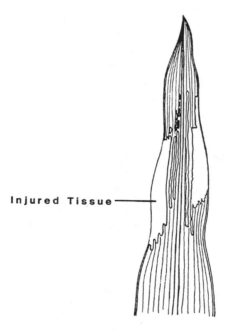

Figure 6.5   Sulfur dioxide injury on grain sorghum.

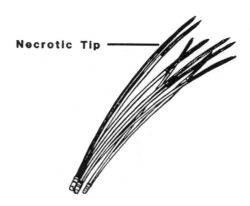

Necrotic Tip

Figure 6.6   Sulfur dioxide injury on scotch pine.

chlorotic. A variation of this symptom is called stipple. In stipple, the injured palisade cells may become intensely pigmented, giving the leaf surface a red-purple to black-brown appearance. The collapse of the overlying epidermal cells is generally not observed with the stipple symptom. Although flecking may be considered to be a more severe reaction than stipple, it is apparent that these symptoms are more related to plant species than to O_3 dose. In both the fleck and stipple symptom patterns, injury is confined to the interveinal tissues, the veins being uninjured. In narrow-leaved species the most common symptom is small necrotic or chlorotic bifacial areas (flecks) between the veins which are most severe in the bend of the leaf blade.

Ozone injury on coniferous species such as pine is common. It is the causal factor in emergence tipburn of eastern white pine, a disease observed on white pines since the early 1900s. The disease syndrome is characterized by the death of tips of young elongating needles in spring and early summer. The killed needle

Table 6.1   Relative Sensitivity of Plants to SO_2 Exposures

Sensitive	Intermediate	Resistant
Flowering shrubs/ornamental plants		
Sweet pea	Gladiolus	Chrysanthemum
Cosmos	Tulip	
Aster	Zinnia	
	Trees/shrubs	
White pine	Austrian pine	Red cedar
White birch	Alder	Maple
Trembling aspen	Douglas fir	Linden
	Field/garden crops	
Spinach	Cucumber	Corn
Turnips	Tomato	Potato
Beets	Lettuce	Cabbage
Alfalfa	Wheat	Broccoli
Oats	Beans	

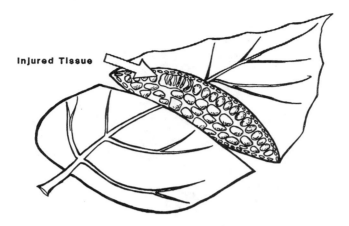

Figure 6.7 Typical tissue injury associated with O₃ exposure.

tips become reddish-brown. Ozone apparently not only produces tip necrosis, but also symptoms ranging from chlorotic flecks through chlorotic mottling (small chlorotic patches). In general, chlorotic flecking and mottling occur on less sensitive trees whereas tip necrosis is observed on trees which are more sensitive. On O_3-affected trees, the older needles become prematurely aged and discolored and are cast early. As a result, the tree may only possess the current year's needles.

Ozone, along with SO_2, may be responsible for the chlorotic dwarf disease of eastern white pine which has been observed in the Northeast for nine decades. In young, sensitive trees, current year needles become spotted with chlorotic flecks and mottling. Older needles become prematurely chlorotic and are shed before the new set of needles is fully developed. These trees are severely stunted and usually die before they reach 15 years of age.

In the San Bernadino Mountains of southern California, large numbers of ponderosa pines have been affected by ozone needle mottle. The earliest symptoms are small chlorotic mottles which develop from the tip to the base in older needles. Needle tip necrosis may also occur. Chronic O_3 exposures result in the premature aging and casting of affected older needles. As a result, O_3-injured trees may possess only 1-year-old needles during the summer months rather than the 3- to 5-year normal complement.

Ozone injury to representative plants is illustrated in Figures 6.8 to 6.10. The relative sensitivity of plants to O_3 is summarized in Table 6.2. As can be seen from Table 6.2, O_3 sensitivity also varies from species to species and from variety to variety within species. Ozone is more phytotoxic than SO_2. It is not uncommon to observe symptoms on sensitive plants from exposures in a concentration range of 0.10 to 0.30 ppmv for a few hours. Indeed some plants, such as a tobacco variety called BEL-W3, show acute injury from O_3 exposures as low as 0.05 ppmv for a few hours.

Ozone affects a variety of physiological processes in plants. Among the more important of these is suppression of photosynthesis; the higher the O_3 concentra-

Figure 6.8 Ozone injury on white ash.

tion, the greater the reduction in photosynthesis. This effect on photosynthesis may occur in the absence of visible injury and may be responsible in part for growth reduction observed in crop plants and forest species.

Ozone can damage cell membranes. Such damage may result in severe nutrient loss. Nutrient imbalances associated with atmospheric pollutants have been implicated as contributing to the severe injury to forest trees observed in central Europe. Ozone is considered by plant scientists to be the most important phytotoxic air pollutant. It is estimated to cause more than 90% of the plant injury associated with atmospheric pollutants in the U.S.

Peroxyacyl Nitrate. Peroxyacyl nitrate (PAN) is a secondary pollutant formed in photochemical smog. It is the causal agent of smog injury observed on vege-

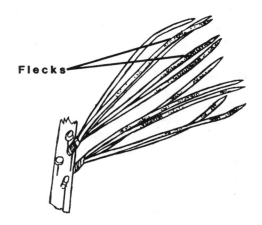

Figure 6.9 Ozone injury on eastern white pine.

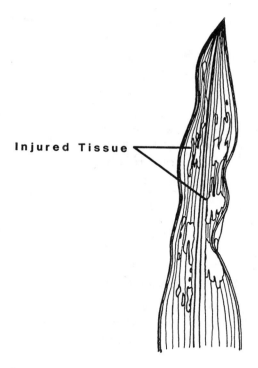

Injured Tissue

Figure 6.10 Ozone injury on corn.

table crops in the Los Angeles Basin, where it has been a major problem since the mid-1940s. Peroxyacyl nitrate injury has also been observed in other western cities but to a far lesser degree. Vegetable injury by PAN has also been observed in New Jersey and southern Canada.

As PAN enters the leaf, it reacts with spongy mesophyll adjacent to the stomates, causing their collapse (Figure 6.11). Large air pockets are formed that are responsible for the glazing or bronzing symptom which is observed on the lower leaf surface of many dicot species. In some plants, such as petunia, however, total tissue collapse and necrosis of sensitive tissue may occur, resulting in bifacial necrosis.

A distinct pattern of banding is usually observed with PAN exposures. These bands occur in the same location on leaves of the same physiological age. Peroxyacyl nitrate injury often appears as bands at the apex of the youngest sensitive leaf, the middle of an intermediate-aged leaf, and the base of the oldest sensitive leaf. Young, rapidly developing leaves on young, rapidly growing plants are most sensitive to PAN.

Glazing and bronzing rarely occur in monocots such as grasses. Peroxyacyl nitrate injury usually appears as distinct bands several millimeters to 2 cm wide across the leaf blade. If the injury is slight, the injured tissue may be chlorotic. Severe injury results in tissue collapse and bleaching of the necrotic tissue.

Table 6.2   Relative Sensitivity of Plants to O_3 Exposures

Sensitive	Intermediate	Resistant
	Ornamentals	
Coleus	Begonia	Chinese azalea
Chinese lilac	Common privet	Gladiolus
Snow azalea	Lynwood gold forsythia	American holly
Petunia	Common lilac	Laland's firethorne
Bridalwreath	Sweet mock orange	Mountain laurel
Spreading cotoneaster	Linden viburnum	Trailing mahonia
	Deciduous trees/shrubs	
White ash	Box elder	American linden
Quaking aspen	Eastern redbud	Black locust
American sycamore	Pin oak	Sugar maple
Honey locust	Black oak	Black walnut
White oak	Rhododendron	Red maple
Tulip poplar		
	Coniferous trees/shrubs	
Jack pine	Eastern white pine	Red pine
Loblolly pine	Scotch pine	Douglas fir
Austrian pine	Short leaf pine	Colorado blue spruce
Ponderosa pine	Incense cedar	Norway spruce
	Field/garden crops	
Alfalfa	Cabbage	Beet
Potato	Field corn	Cotton
Radish	Cucumber	Rice
Tobacco	Pea	Strawberry
Tomato	Turnip	Sweet potato

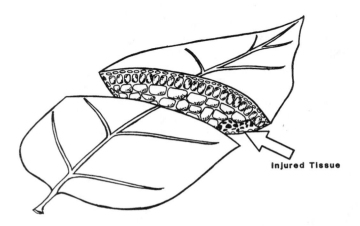

Figure 6.11 Typical tissue injury associated with PAN exposure.

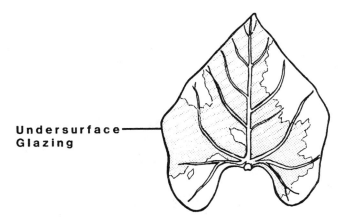

Undersurface Glazing

Figure 6.12 Peroxyacyl nitrate injury on lower surface of pinto bean.

Peroxyacyl nitrate is not known to cause injury in conifers or, for that matter, in woody plants.

Peroxyacyl nitrate injury is illustrated in Figures 6.12 and 6.13. Peroxyacyl nitrate is more phytotoxic than O_3. Sensitive plants such as petunia, tomato, and romaine lettuce may be injured by exposures to as little as 15 to 20 ppbv for a few hours. The relative sensitivity of plants to PAN is summarized in Table 6.3.

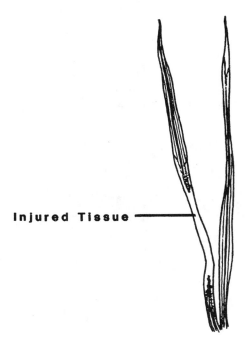

Injured Tissue

Figure 6.13 Peroxyacyl nitrate injury on oats.

Table 6.3   Relative Sensitivity of Plants to PAN
             Exposures

Sensitive	Intermediate	Resistant
Flowering shrubs/ornamental plants		
Aster	Carnation	Lily
Petunia	Coleus	Periwinkle
Mint	Lilac	Orchids
Weeds		
Chickweed	Cheeseweed	
Pigweed	Lamb's quarters	
Jimsonweed		
Trees/shrubs		
	Apple	Norway maple
	Green ash	White oak
	Basswood	Austrian pine
	Lilac	Blue spruce
Field/garden crops		
Bean	Alfalfa	Cabbage
Tomato	Spinach	Corn
Lettuce	Soybean	Radish

Fluorides. Fluoride injury may result from the uptake of gaseous HF through the stomates or from soluble particulate fluorides absorbed through the leaves and/or roots. Whatever the route of entry, fluorides enter the veins and are transported to leaf margins and/or the leaf tip, where they accumulate. Fluoride injury generally results from the accumulation of small quantities of fluoride over a period of weeks to months. The severity of injury is directly related to level of fluoride in leaf tissue.

In dicot species, fluoride injury appears as a marginal or tip necrosis. The injury occurs in both mesophyll and epidermal cells (Figure 6.14). The necrotic tissue is often separated from healthy tissue by a narrow, sharply defined reddish band. The necrotic tissue may have a characteristic wavy color pattern due to

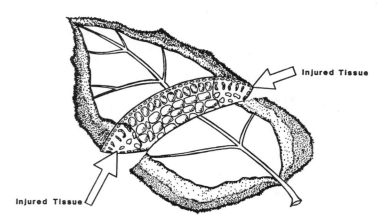

Figure 6.14 Typical tissue injury associated with fluoride exposure.

successive fluoride exposures. The amount of color contrast, as well as the color of necrotic tissue, varies from species to species. The necrotic tissue may break away from uninjured tissue near the reddish band, giving the leaf a ragged appearance.

Injury on monocot species is characterized by a tip necrosis which extends in irregular streaks down the leaf. The necrotic tissue may vary in color from ivory to various shades of brown. A band of dark brown tissue sharply demarks injured from healthy tissues. Symptoms may consist of chlorotic mottle at the margins and leaf tips in species such as corn and sorghum. The small irregular chlorotic patches form between the veins and merge to form continuous chlorotic bands as injury becomes extensive. Necrosis of the chlorotic tissue may occur if the injury is severe.

Tip necrosis is the characteristic symptom of fluoride injury in conifers. The necrosis begins at the tip of current year needles and progresses downward. The border between living and dead tissue may be sharply delimited, with a band of darker tissue adjacent to the unaffected tissue. The necrotic tip is usually reddish-brown. Conifer needles are most sensitive when they are in the early stages of development in the spring, becoming more resistant as they become older.

Typical fluoride-induced injury on selected species is illustrated in Figures 6.15 to 6.17. The relative sensitivity of plants to fluoride is summarized in Table 6.4. Some sensitive species can be injured by fluoride exposures of only a few ppbv for 1 week.

Ethylene. Plant injury induced by ethylene was first reported when ethylene was a component of illuminating gas used in greenhouses at the turn of the century. Today, ethylene is a ubiquitous pollutant found in urban environments,

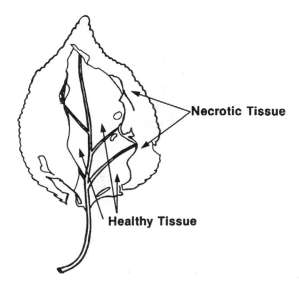

Figure 6.15 Fluoride injury on apricot.

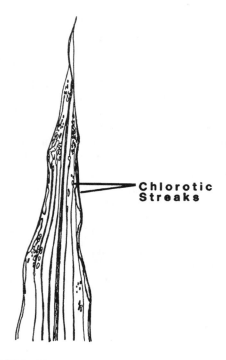

Figure 6.16 Fluoride injury on corn.

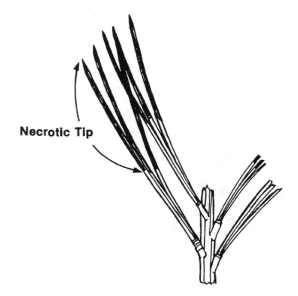

Figure 6.17 Fluoride injury on ponderosa pine needles.

Table 6.4   Relative Sensitivity of Plants to Fluoride Exposures

Sensitive	Intermediate	Resistant
Flowering shrubs/ornamental plants		
Gladiolus	Azalea	Petunia
Jerusalem cherry	Rose	Bridalwreath
Iris	Dahlia	Chrysanthemum
Tulip	Lilac	Privet
Trees/shrubs		
Ponderosa pine	Grand fir	Lodgepole pine
Blue spruce	White spruce	Arborvitae
Oregon grape	Quaking aspen	Sycamore
Box elder	Black walnut	Dogwood
Serviceberry	Silver maple	Black locust
Field/garden crops		
Sweet corn	Field corn	Tobacco
Sweet potato	Tomato	Cabbage
Wheat (young)	Alfalfa	Cucumber
Oats (young)	Spinach	Summer squash

as it is one of the major substances found in automobile exhaust. Concentrations of ethylene in urban areas may be sufficiently high to cause significant effects on plants. In addition, plants may produce increased ethylene levels when they are under stress from biotic or abiotic factors. Some symptoms associated with O_3 injury such as epinasty (abnormal curving of the leaf stem), advanced senescence (aging) of affected leaves, and decreased plant growth, may be the result of ozone-induced stress ethylene production.

One of the major normal functions of ethylene is to act as a maturation hormone controlling the ripening and senescence of many fruits. The presence of ethylene at elevated levels at other development stages may also have significant effects. Ethylene produced by ripening apples may be sufficient to accelerate senescence and decrease the keeping quality of cut flowers stored in the same environment. Indeed, the saying that "one bad apple spoils the bunch" is literally true; the elevated ethylene levels of an overly mature or rotting apply hormonally induces other apples to mature more rapidly.

The effects of ambient levels of ethylene on plants growing in the urban environment are potentially great, because ethylene has significant biological activity at very low concentrations. Anthropogenically produced ethylene-induced changes are difficult to detect, as they mimic natural processes. The acceleration of maturation induced by ambient ethylene exposures can, however, be expected to significantly affect plant growth.

One of the best-documented effects of ethylene injury is "dry sepal" of orchid associated with the use of ethylene illuminating gas in greenhouses. Dry sepal has also been reported in urban areas, presumably from elevated levels of ethylene resulting from automobile emissions.

Studies by researchers at Beltsville, MD have shown that elevated ethylene levels measured in the Washington, DC metropolitan area are capable of causing plant injury. Comparable exposures under laboratory conditions resulted in reduced plant growth, premature senescence, and reduced flowering and fruit production. Similar effects were observed on greenhouse plants that were exposed to ethylene in ambient air.

Flowering plants are particularly sensitive to ethylene. Exposures may cause premature bud break (bolting), inhibition of flowering (bullheading), and accelerated flower aging. Because of these effects, ethylene may pose some economic threat to commercial growers where ambient levels are elevated from automobile emissions.

Because of its hormonal properties, ethylene has the potential for causing widespread deleterious effects to vegetation. However, field identification of injury resulting from such exposure may be nearly impossible, as injury may mimic natural senescence and changes induced by a variety of environmental stresses. An interesting example of this problem is the observed premature pigmentation of tree and shrub leaves along many of our highways. Is this effect due to ethylene or some other atmospheric pollutant, or is it due to other factors such as deicing salt and/or thermal emissions from motor vehicles and road surfaces?

Acidic Deposition. Acidic materials deposited on plants and/or the plant soil medium have the potential for causing injury or measurable changes in plants. Laboratory and greenhouse studies with simulated acidic rain events have shown that a wide variety of plants can be injured by exposure to solutions with a pH of approximately 3.0. Symptoms reported include (1) pitting, curl, shortening, and death in yellow birch; (2) small (<1 mm) necrotic lesions, premature abscission in kidney bean, soybean, loblolly pine, eastern white pine, and willow oak; (3) cuticular erosion in willow oak; (4) chlorosis in sunflower and bean; (5) wrinkled leaves, excessive adventitious budding, and premature senescence in bean; (6) marginal and tip necrosis in bean, poplar, soybean, ash, birch, corn, and wheat; and (7) galls, hypertrophy, and hyperplasia in hybrid poplar.

The most common symptom of acid injury on plants observed in a wide variety of studies is small necrotic lesions. These appear to be a result of the deposition and retention of acidified water on the leaf surface and subsequent evaporation once lesions form. Reported exposures to simulated acid rain events result in an increased number of lesions on affected leaves. The injury begins on the upper surface and is frequently followed by the collapse of tissue so that both surfaces are affected. Fully expanded leaves are the most sensitive to the direct effects of acid rain.

Other effects (not manifested as visible injury) include changes in yield and leaching of a variety of substances from plant surfaces. Simulated acid rain effects on yield have resulted in a number of outcomes. For example, the yield of crops such as tomatoes, green peppers, strawberries, alfalfa, orchard grass, and timothy were observed to be stimulated, whereas yields of broccoli, mustard greens, carrots, and radishes were inhibited. In other crop plants, yields were unaffected.

Leaching of a variety of substances from the surfaces of plants appears to be a common outcome of exposure to acidic species deposited in a wet form on plants. Leachates have been reported to include amino acids, free sugars, organic acids, vitamins, a variety of essential mineral nutrients (nitrogen, calcium, phosphorous, magnesium), alkaloids, growth regulators, and pectic substances.

The potential for acidic deposition to cause significant direct effects to agricultural crops has been demonstrated in simulated acidic deposition experiments under controlled exposure conditions. Foliar injury has been observed under laboratory conditions at pH levels associated with recorded severe ambient acidic deposition events. There have, however, been no confirmed reports that atmospheric deposition has caused similar symptoms on vegetation under ambient (field) acidic deposition conditions.

A variety of studies have attempted to determine the potential impact of acidic deposition on forests. The evidence available indicates that rainfall acidity does not visibly damage foliage or reduce tree growth. However, there remain significant questions about the potential for direct injury of tree tissues by other mechanisms including: (1) acid toxicity to reproductive structures and sensitive life stages, and (2) erosion of the leaf cuticle by long-term exposures and foliar injury caused by highly acidic solutions deposited as aerosols.

A variety of potential impacts of acidic deposition on forest element cycles and tree nutrition have been proposed. These include: (1) accelerated leaching of base cations from foliage and soils, which may result in nutrient imbalance, (2) increased mobilization of aluminum and other metals, which may lead to nutrient deficiency, root damage, and reduced drought tolerance, (3) inhibition of soil microbiological processes which may lead to reduced decomposition of organic matter, damage to mycorrhizae, nutrient deficiency and altered pathogen–host relationships, and (4) increased bioavailability of nitrogen, leading to accelerated organic matter decomposition and increased susceptibility to natural stresses such as freezing temperatures, drought, etc.

The significance of these mechanisms in altering tree and forest growth is uncertain. They represent, however, areas of major scientific concern and research activity.

Forest Declines

In previous sections, symptoms associated with exposures to known phytotoxic pollutants were described. Symptoms were mainly acute in nature. Both acute and chronic effects on plants have been observed in controlled exposure studies and under field conditions. Causal relationships have in many cases been well established.

In many real-world cases of plant injury, particularly in forest ecosystems, the relationship between observed injury and exposures to atmospheric pollutants has not been clearly established. This has been particularly the case of the many so-called forest declines.

The term decline is used to describe the process by which large numbers of trees die. In a decline, tree death occurs progressively, that is, trees are weakened, become less vigorous and eventually die.

Declines may occur as a result of exposures to a variety of natural and/or anthropogenic stress factors. Natural phenomena which may initiate forest declines include drought, insects, and frost. Weakened trees may succumb to a variety of other factors including root rot, insects, and disease. In most cases, no single factor can explain the observed death of trees.

A number of forest declines have been observed in North America and central Europe in the past three decades. In some cases, these declines have been definitively linked to atmospheric pollutants. In other cases, one or more atmospheric pollutants are suspected by forest scientists as at least contributing to the observed decline.

California Pines. Declines in ponderosa and Jeffrey pines were first observed in the San Bernardino National Forest east of Los Angeles in the 1950s. Older needles on affected trees lost their chlorophyll and prematurely died. Injured pines had reduced radial growth and tolerance to the western pine beetle and other stresses. Similar declines were later observed in the Laguna Mountains east of San Diego, and in the Sierra Nevada and San Gabriel Mountains. Both ponderosa and Jeffrey pine are very sensitive to elevated O_3 levels, and research scientists have been able to definitively link this decline of California pines to O_3.

Eastern White Pine. Sensitive white pine genotypes have shown evidence of decline throughout their range in eastern North America for decades. This decline has been a selective one, with only the most sensitive trees showing evidence of foliar injury as well as reduced height and diameter growth. These effects are most pronounced in high O_3 regions, and a causal connection between elevated O_3 levels and decline of eastern white pine appears evident.

Red Spruce and Fraser Fir. Declines in red spruce forests have been reported in the eastern mountains from New England to North Carolina. Radial growth has decreased sharply with reductions in the Northeast observed as early as the 1960s. The most dramatic diebacks have occurred at elevations above 800 m. The number of trees has declined by 50% or more in the past 25 years. Symptoms observed include the dieback of branch tips from the top and inward from terminal shoots and yellowing of the upper surfaces of older needles.

As a general rule, the percentage of severely damaged and dying trees increases with elevation. On three New England mountains, approximately 60% of red spruce were dead or dying at 1000 m compared to 20% at 700 m. Less severe damage has been reported to occur at lower elevations as well, including foliar symptoms, reduced growth, and tree death.

In the mountains of North Carolina, forest declines are occurring which include both red spruce and Fraser fir. Growth of red spruce and to a lesser extent Fraser fir began to decrease in the 1960s. By 1987, most red spruce and Fraser fir trees on Mt. Mitchell's western-facing slope had died. High mortality has also been observed in similar forests in Tennessee and Virginia.

In the high elevation mountains where these declines are occurring pollution levels are relatively high compared to those at lower elevations. This is particularly true for O_3. In addition, deposition rates of H^+ may be ten times higher than at lower elevations. In elevations where forest damage is the greatest, trees may be exposed to highly acidic, high O_3 fogs/clouds for up to 1000 hour/year.

Pines in the Southeast. Recent surveys conducted by the U.S. Forest Service in the southeastern states of Virginia, North and South Carolina, Georgia, and Florida have shown that average radial growth rates for loblolly and slash pines under 16 in. (6.3 cm) diameter have decreased by 30 to 50% over the past 30 years. Pine mortality was observed to increase from 9 to 15% from 1975 to 1985.

Hardwoods in the Northeast. Crown dieback and increased mortality rates in sugar maple was reported in southeastern Canada in the late 1970s. Similar dieback symptoms have been reported for sugar maples in Pennsylvania, New York, Massachusetts, and Vermont. Yellow birch, American beech, and white birch have also been reported to be showing decline symptoms.

Declines in Central Europe. Multispecies forest declines were first reported in central Europe on low mountains in the 1970s. Silver fir in the Black Forest lost their needles from the inside of branches outward and from the bottom upward toward the crown and died in large numbers. Within a few years they began to show yellowing of needles (which was clearly linked with nutrition deficiencies) and defoliation of older trees. Pines were next to show symptoms, followed by beech and oak. Symptoms of discoloration and premature leaf fall have now been reported for all major forest species in western Germany.

Though damage to trees was first reported to occur above 800 m elevation, it now occurs at lower elevations as well. Older trees on west-facing slopes which face prevailing winds are most affected.

A survey conducted in 1983 throughout what was then West Germany showed that approximately 34% of the country's trees were affected; by 1986, 54%. Damage was greatest on firs, with 83% showing some symptoms and more than 60% with moderate-to-severe damage. Substantial injury to spruce, beech, and oak was also observed.

Similar injury to forest trees has now been reported throughout central Europe at all elevations and on all soil types. In surveys conducted between 1983 and 1986, 14% of Swiss forests showed symptoms of decline, 22% of Austrian forests, 22% of coniferous and 4% of deciduous trees in France, and 24% of forest areas in Holland.

Research conducted in Germany and other countries of central Europe indicates that air pollutants, through mechanisms which vary by site, are a significant contributing factor in the severe forest declines observed. A link to air pollution is suggested by the large number of species affected, the rapid onset of symptoms, the large geographical area, and wide range of associated climates and soil conditions involved. The scientific consensus on this widespread decline is a combination of direct foliar damage and nutrient imbalance, both due to exposure to atmospheric pollutants. Ozone, along with acid fogs, has been proposed as the principal cause of tree damage in central European forests.

Problems of Injury Recognition

In the preceding pages it is apparent that, to a degree, the individual major phytotoxic pollutants produce rather specific acute injury patterns which are distinctive. The actual diagnosis of air pollution injury in the field is not, however, an easy task, as injury by biotic and abiotic factors may mimic air pollution-induced injury. Indeed, injury caused by minor pollutants such as Cl_2 and HCl may mimic those produced by SO_2, fluorides, PAN, and O_3. Confounding symptom patterns may be produced by drought, late and/or early frosts, winter injury, nutrient disorders, deicing salts, pesticides, insect pests, and plant diseases.

Interaction Effects

Pollutants do not occur in the atmosphere singly. Therefore it is possible, even probable, that simultaneous, sequential, or intermittent exposures to several phytotoxic pollutants may result in plant injury. Available evidence indicates that simultaneous exposures to gaseous mixtures can produce synergistic, additive, or antagonistic effects.

Studies of plant exposures to mixtures of $SO_2 + O_3$ and $SO_2 + NO_2$ have reported decreased injury thresholds. Injury produced from exposures to gaseous mixtures was more severe than the sum of injuries caused when plants were exposed to single pollutants (a synergistic response). In other studies, exposures to gaseous mixtures report synergistic, additive, or antagonistic effects. The type of interaction response appears to be related to plant species exposed, pollutant concentrations, duration of exposure, and amount of injury resulting when plants are exposed singly. Antagonistic responses are generally observed when injury caused by pollutants applied singly is severe; the effect of mixtures is to reduce the severity of injury. Visible symptoms produced by gaseous mixtures are usually characteristic of a single pollutant.

Economic Losses

Numerous instances of vegetation injury have been reported in this century. Excluding the photochemical oxidant problem in southern California and some areas of the Northeast, most reports of air pollution-induced plant injury have been associated with point sources, with injury being localized. With the exception of primary metal smelters, economic losses associated with point sources have often been insignificant. Many offending point sources in the U. S. have since been controlled. Nevertheless, because of photochemical oxidants such as O_3 and PAN, air pollution injury to vegetation is still widespread. These photochemical oxidants continue to cause significant injury to plants and economic losses in many regions of the country.

Several attempts have been made to estimate the annual economic losses to crops. Some have been gross ball park estimates, while others have been based on actual field studies. Surveys of vegetable, citrus, and grape growers in southern

California in the early 1970s estimated crop losses to be between $40 and $50 million per year. These loss estimates did not take into account the widespread damage to ponderosa pine and sensitive trees and ornamental species. From the California survey, a $135 million annual loss to crops has been projected for the U. S.

Scientists who conduct research on the toxic effects of air pollutants on plants generally agree that O_3 causes 90% or more of the air pollution damage to crops in the U.S. This recognition has led to the establishment of a National Crop Loss Assessment Network (NCLAN). From summaries of O_3 monitoring data, determination of the O_3 sensitivity of major crop plants such as corn, soybeans, wheat, cotton, grain sorghum, and barley, and economic data, NCLAN participating scientists have modeled the economic impact of O_3 on U.S. agriculture. They have estimated that a 25% reduction in current ambient O_3 levels would result in a $1.71 billion annual increase in agricultural production, and a 40% reduction would result in a $2.52 billion annual increase.

The quantification of economic loss is a difficult task. It requires surveys of suspected air pollution injury, confirmation of the causal factor, and an estimation of the dollar costs. Loss estimates may be confounded by several factors. For example, the presence of visible injury on plants may not be translatable into economic loss. In many cases, plants with severe foliar injury may quickly recover, replacing injured leaves with younger healthy ones with no apparent lasting effects. In other instances, even a slight amount of injury can decrease the marketability of a crop. To deal with this problem, plant scientists use the concept of plant damage rather than plant injury when it relates to potential economic losses. It is implicit in the damage concept that economic losses have resulted. Plant injury does not have this connotation.

It is reasonable to expect that when a phytotoxic pollution problem continues for a period of time, economic losses will decline, as no grower can sustain such losses indefinitely and continue to operate. The grower may discontinue agricultural production altogether or utilize pollution-resistant varieties or other pollution-tolerant crops.

DOMESTICATED ANIMALS

Retrospective studies of air pollution disasters, such as that which occurred in Donora, PA in 1948, indicate that in addition to death and illness observed among the human population, domesticated animals such as dogs and cats were also affected. From epidemiological studies it is evident that humans chronically exposed to high community air pollution levels have a higher incidence of respiratory disease. Although similar studies have not been made on domesticated animals, animals such as dogs and cats, which share the same basic environment as humans, would also be expected to be affected by air pollution exposures. Of course animals don't smoke, nor do they experience the same stresses as humans.

Fluoride

Fluoride has caused more confirmed air pollution injury to domesticated animals than any other air pollutant. Most cases of fluoride toxicity (fluorosis) have resulted from the contamination of forage. It has been forage-consuming livestock animals such as cattle, sheep, horses, and pigs that have been the most commonly poisoned by fluoride.

In the U. S., fluorosis had been prevalent in Florida, Tennessee, Utah, Washington, and Oregon. It was often diagnosed in livestock in the vicinity of phosphate fertilizer plants and primary aluminum smelters. Fluoride emissions also result from the manufacture of steel, brick and tile products, elemental phosphorus, phosphoric acid, glass, enamel, phosphate animal feed, and nonferrous metals, as well as the combustion of coal. Fluorides may be emitted either in the gaseous or particulate form.

Fluoride is a ubiquitous, naturally occurring element found in varying amounts in soil, water, air, vegetation, and animal tissues. It may be beneficial when ingested in small quantities. Adverse effects result when fluoride intake is excessive. Fluoride ingestion associated with fluoride sources may result from the accumulation of fluoride in the leaves of forage plants exposed to gaseous fluoride, contamination of foliage by fluoride-containing dusts, and the uptake of fluorides from soils contaminated by particulate fluoride dusts.

Fluorosis. Fluorosis can be either acute or chronic. Acute fluorosis is rare because livestock will not voluntarily consume heavily contaminated forage. Acute fluorosis, when it is observed, is usually associated with an accidental ingestion. Chronic fluorosis was, however, commonly observed in livestock which ingested fluoride-contaminated forage over a period of time.

As fluoride interferes with the normal metabolism of calcium, chronic fluoride toxicity is characterized by dental and skeletal changes. One of the earliest signs of fluorosis is white chalky patches or mottling of dental enamel. Chronic fluoride exposures may result in the incomplete formation of the enamel, dentine, or the tooth itself. The teeth may be softened so that excessive dental wear occurs, interfering with the proper mastication of food. These dental changes occur during the period of tooth development and are not reversible. Dental changes are generally not observed if fluoride exposure occurs after the permanent teeth are formed. Therefore, the age of the animal at the time of fluoride ingestion is the most important determinant in the development of dental changes.

Skeletal changes are usually manifested as a diffuse thickening of the long bones and calcification of the ligaments. Stiffness and lameness result when these changes occur in the joints of affected animals. Lame animals eat less because they have difficulty standing for any length of time. Affected animals become emaciated as a result of the reduced food intake and difficult mastication. In dairy cattle this results in decreased milk production.

Fluorosis of livestock resulting from air contamination is no longer as serious a problem in the U.S. as it was in the 1950s and 1960s. This is a result of stringent control requirements imposed on fluoride sources in the past three decades.

Lead

The second most commonly observed air pollution-related effects on live-stock were those associated with lead emissions. Lead poisoning may, however, result from other exposures as well, as lead is present in numerous products including paint, storage batteries, metal containers, and pesticides. Lead poisoning in livestock was common in the early part of this century when lead arsenate insecticide was being used as an orchard spray.

Lead poisoning resulting from air pollution has been reported in the vicinity of primary and secondary lead smelting operations. In most instances such poisoning was due to the ingestion of lead dust-contaminated forage. Lead poisoning may be acute or chronic depending on the exposure level. In cattle, the early stages of acute lead poisoning are characterized by excessive thirst, salivation, loss of appetite, constipation, delirium, and reduced milk production. If lead exposure is high, affected animals may die. Chronic lead poisoning may result in a spectrum of symptoms including diarrhea, colic, nervous disorders, swollen joints, lethargy, incoordination, bellowing stupor, rough coat, and emaciation, as well as a variety of metabolic changes similar to those observed in humans chronically exposed to lead. Lead-poisoned animals may not exhibit all of these symptoms at one time. Because of this and the fact that symptoms are nonspecific, lead poisoning is often difficult to diagnose, particularly in its early stages.

Lead air pollution may not only cause poisoning in livestock but in other animals as well. For example, lead poisoning of animals in New York zoos was reported in the early 1970s — lead aerosols from automobile emissions were suggested as the potential causal agent.

In most cases, when air pollution injury to livestock has been confirmed, such injury has been associated with stationary sources that are processing minerals. These minerals may contain biologically active elements that, when ingested in excessive amounts, produce toxic effects. Such effects have been described for fluoride and lead. Other elements in mineral dusts such as arsenic, selenium, and molybdenum may also poison livestock and other animals.

MATERIALS

Gaseous and particulate air pollutants are known to significantly affect non-biological materials. In the U. S., these effects cause economic losses estimated to be in the billions of dollars each year. Of particular importance are effects on metals, carbonate building stones, paints, textiles, fabric dyes, rubber, leather, and paper. Significant effects on these materials have been observed in other industrialized nations as well. In western Europe, which is a repository for many monuments of history and fine works of art, air pollutant-induced damage has been incalculable. Because these cultural treasures are irreplaceable, their preservation from the destructive effects of airborne contaminants poses a significant challenge to their present guardians. In a sense, the destruction of western European antiquities symbolizes the seriousness of the material damage problem.

Materials can be affected by both physical and chemical mechanisms. Physical damage may result from the abrasive effect of wind-driven PM impacting on surfaces and the soiling effect of passive dust deposition. Chemical reactions may result when pollutants and materials come into direct contact. Absorbed gases may act directly on the material, or they may first be converted to new substances that are responsible for observed effects. The action of chemicals usually results in irreversible changes. Consequently, chemical damage to materials is a more serious problem than the physical changes caused by particulate dusts.

Metal Corrosion

The corrosion of metal in industrialized areas represents one of the most ubiquitous effects of atmospheric pollutants. Because corrosion is natural, we tend not to recognize the role that pollutants play in accelerating this process. Because the ferrous metals, iron and steel, account for about 90% of all metal usage, pollution-induced corrosion on these metals is of particular significance. When ferrous metals corrode, they take on a characteristic rusty appearance. Corrosion is at first rapid, then slows as a partially protective film develops. It is apparent that the acceleration of corrosion in industrial environments is associated with SO_2 and PM pollution. The probable agent of corrosion in both instances is sulfuric acid produced from the oxidation of SO_2. Interestingly, the corrosive effect of these substances is reduced in atmospheres where relatively high oxidant levels are also observed. It has been suggested that oxidants, such as O_3, inhibit the effects of SO_x by producing a more corrosion-resistant product.

Nonferrous metals can also experience significant pollution-induced corrosion. For example, zinc, which is widely used to protect steel from atmospheric corrosion, will itself corrode when acidic gases destroy the basic carbonate coating that normally forms on it. The reaction of SO_x with copper results in the familiar green patina of copper sulfate that forms on the surface of copper materials. This coating of copper sulfate makes the metal very resistant to further corrosion. The green discoloration of SO_x-induced copper corrosion is most commonly observed on bronze statues in city parks and squares. Copper may also be discolored by exposures to H_2S, which blackens it.

Because nonferrous metals are used to form electrical connections in electronic equipment, corrosion of such connections by atmospheric pollutants can result in serious operational and maintenance problems for equipment users. An example is the stress corrosion and breakage of nickel-brass spring relays in the central offices of telephone companies in Los Angeles, New York City, Philadelphia, and other cities in California, Texas, and New Jersey. The breakage occurred when nickel-brass connections were under moderate stress and positive electrical potential. Stress corrosion and breakage of the nickel-brass relay springs was caused by the high nitrate concentrations in dust which accumulated on these metal components. In new equipment the problem is avoided by using copper-nickel spring relays which are not susceptible to stress corrosion caused by airborne dusts.

Pollutants may also damage low-power electrical contacts used in computers, communications, and other electronic equipment by forming thin insulating films over contacts. Such films may result in open circuits, causing equipment to malfunction. These films are produced from chemical reactions with the contact metals and contaminants such as SO_2 and H_2S. Equipment malfunction may also result from the contamination of electrical contacts by PM. Particles may physically prevent contact closing, or they may result in chemical corrosion of contact metals.

Building Materials

Unlike the corrosion of metals whose true cause is difficult to discern, pollutant effects on building surfaces are far more self-evident. The soiling and staining of buildings in urban areas which have or have had high atmospheric aerosol levels is a particularly striking example of this. The dirtiness associated with many cities is due to the unsightly coatings of sticky particulate dusts which have been deposited by the smoke of industry and coal-fired space heating over many years. The cleaning of soiled buildings requires sand blasting, which for a single large building may cost tens of thousands of dollars.

In addition to soiling, building materials such as marble, limestone, and dolomite may be chemically eroded by weak to strong acids. This erosion may be caused by exposures to acidic gases in the presence of moisture, acid aerosols, or acid precipitation. The reaction of sulfuric acid with carbonate building stones results in the formation of $CaSO_3 \cdot 2H_2O$ and $CaSO_4 \cdot 2H_2O$ (gypsum), both of which are soluble in water. These erosion effects are not limited to the surface, as water transports acids into the interior of the stone. The soluble salts produced may precipitate from solution and form incrustations, or they may be washed away by rain.

The chemical erosion of priceless, irreplaceable historical monuments and works of art in western and southern Europe can be described as no less than catastrophic. Although it is primarily a European problem, similar destruction is occurring on monuments in the U.S. In the late 19th century, the obelisk Cleopatra's Needle was moved from Egypt to a park in New York City. The monument, which withstood the ravages of desert heat and sands for thousands of years without significant deterioration, has had hieroglyphics obliterated on two of its faces in the past half century. The other two faces have been eroded less, because they were away from the direction of prevailing winds.

A significant concern is the Acropolis, located downwind of the city of Athens, Greece. Although the chemical erosion of the marble structures in the Acropolis has been occurring for more than 100 years, the rapid population growth and use of high-sulfur heating oils since World War II has greatly accelerated their destruction. Unless significant control of sulfur emissions is achieved in Athens and acceptable preservation methods are developed and implemented, many of the sculptured forms in the Acropolis buildings will soon be unrecognizable.

In addition to the Acropolis, the Colosseum in Rome and the Taj Mahal are in various stages of dissolution. In many European cities, marble statues have had to be moved indoors. Cities such as Venice have enacted legislation to preserve and restore hundreds of damaged plazzi, churches, and historical buildings.

Paints

Paints consist of two functional components: vehicles and pigments. The vehicle consists of a binder and a variety of additives which hold the pigment to a surface. The principal materials used as vehicles include oils, such as linseed oil, and resins, which may be natural, modified natural, or synthetic. The pigments, which may be classified as either white or colored, provide aesthetic appeal, hiding power, and durability. White lead, titanium dioxide, and zinc oxide were/are commonly used white pigments. Colored pigments include a variety of mineral, metal, and organic compounds. Paints provide an aesthetically appealing surface coating that protects the underlying material from deterioration.

The appearance and durability of paints is affected to a significant degree by natural weathering processes. Paint weathering is accelerated by environmental factors such as moisture, temperature, sunlight (particularly ultraviolet light), and fungi. Paint appearance and durability may also be affected by air pollutants such as PM, H_2S, SO_x, NH_3, and O_3.

Pollutant effects on paints may include soiling, discoloration, loss of gloss, decreased scratch resistance, decreased adhesion and strength, and increased drying time. If the paint coating is significantly damaged, the underlying surface may be open to attack by pollutants and natural weathering processes. This, of course, defeats the purpose of applying a protective coating of paint.

One obvious effect on paints is the soiling or dirty appearance that results from the accumulation of aerosol particles. These particles may also cause chemical deterioration of freshly applied paints that have not completely dried. Particles may also serve as wicks that allow chemically reactive substances to reach the underlying material, resulting in corrosion if the underlying material is metallic.

Exterior paints pigmented with white lead may be discolored by reaction of lead with low atmospheric levels of H_2S. The blackening of the paint surface by the formation of lead sulfide can occur at H_2S exposures as low as 0.05 ppmv for several hours. The intensity of the discoloration is related not only to the concentration and duration of the exposure but also to the lead content of the paint and the presence of moisture.

Textiles and Textile Dyes

Textiles consist of a basic fiber component and additives such as dyes, water repellents, and finishes. Each of these is subject to attack from pollutants.

Exposures to atmospheric pollutants may result in significant deterioration and weakening of textile fibers. Fabrics such as cotton, hemp, linen, and rayon which are composed of cellulose are particularly sensitive to acid aerosols and acid-forming gases. Acids chemically break the cellulose chain, producing water-soluble products that have little tensile strength. Synthetics such as nylon may also incur significant acid damage. Nylon polymers may also undergo oxidation by NO_2, which reduces the affinity of nylon fibers for certain dyes.

In addition to the chemical action of pollutants on textiles, soiling by PM may occur. Although this soiling may not cause direct damage to the textile itself, it requires more frequent washing, which accelerates fabric deterioration.

Air pollutants may react with fabric dyes, causing them to fade. Such fading has been associated with SO_2, PM, NO_2, and O_3. Fading of textile dyes associated with NO_x exposure, particularly NO_2, has a long history. For example, NO_x emissions from open electric-arc lamps and incandescent gas mantles were reported to cause fading of stored woolen goods in Germany prior to World War I. Nitrogen oxide emissions from gas heaters were also believed responsible for a serious fading problem on acetate rayon fabrics during the 1930s. During the 1950s, NO_x emissions from home gas dryers were reported to have caused fading of colored cotton fabrics. Nitrogen dioxide-induced fading has also been reported for fabrics exposed in urban environments. Such exposures resulted in a yellow discoloration of undyed white or pastel-colored fabrics. Investigation of the problem by textile industry scientists showed that the observed yellowing was due to the effect of NO_2 on textile additives such as brighteners, softeners, resinous processing agents, and antistatic and soil release finishes. The problem was resolved by using resistant, more expensive additives.

Reports of NO_2-induced fading of textile materials from ambient NO_2 exposures were soon followed by observations that ambient O_3 levels were also a major cause of fading. In the early 1960s, O_3 fading was reported for polyester-cotton permanent press fabrics and nylon carpets. In California, Texas, and Tennessee, fading was reported on permanent press slacks stored in warehouses or on retail shelves. Some manufacturers suffered significant economic losses. Fading of nylon carpets was a problem along the Gulf Coast from Florida to Texas. Blue dyes were particularly sensitive to a combination of ambient O_3 exposure and high humidity. This "Gulf Coast fading" was overcome by using O_3-resistant dyes and modification of the nylon fibers to decrease the accessibility of O_3.

Pollution-induced fading of fabric dyes has posed a significant problem for textile manufacturers. Consequently, the industry has had to develop resistant dyes and chemical inhibitors to mitigate the problem. Neither of these has been completely satisfactory. Resistant dyes, in addition to being more expensive than sensitive dyes, tend to have poor dyeing properties requiring slower processing and more care in their application. Chemical inhibitors, which selectively react with pollutants, can, on the other hand, reduce the fading of sensitive dyes without the processing problems of resistant dyes. The effect of inhibitors is, however, only temporary, because they are eventually consumed when exposed to pollutants over a period of time.

Paper and Leather

Papers that have come into use after 1750 are very sensitive to SO_2. This sensitivity is apparently due to the catalytic conversion of SO_2 to sulfuric acid by metallic impurities. The sulfuric acid causes the paper to become brittle, decreasing its service life. This embrittlement is of concern to libraries and museums in environments with high SO_2 levels, as it makes the preservation of historical books and documents much more difficult.

The preservation of historical books involves more than the preservation of the paper component. Air pollution damage to leather-bound books is also a serious problem. Damage to leather is initially characterized by cracking. The cracking results in more leather being exposed, accelerating the deterioration. The leather eventually loses its resiliency and turns to a reddish-brown dust. Pollution damage to leather is similar to that of paper, as both are apparently caused by sulfuric acid produced from SO_2 by the catalysis of metallic impurities.

Rubber

Ozone can induce cracking in rubber compounds that are stretched or under pressure. The depth and nature of this cracking depends on the O_3 concentration, the rubber formulation, and the degree of rubber stress. Unsaturated natural and synthetic rubbers such as butadiene–styrene and butadiene–acrylonitrile are especially vulnerable to O_3 cracking. Ozone attacks the double bonds of such compounds, breaking them when the rubber is stressed. Saturated compounds such as thiokol, butyl, and silicon polymers resist O_3 cracking. Such resistance is necessary for rubber in such important products as automobile tires and electrical wire insulation.

Rubber cracking was one of the first effects of smog observed in the Los Angeles area. During the 1950s, it was used to monitor O_3 concentrations in the atmosphere. Under standardized conditions, O_3 concentrations could be directly related to the depth of the O_3-induced cracks.

Glass

With the exception of fluoride, glass is highly resistant to most air pollutants. Fluoride reacts with silicon compounds to produce the familiar etching pattern. Etching of glass has been observed in the vicinity of major fluoride sources such as phosphate fertilizer mills, aluminum smelters, enamel frit plants, and even steel mills. Continuous exposure to fluoride may render glass opaque. The etching of glass, where it has occurred, has generally been considered to be of minor importance relative to such problems as fluoride injury to vegetation and fluorosis in livestock.

Economic Losses

Economic losses due to air pollution-induced damage to materials are difficult to quantify because one cannot easily distinguish what is due to the natural deterioration of materials and what is caused by air pollutants. In addition, it is difficult to assess indirect costs of early replacement of materials that may be worn out by excessive cleaning.

Table 6.5   Estimated Annual Losses
              Resulting from Air Pollution
              Damage to Materials in the
              U.S. in the Early 1970s

Materials	Losses ($)
Textiles and fabrics	2.0 billion
Metals	1.5 billion
Paints	700 million
Rubber	500 million
Electrical components	65 million

From Liu, B. and E. Xu. 1976. *Physical and
Economic Damage Functions for Air Pollutants by Receptors*. EPA/600/5-76-011.

Costs of air pollution damage to materials have only occasionally been estimated and reported. Estimates based on a study conducted in the early 1970s are summarized in Table 6.5. Despite the fact that these estimates are relatively crude and somewhat dated, they nevertheless indicate the relative magnitude of the pollution/materials damage problem that this nation faced when it seriously began cleaning up the air of our cities and towns.

ODOR POLLUTION

The welfare effects described in this chapter do not impact people directly. Odor, on the other hand, does. Though unpleasant odors may cause symptoms in some individuals, the problem of odor or odor air pollution is usually viewed as being one of annoyance. Within this context, odor, from a regulatory standpoint, is seen as a welfare and not a health issue.

The term odor as commonly used in everyday speech generally carries with it an unpleasant connotation. The term as defined by scientists is, of course, neutral; it connotes only that some volatile or semivolatile substance is being sensed by the human olfactory system.

The ability to sense or to smell an odor is widespread, if not the norm, in the animal kingdom. In many species, the olfactory sense plays a significant role in locating food, attracting individuals of the opposite sex for purposes of breeding, and in many cases, warning of approaching danger. In humans, the olfactory sense plays a lesser and apparently less obvious role.

The olfactory function in humans consists of two different organs found in the nose. The olfactory epithelium (found in the highest part of the nose) consists of millions of yellow-pigmented bipolar receptor cells which connect directly to the olfactory bulbs of the brain. The free endings of trigeminal nerves distributed throughout the nasal cavity serve as another olfactory organ or organ of smell. Trigeminal nerves respond to odoriferous substances that cause irritation, tickling, or burning. The chemical senses that correspond to these two olfactory organs are not easily separated, with many odoriferous substances or odorants stimulating both systems. They stimulate different parts of the brain and, therefore, may have different effects. The major function of the trigeminal nerve system is to stimulate reflex actions such as sneezing or the interruption of breathing when the body is being exposed to potentially harmful odor-producing substances.

Odor Measurement

The perception of odor is a physiopsychological response to the inhalation of an odoriferous chemical substance. Odors cannot, therefore, be chemically measured. However, some of the sensory attributes of odors can be measured by exposing individuals under controlled conditions. Elements of odor subject to measurement are detectability, intensity, character (quality), and hedonic tone (pleasantness, unpleasantness).

The limit of detection is called the odor threshold. It may be characterized in one of two ways. The threshold may be that concentration of an odoriferous substance where there is a detectable difference from the background. Alternatively, the threshold may be defined as the first concentration at which an observer can positively identify the quality of the odor. The former would be best described as a detection threshold, the latter as a recognition threshold.

Though the olfactory sense in humans is not as acute as it is in many animals, it nevertheless has the ability to detect many substances at very low concentrations. Odor thresholds for a variety of chemicals are summarized in Table 6.6. Note that humans can detect H_2S at approximately 0.5 ppbv.

Odors differ in their character or quality. This odor parameter allows us to distinguish odors of different substances by prior odor associations. The characters of a variety of selected chemicals are summarized in Table 6.6. For example, dimethyl amine is described as fishy, phenol as medicinal, paracresol as tar-like, etc.

Table 6.6   Odor Thresholds and Characteristics of Selected Chemical Compounds

Chemical	Odor threshold (ppmv)	Odor character
Acetalydehyde	0.21	Green, sweet
Acetone	100.0	Chemical sweet, pungent
Dimethyl amine	0.047	Fishy
Ammonia	46.8	Pungent
Benzene	4.68	Solvent
Butyric acid	0.001	Sour
Dimethyl sulfide	0.001	Vegetable sulfide
Ethanol	10.0	Sweet
Ethyl mercaptan	0.001	Earthy, sulfidy
Formaldehyde	1.0	Hay, straw-like, pungent
Hydrogen sulfide	0.00047	Eggy sulfide
Methanol	100.0	Sweet
Methyl ethyl ketone	10.0	Sweet
Paracresol	0.001	Tar-like, pungent
Perchloroethylene	4.68	Chlorinated solvent
Phenol	0.047	Medicinal
Sulfur dioxide	0.47	Sulfury
Toluene	2.14	Moth balls, rubbery

Related to the character or quality aspects of an odor is hedonic tone or the degree of pleasantness or unpleasantness. The character of an odor may, in many cases, be the primary determinant of whether it is pleasant or unpleasant. The olfactory sense is highly subjective, with some odors being pleasant to some and unpleasant to others. The difficulty of defining odor problems lies in part with the lack of unanimity as to what is an unpleasant odor.

Unpleasantness may also be related to odor intensity. Some odors may elicit a pleasant sensation in an individual in relatively low-to-moderate concentrations. The olfactory response to an odorant decreases as the concentration decreases. This decrease is nonlinear. For a substance such as amyl butyrate, the perceived odor intensity decreases by 50% for a tenfold reduction in concentration. This logarithmic relationship is common for most other odorants as well.

The perceived intensity of an odor rapidly decreases after initial exposure. Generally within a few minutes, the odor may not be perceived at all by those who are exposed. This phenomenon is called olfactory fatigue. Within minutes after the exposure ceases, sensitivity to the odor is restored.

A person may also become habituated to an unpleasant odor ("get used to it" or become "more tolerant"). Habituation operates over much longer time periods than olfactory fatigue.

Unpleasant odors may affect our sense of well-being. Responses to a variety of malodors can include nausea, vomiting, headache, coughing, sneezing, induction of shallow breathing, disturbed sleep, appetite disturbance, sensory irritation, annoyance, and depression. Effects may be physiological, psychological, or both.

Odor Problems

It is likely that malodors emanating from a nearby source result in more complaints to regulatory agencies than any other form of air pollution. Particularly notable sources of malodors (and, in many instances, citizen complaints) are rendering plants, soap-making facilities, petrochemical plants, refineries, pulp and kraft paper mills, fish-processing plants, diesel exhaust, sewage treatment plants, and agricultural operations including feedlots, poultry houses, and hog confinements. Malodors associated with such sources include a variety of amines, sulfur gases (such as H_2S, methyl and ethyl mercaptan, and carbon disulfide), phenol, ammonia, aldehydes, fatty acids, etc.

In addition to the annoyance caused by malodors, the presence of a continuous source may result in a decrease in property values. Individuals may not want to move into a neighborhood with a known malodor problem. Increasingly, however, many individuals are building homes in rural areas where malodorous agricultural operations already exist. It is ironic that some individuals move to the country and then conclude that one of the most common smells of the country, the barnyard, is an unacceptable form of air pollution. To the farmer, the stench of the barnyard is but a long-standing part of life.

READINGS

1. Adams, R.M., S.A. Hamilton, and B.A. McCord. 1985. "An Assessment of Economic Effects of O_3 on U.S. Agriculture." *JAPCA* 35:938–943.
2. Graedel, T.E. and R. McGill. 1986. "Degradation of Materials in the Atmosphere." *Environ. Sci. Technol.* 20:1093–1100.
3. Heck, W.W. and C.S. Brandt. 1977. "Effects on Vegetation: Native Crops, Forests." 158–220. In: Stern, A. C. (Ed.). *Air Pollution. Vol II. The Effects of Air Pollution.* 3rd ed. Academic Press, New York.
4. Hellman, T.M. and F.H. Small. 1974. "Characterization of the Odor Properties of 101 Petrochemicals Using Sensory Methods." *JAPCA* 24:979–982.
5. Jacobson, J.S. and A.C. Hill (Eds.). 1970. *Recognition of Air Pollution Injury to Vegetation: A Pictorial Atlas.* Air Pollution Control Association, Pittsburgh, PA.
6. Liu, B. and E. Xu. 1976. *Physical and Economic Damage Functions for Air Pollutants by Receptors.* EPA/600/5-76-011.
7. Lucier, A.A. and S.G. Haines (Eds.). 1990. *Mechanisms of Forest Response to Acidic Deposition.* Springer-Verlag, New York.
8. MacKenzie, J.J. and M.T. El-Ashry. 1988. *Ill Winds: Airborne Pollution's Toll on Trees and Crops.* World Resources Institute, Washington, DC.
9. MacKenzie, J.J. and M.T. El-Ashry (Eds.). 1989. *Air Pollution's Toll on Forests and Crops.* Yale University Press, New Haven, CT.
10. National Research Council, Committee on Biological Effects of Atmospheric Pollutants. 1971. *Fluorides.* National Academy of Sciences, Washington, DC.
11. Schulze, E.D. 1989. "Air Pollution and Forest Decline in a Spruce (*Picea abies*) Forest." *Science* 244:776–783.
12. USEPA. 1982. *Air Quality Criteria for Particulate Matter and Sulfur Oxides.* EPA/600/8- 82-029a-c.
13. Yocum, J.E. and R.A. Duffee. 1970. "Controlling Industrial Odors." *Chem. Eng.* June 15, 160–168.
14. Yocum, J.E. and J.B. Upham. 1977. "Effects on Economic Materials and Structures." 65–110. In: Stern, A. C. (Ed.). *Air Pollution. Vol. II. The Effects of Air Pollution.* 3rd ed. Academic Press, New York.
15. Yocum, J.E. 1979. "Air Pollution Damage to Buildings on the Acropolis." *JAPCA* 29:333- 338.

QUESTIONS

1. Ozone is the most important phytotoxic air pollutant in North America. Why is this the case?

2. What effect do environmental factors have on symptom development in plants exposed to toxic doses of pollutants?

3. Indicate pollutants responsible for the following effects on materials: (1) cracking of rubber, (2) oxidation of fabric dyes, (3) soiling, (4) darkening of lead-based paint, (5) disintegration of paper and leather, and (6) erosion of building stone.

4. Indicate pollutants responsible for the following symptoms in plants: (1) marginal/tip necrosis, (2) interveinal necrosis, (3) stipple/chlorotic mottle, (4) premature maturation, and (5) undersurface glazing.

5. What kind of phytotoxic exposures result in (1) necrosis, (2) chlorosis, and (3) growth changes.

6. Why is it difficult to diagnose ethylene injury in the field?

7. Describe three ways acidic deposition may adversely affect the growth of plants such as trees.

8. Significant declines in forests/forest species have been reported in North America and Europe. What specific factors may be responsible?

9. Livestock (in the U.S.) have in the past been poisoned by various pollutant exposures, including fluoride and lead. Describe the nature and the effects of such exposures.

10. Malodors represent a significant air-quality problem. Why?

11. Considering the various welfare effects, what pollutant is likely to cause the most environmental harm? Why?

12. Relative to pollutant effects on plants, what is the conceptual difference between injury and damage?

7

AIR QUALITY AND EMISSIONS ASSESSMENT

In previous chapters the quality of ambient air was discussed in the context of major pollutants, sources, effects, and atmospheric dispersion processes. Air quality, of course, is ever changing. It varies greatly over the course of the year or over 24 hours and from one location to another. The quality of ambient air for a specific time and place depends on a combination of factors including the magnitude of emissions from individual sources, source emission density, topography, and the state of the atmosphere.

How can air quality be defined? Obviously, in the disastrous episodes in the Meuse Valley, Donora, and London, air quality could be defined in terms of health because of the illness and deaths associated with those events. Air quality could also be defined as a function of visibility or some other pollutant-induced effect. Any definition of air quality should be, in whole or in part, based on its effects. In many instances effects are only recognized or measured after the fact. Due to confounding variables, health effects associated with pollutant exposures are often difficult to demonstrate and quantify. Because of the *a posteriori* recognition of most pollutant effects and the difficulty of identification due to confounding variables, the qualitative or quantitative use of pollutant effects to determine ever-changing air quality is not practical. It is more desirable to define air quality in terms of some real-time or average measurable condition of the atmosphere. Such a definition can be based on concentrations of specific atmospheric pollutants measured by chemical or physical techniques. These concentrations, when related to effects data, serve as indicators of potential pollutant effects and therefore air quality.

The protection of the health of individuals in our communities from excessive exposures to atmospheric pollutants is the primary goal of all air pollution control programs. The success of pollution control efforts depends on the availability of good data on ambient concentrations of major pollutants and emissions from individual sources and/or groups of sources. Consequently, programmatic efforts that attempt to determine pollutant concentrations both temporally and spatially in ambient air, and characterize and quantify emissions from major sources are vital to the success of all pollution control programs. Such efforts typically include ambient air-quality monitoring, source emissions assessment, and modeling. Modeling is used to predict the impact of new or existing sources on ground-level concentrations under various atmospheric conditions.

AIR-QUALITY MONITORING

Air-quality monitoring may be defined as a systematic, long-term (usually over a period of years) assessment of pollutant levels in our communities, less polluted rural areas, and for some pollutants, the global atmosphere. Monitoring activities are intensively conducted by state/provincial and local agencies to determine the compliance status of air quality control regions relative to air-quality standards. Pre-control air-quality monitoring is necessary to determine base levels of pollutants and pollutants which need to be controlled. Routine monitoring is required to determine progress toward compliance for pollutants which exceed air-quality standards and changes in the status of regions that may already be in compliance. Such monitoring also provides the short-term data needs of episode control plans associated with stagnating anticyclones (subsidence inversions).

Though air-quality monitoring activities are primarily associated with compliance efforts, they serve a variety of other vital functions as well. They are especially valuable in the conduct of epidemiological studies which seek to assess the relationship between pollutant levels and various parameters of health among exposed and less exposed populations. Air-quality monitoring is of considerable importance in assessing trends in atmospheric levels of various regulated and unregulated pollutants. These include tropospheric and stratospheric O_3, stratospheric O_3-depleting substances such as chlorofluorocarbons (CFCs) and nitrogen oxides (NO_x), various volatile halogenated hydrocarbons (HCs), and greenhouse gases such as carbon dioxide (CO_2), methane (CH_4), and nitrous oxide (N_2O). Many of these are global concerns with monitoring efforts supported by a number of federal governments.

Air-quality monitoring can be viewed as an activity in which many measurements are made over time and space. Each discrete measurement involves the collection and analysis of individual air samples.

Sampling

Because of the vastness and dynamic nature of the atmosphere, it is not possible to determine the infinite number of values that are typical of pollutants in the atmosphere. As in other cases where the population of values is large, one must take samples from which the characteristics of the whole population can be inferred. Thus, we take or collect samples at fixed locations and infer the general air quality in the area of the sampling station and for the period sampling was conducted.

Concentrations are determined by using sampling methods to collect pollutants in or on a sampling medium or bring them into the vicinity of a sensor. In the former case, sampling and analysis are discrete events, with sample analysis occurring days to weeks after sample collection. In the latter case, sampling and analysis are simultaneous or near-simultaneous events.

Sampling Considerations

In methodologies where sampling and analysis are separated in time, a sufficient quantity of contaminant must be collected to meet the lower detection limit (LOD) requirements of the analytical procedure used. The quantity of collected contaminant will depend on the atmospheric concentration and the sample size (the volume of air which passes through or comes in contact with the sampling medium). The sample size will be a direct function of the sampling rate and duration.

Sampling rates used for individual pollutants and instruments are governed by the collection efficiency of the technique and instrument limitations. Collection efficiency is the ratio of pollutant collected to the actual quantity present in the air volume sampled. Collection efficiency depends on the sampling technique employed and the chemical and/or physical properties of individual pollutants. For gases, collection efficiency decreases with increasing flow rate above the optimum value. Optimum collection efficiencies are often achieved at relatively low sampling rates (<1 L/min). Because of these low flow rates, sampling duration must often be extended to collect a sufficient quantity of pollutant for analysis. For some substances, refrigeration and/or the use of preservatives is necessary to minimize sample loss prior to analysis.

In continuous real-time sampling, the analytical technique is usually sensitive enough that the sample size or volume needed to accurately detect and quantify specific pollutants is relatively small (typically milliliters). As a result, flow rates used in such instruments are also relatively low (ml/min) and optimal rates of flow are used to achieve the sensitivity and accuracy that the instruments are capable of.

Averaging Times

Collection and analytical limitations associated with intermittent sampling methodologies often require extended sampling durations which produce concentration values that are integrated or averaged over the sampling period. The averaging time used for pollutant concentrations is often determined by the duration required to collect the sample. It is also determined by the intended use of the data collected. For example, 24-hour averages are appropriate for long-term trends. For pollutants such as O_3 where peak levels are limited to a few hours in early afternoon, a 1 hour or 8 hour averaging time may be desirable.

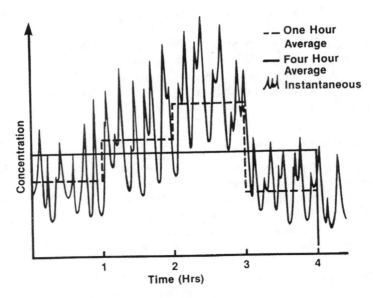

Figure 7.1 Averaging times for gaseous pollutants.

Real-time air-quality monitoring instruments widely used in the past two decades provide a continuous record of fluctuating concentrations. Though such data have been deemed to be more useful than the integrated data obtained from intermittent samples, real-time data is so voluminous that it is mostly uninterpretable. As a result, data from real-time air-quality monitoring instruments are integrated to provide hourly average concentrations or concentrations reflective of the needs of air-quality standards. Differences in average concentrations associated with real-time data can be seen in Figure 7.1.

Averaging times for different pollutants are specified in ambient air-quality standards. For example, the averaging time for PM_{10} is 24 hours and annual; CO, 1 hour and 8 hours; O_3, 1 hour; HCs, 3 hours; SO_2, 3 hours, 24 hours and annual; NO_x, annual; and lead, 3 months. Concentrations averaged over the entire year may be based on a continuous record of measurements or in the case of PM_{10} and lead, averages of a more limited number of samples.

Sampling Techniques

The principal objective of sampling is to collect a contaminant or contaminants for subsequent analysis and/or provide a sensing environment for real-time measurements. Both require a system whereby gases or particles are drawn to the surface of a collecting medium or into a sensing environment. These functions are accomplished by sampling trains which may include a vacuum pump, flow regulator, and a collecting device or sensing unit. Sampling trains for gases may also utilize filters to prevent particles from entering the collection unit.

Based on the type of information desired as well as collection and analytical limitations, sampling may be conducted by static, grab, intermittent, or continuous procedures. Such sampling procedures provide air-quality data representing a range of averaging times, from the instantaneous values of continuous systems to the 30-day average used for some static samplers.

Static or passive sampling has involved the collection of contaminants by the diffusion of gases to a collection medium, the sedimentation of heavy particles into a container, or the impaction of particles on sticky paper. Before 1965, static sampling was commonly used for air-quality monitoring, because it had the advantages of simplicity and low cost. However, because of the long averaging times (typically 30 days) and often semiquantitative nature, static sampling data were limited in their usefulness for ambient air pollution control efforts. Interestingly, relatively recent developments in static or passive sampling technology have made passive sampling a popular and widely used methodology for personal monitoring in occupational environments, measurement of indoor air quality parameters such as formaldehyde and radon, and the conduct of epidemiological studies.

Grab sampling is not commonly employed to monitor air quality on a day-to-day basis; rather, it is widely used for control personnel and researchers to identify problem pollutants. In grab sampling a small volume of air is collected in a matter of seconds or minutes. Grab samples can be collected by a variety of techniques. Evacuated bottles, gas syringes, and bags made of synthetic materials have received widespread use. Gas syringes can be used dry, sealed, and returned to the laboratory for analysis.

Intermittent sampling using dynamic sampling methods came into widespread use in air-quality monitoring networks in the U.S. in the 1950s through early 1970s both for PM and gases such as SO_2 and NO_x. Intermittent sampling has the advantage of relatively low cost and reasonably useable data. It has the disadvantage of relatively long averaging times (typically 24 hours) which has been historically viewed by regulators as being undesirable for gaseous pollutants. Intermittent sampling continues to be the sampling method of choice for particulate matter (PM) and in acid rain sampling networks. Because of its low costs, intermittent sampling can be used to supplement continuous monitoring networks. Intermittent collection of samples at multiple locations may provide data of more usefulness than a continuous record at a few locations.

Continuous monitoring devices provide instantaneous sampling results. Because of real-time data acquisition, continuous monitors are the systems of choice in air-quality monitoring programs. Their major disadvantages are high capital equipment costs and a need to reduce data to a manageable form. A continuous monitoring instrument is illustrated in Figure 7.2.

Gases

Gas collection can be accomplished by utilizing the principles of absorption, adsorption, and condensation. Of these, absorption is the most widely used. In absorption systems, contaminated air is drawn through a liquid reagent. The

Figure 7.2 Continuous O_3 monitoring instrument. (Courtesy of DASIBI Corp.)

contaminant is removed from the sampled airstream by being dissolved in absorbing reagent contained in a bubbler or impinger. The sampled gas enters the medium through a dispersion tube which is below the surface of the liquid. The gas forms bubbles and is dispersed in the liquid collection medium where it is absorbed.

Adsorption is a process that collects gaseous materials on a solid surface. In adsorption, gases are physically attracted to a solid which has a large surface area, such as activated carbon, silica gel, molecular sieve, and others. The collected gases may be desorbed from the collecting medium and subsequently analyzed. Adsorption sampling is common for a variety of organic gases. Because adsorbents in general are not very specific, a mixture of gases is usually collected, making analysis more difficult. In addition, some collected gases may react with each other.

Gases may also be collected by condensation or freezing. In such sampling, air is drawn through a collection vessel maintained at subambient temperatures. The low temperatures cause some vapors to condense or freeze and be retained in the collecting vessel. The sample must be removed and an analysis subsequently performed. Such cryogenic or freeze-out sampling has been used to condense and isolate various organic gases. It is the most difficult to use of the gas sampling techniques available.

Particulate Matter

Particles may be collected by a variety of techniques including gravitational settling, filtration, impaction, and electrostatic and thermostatic precipitation. Of these, filtration and impaction are the most widely used for sampling ambient PM.

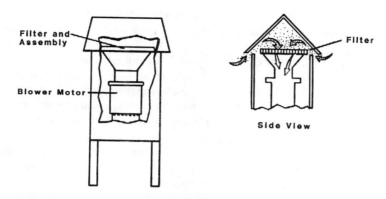

Figure 7.3 Schematic of a high volume (hi-vol) sampler.

Hi-vol Sampling. The high volume sampler (Figure 7.3) was widely used in ambient air-quality monitoring programs in the U.S. until the air-quality standard for PM was changed in 1988. It is still used for some applications (sampling lead levels around emission sources) in the U.S. and is used in a variety of countries where its low cost makes it affordable. The collecting glass fiber filter is located upstream of a heavy-duty vacuum cleaner type motor which is operated at a high air flow rate (1.13 to 1.7 m³/min or 40 to 60 ft³/min). The sampler is mounted in a shelter with the filter parallel to the ground. The covered housing protects the glass fiber filter from wind and debris, and from the direct impact of precipitation. The hi-vol collects particles efficiently in the size range of 0.3 to 100 μm. The mass concentration of total suspended particles (TSP) is expressed as micrograms per cubic meter for a 24-hour period. The hi-vol is an intermittent sampling device. It is normally operated on a 6-day sampling schedule, with a 24-hour average sample collected every sixth day.

Paper Tape Samplers. Paper tape samplers draw ambient air through a cellulose tape filter. After a 2-hour sampling period, the instrument automatically advances to a clean piece of tape and begins a new sampling cycle.

Advanced paper tape samplers are equipped with densitometers for optical density measurements during the sampling period. These instruments record changes in light transmission which can be converted into coefficient of haze (COH) units.

Coefficient of haze units are based on light transmission through the soiled filter area. The higher the ambient particle loading, the more soiled the filter. The subsequent increase in optical density can in most instances be directly related to mass concentrations.

In the early 1970s, paper tape samplers were widely used for continuous monitoring of PM concentrations. Because of difficulties in relating data acquired by this optical method to the gravimetric data of the hi-vol reference method, most paper tape sampling was discontinued. Paper tape samplers may still be used on a standby basis in metropolitan areas where rapid data acquisition is essential to implement episode control plans during stagnating anticyclones.

Size Selective Samplers. A variety of sampling devices are available that fractionate collected suspended PM into discrete size ranges based on their aerodynamic diameters. These particle samplers may employ one or more fractionating stages. The physical principle by which particle fractionation occurs is inertial impaction. Therefore, most such devices are called impactors.

In impactors, air is drawn through the unit and deflected from its original path of flow. The inertia of suspended particles causes them to strike or impact a collecting surface. The size range of particles collected on the impaction surface depends on (1) gas velocity, (2) particle density and shape, (3) air flow geometry, (4) gas viscosity, and (5) the mean free path of the gas.

Multistage or cascade impactors can fractionate suspended particles into six size fractions. In theory, each stage collects particles above a certain "cutoff" diameter which is smaller than the previous stage.

Other impactors have been developed to fractionate suspended particles into two size fractions, i.e., coarse (from 2.5 to 10 µm) and fine (< 2.5 µm). Although these virtual or dichotomous impactors operate like a typical inertial unit, large particles are impacted into a void rather than an impervious surface. Both particle fractions are collected on individual membrane filters.

In 1987, the primary air-quality standard for PM was changed from mass values that included particles up to 100 µm to a PM_{10} standard with an upper cutoff diameter of 10 µm. Although a number of sampling devices, including the cascade and virtual impactors, can meet the performance requirements of the standard, the most widely used device in ambient air-quality monitoring networks is a modified hi-volume sampler with a size-selective unit (Figure 7.4). The size selective unit inertially removes particles that have aerodynamic diameters greater than 10 µm by impaction on a surface with the smaller fraction ($\leq$10 µm) collected on a glass fiber filter.

Accuracy and Precision

The accuracy of sampling and analytical techniques must be known before they can be accepted for use in air-quality monitoring programs. Accuracy is defined as the relative closeness of a measured value to the true value. Sampling and analytical techniques used to determine the concentration of air contaminants vary in their degree of accuracy. Accuracies greater than 90% are generally considered to be acceptable. In addition to accuracy, a sampling and analytical procedure should be precise. Precision is a measure of a method's reproducibility. It is usually represented as a ±% value around the population mean of multiple sample results from a standard atmosphere using the same test procedure. Precision values of ±10% are considered to be very good, and for some sampling and analytical procedures (notably used in occupational measurements), a precision of ±25% is acceptable. Unfortunately, in common usage, the terms accuracy and precision are often used interchangeably. Differences in accuracy and precision

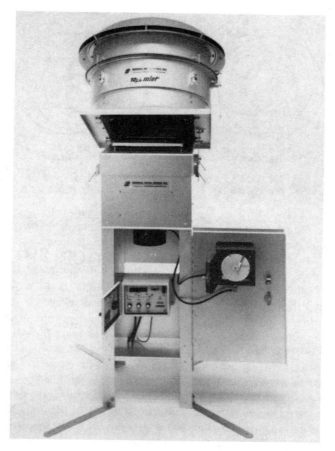

Figure 7.4   PM$_{10}$ sampler with a size-selective inlet. (Courtesy of General Metal Works —
                 Anderson Graesby.)

are illustrated in Figure 7.5. Ideally, sampling and analytical procedures for atmospheric pollutants should have high accuracy and should be relatively precise.

Calibration and Quality Assurance

Both real-time continuous and intermittent air sampling instruments must be
periodically calibrated to ensure the validity of monitoring data. Of particular
importance is the calibration of instrument air flow rates and, in continuous
monitors, pollutant concentrations. In calibration, measured values are compared
to standard values.

Air-flow calibrations are typically conducted using a variety of primary and
secondary standards. A primary standard is traceable directly to a standard determined by the National Institute of Standards and Technology (NIST) and a
secondary standard has been calibrated against a primary standard. An example

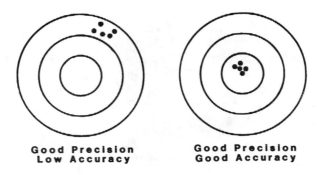

Good Precision   Good Precision
Low Accuracy   Good Accuracy

Figure 7.5 Schematic illustration of accuracy and precision.

of a primary standard for air flow calibration is a bubble meter (based on using a gas burette and soap bubbles); secondary standards may include wet test meters, dry test meters, and rotometers. Air-flow calibration devices which have been themselves calibrated against primary standards are widely used in air-quality monitoring because of their suitability for field calibration of instruments.

Calibration is also essential to ensure the validity of gas concentrations measured and recorded by real-time continuous monitoring instruments. Such calibrations may be conducted by the use of standard gas mixtures, dynamic calibration systems which use permeation tubes containing the calibration gas (which diffuses through a membrane at a known rate), and, in some cases, comparisons to reference methods wherein samples are collected in impingers with subsequent wet chemical analysis of the sample reagent.

All air-quality monitoring programs must have a strong quality control program. These programs systematically ensure that data of high quality are being collected and used. Quality assurance requires periodic calibration and collection of data according to a standardized protocol. Deviations from monitoring and quality assurance protocols may require the invalidation of collected data.

ANALYTICAL PROCEDURES

A variety of chemical and physical techniques are used to determine concentrations of pollutant gases. Some of these are manual methods which require the collection of gases with subsequent analysis conducted in the laboratory. Most of those discussed below are utilized in continuous-flow, instantaneous-response instruments.

Wet Chemistry

Wet chemical techniques are primarily used to determine concentrations of gaseous pollutants collected by bubblers or impingers. Most of the numerous wet chemical procedures available employ absorbing reagents to collect the desired gas or gases. The reagent or pollutant reaction product is analyzed in the laboratory, and atmospheric concentrations are calculated. The most commonly used analytical principle is colorimetry, which is the formation of colored solutions whose light absorbance can be determined by spectrophotometric instruments.

Although wet chemical techniques are not as widely used for routine ambient air-quality monitoring as they were in the past, they are still useful in the determination of various gases for which instrumental analysis is unavailable. They can also be used to calibrate instrumental methods.

Chemiluminescence

Chemiluminescence is the emission of light as a result of chemical reactions. Chemiluminescence is commonly used in continuous air monitoring instruments for determining O_3 and NO_x. In most chemiluminescent analyzers, sampled air is introduced into a chamber with a reactant gas. Emitted light is sensed by a photomultiplier tube and amplified. Chemiluminescence methods have high specificity and sensitivity, and require no liquid chemicals.

Electrochemistry

Electrochemical analyzers are widely used to measure gases such as SO_2. Prior to the development of chemiluminescent techniques, they were used to measure total oxidants, which were assumed to be primarily O_3. In electrochemical analyzers, sampled air passes through an aqueous solution, where a chemical reaction between the pollutant and the reagent liberates electrons, creating a small current flow. Current flow can be calibrated to indicate the concentration of pollutants being measured. A major drawback of such methods is the interference resulting from the reaction of the electrolyte with other contaminants in the air sample. This problem may be reduced by removing interferent gases before they reach the reaction chamber.

Infrared Analysis

Nondispersive infrared (NDIR) analysis is primarily used for the ambient monitoring of CO concentrations. The principle employed is the selective absorption of infrared energy by certain gases. Infrared analyzers generally consist of sample and reference cells with split- or dual-beam infrared sources. The absorption of infrared energy by gases such as CO results in differential heating and expansion between sample and reference cells, causing a diaphragm between them to move with a frequency related to the concentration of the absorbing gas. Sampled air must usually be dried prior to analysis, because H_2O interferes with CO analysis.

Ultraviolet Absorption

Depending on their molecular structure, a variety of chemical substances may absorb UV light energy to a significant degree. The selective absorption patterns can be used to identify UV-absorbing gases and their concentration. Ultraviolet absorption instruments are very sensitive to O_3.

Photometry

In photometric analysis a pollutant in sampled air absorbs some or all wavelengths from a light beam. In wavelength-selective instruments, pollutant concentrations are usually measured by analyzing the beam at two different wavelengths in separate sample and reference cells.

Chromatography

In gas chromatography, molecules are adsorbed on a column of granular packing material. The collected gases are desorbed from the chromatographic column by heating. Because of differences in the strength of adsorption, each gas will be released at distinct intervals.

Gases segregated by their differential desorption rates pass through a detector where the relative concentration of each is determined. A variety of detector systems are used in gas chromatography. Among the more widely used are those based on measurements of changes in heat capacity (thermal conductivity), differences in ionization due to the combustion of sample gases (flame ionization), and current flow between two electrodes caused by ionization of gases by radioactive isotopes (electron capture). Thermal conductivity detectors are used to identify and quantify gases such as N_2, O_2, CO_2, CO, and CH_4; flame ionization is used for a large variety of paraffinic, olefinic, and aromatic HCs and their O_2 derivatives; and electron capture is commonly used to quantify a variety of chlorinated HCs.

The gas chromatograph is a widely used instrument. Because of its size and the complexity involved in identifying and quantifying individual components of complex gas mixtures, gas chromatographs are, for the most part, laboratory instruments with limited field applications. Flame ionization detectors commonly used in gas chromatographs are used to determine total nonmethane HCs (NMHCs) in major metropolitan areas where compliance with the air-quality standard for NMHCs is a concern.

EMISSIONS ASSESSMENT

In addition to ambient air-quality monitoring, air pollution control programs need to assess emissions from sources within area quality control regions. It is insufficient to maintain an elaborate program of ambient air-quality monitoring without also having a good knowledge of sources, especially what and how much they emit. Good quality source emission data are essential for effective control of air pollution. Because of cost factors and the complexities involved in source emissions determinations, this aspect of air surveillance programs has not received the attention it deserves. Recent new permitting requirements promulgated under the 1990 Clean Air Act (CAA) Amendments have significantly increased emissions assessment requirements.

Source Sampling

A variety of approaches may be used to determine emission rates for a given source. The most desirable means is to actually measure emissions by source sampling. Because of regulatory requirements, emissions from some sources are monitored continually with automated instruments. Source monitoring (or more commonly, source sampling) has been for many sources a relatively infrequent occurrence. Source sampling can be a costly, time-consuming activity.

Source sampling in many cases requires that a sampling team collect samples from the emission stream of a source. This usually requires the erection of scaffolding around a smokestack at a sufficient height to collect representative samples. Such work contains practical access limitations, technical difficulties, and an element of physical danger. Such source sampling, usually called stack sampling, is often a one-time activity conducted over several days. From one-time sampling, inferences are made as to emissions which occur on a continuous basis. Such inferences, depending on a variety of plant-operating factors, may be subject to considerable error. Nevertheless, a case can be made that some quantitative data based on actual measurements is better than no data at all or emission assessments based on calculations or generic extrapolation.

Source sampling consists of introducing a probe into a waste gas stream flowing through a smokestack. This probe must then withdraw a sample of waste gas, which will be analyzed in the laboratory. Gaseous pollutants may be collected by absorption in impingers, adsorption on charcoal or other media, or condensation in collecting traps. Gases may also be measured using continuous monitors. Particulate matter may be collected by a variety of techniques including wet scrubbing, filtration, impaction, and electrostatic precipitation. Cascade impactors are used in source sampling when characterization of particle size is desired — a factor which may determine the selection of control equipment.

Source sampling represents unique conditions of air flow and temperature. When emissions are associated with combustion systems, the velocity and temperature may be considerably higher than ambient conditions. Consequently, temperature and pressure must be measured to correct the sample volume to reference conditions (e.g., 20°C and 760 mm Hg). Velocity data determined from pressure measurements utilizing a pitot tube are also necessary to calculate mass loading to the atmosphere (i.e., emission rates).

Because PM comprises a range of inertial and other flow characteristics, such sampling requires that air flow through the sampling probe be at the same rate as that flowing in the waste gas stream. Such sampling is described as being isokinetic.

Source sampling may also be qualitative or semiquantitative. Opacity measurements of smoke plumes utilizing Ringelmann charts (Figure 7.6) represent a semiquantitative form of source sampling. Ringelmann measurements or, more accurately, assessments of plume optical density (degree of plume darkness), have

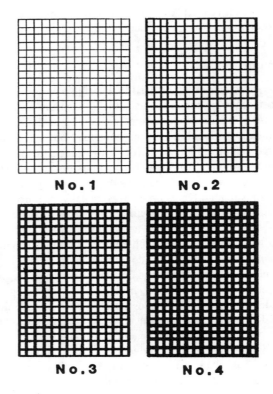

Figure 7.6 Ringelmann plume opacity charts.

been employed by air pollution control agencies for more than 80 years. The testimony of air pollution inspectors certified by Smoke Schools to read plume density is accepted by courts as valid evidence of violations of opacity standards. Ringelmann measurements have the advantages of simplicity, low cost, and a long history of legal acceptability. Ringelmann measurements are used in conjunction with emission standards that limit the opacity of a plume. Opacity measurements may also be determined quantitatively and continuously by projecting a light beam through the waste gas stream onto a photocell prior to emission from a smokestack.

Emission Factors

Emissions may be determined by means other than direct measurement of effluent gases. One common approach is to use emission factors published by the U.S. Environmental Protection Agency (USEPA). Emission factors are a listing of average emission rates that can be expected from individual source processes under given operating conditions. Emission factors may be determined from source sampling data or calculated from levels of contaminants expected from the use of a given raw material or fuel. For fuel, the production and emission of SO_2 can be calculated with reasonable accuracy from measurements of fuel sulfur content. For some particulate dusts, emissions may be calculated by determining the weight loss associated with the processing of a raw material into a finished product.

The assessment of emissions from both stationary and mobile sources is desirable for several reasons. Most importantly, effective air pollution control depends on identifying the major sources of emissions and applying appropriate limitations to them. In most cases, sources comply with emission limitations because it is corporate policy to obey the law. However, when sources do not comply and all reasonable attempts to require compliance have failed, air pollution authorities have no other recourse but to file legal action against violators. Source sampling provides the legal evidence to support the substance of the complaint. Emission assessment programs are also valuable in developing an ambient air-quality monitoring program and in providing adequate emission data for air-quality modeling.

AIR-QUALITY MODELING

Ambient air-quality monitoring is an expensive undertaking; as a result, even large cities may have only a relatively few monitoring stations. How then can air pollution control authorities determine whether emissions from a specific point source are resulting in violations of air-quality standards at some location downwind, a location not being monitored? Additionally, how can one know the degree of emission reduction necessary when results of ambient air-quality monitoring indicate that standards are being violated? The answer to both questions is to use air-quality models to predict ground-level concentrations downwind of point sources. Such models provide a relatively inexpensive means of determining compliance and predicting the degree of emission reduction necessary to attain ambient air-quality standards. Air-quality models are widely used by regulatory agencies as surveillance tools, that is, tools that assess the impact of emissions on ambient air quality. Additionally, under the 1977 CAA Amendments, the use of models is required for the evaluation of permit applications associated with permissible increments under the Prevention of Significant Deterioration (see Chapter 8) requirements.

The object of a model is to determine mathematically the effect of source emissions on ground-level concentrations, and to establish that permissible levels are not being exceeded. Models have been developed to meet these objectives for a variety of pollutants and time circumstances. Short-term models are used to calculate concentrations of pollutants over a few hours or days. They require detailed understanding of chemical reactions and atmospheric processes on the order of minutes to hours. Such models can be employed to predict worst-case episode conditions and are often used by regulatory agencies as a basis for control strategies. Long-term models are designed to predict seasonal or annual average concentrations, which may prove useful in studying potential health effects.

Models may be described according to the chemical reactions involved. So-called nonreactive models are applied to pollutants such as CO and SO_2 because of the simple manner in which their chemistry can be represented. Reactive models address complex multiple-species chemical reactions common to atmospheric photochemistry and apply to pollutants such as NO, NO_2, and O_3.

Models can be classified according to the type of coordinate system used. In grid-based models, the region of interest is subdivided into an array of cells which may range in horizontal size from 1 to 2000 km^2, depending on the scale desired. The average concentration of each pollutant is then calculated for each grid cell. This approach is commonly used by regulatory agencies to determine compliance with air-quality standards. Trajectory models, unlike the fixed-grid approach described above, follow parcels of air as they move downwind. These models are less costly than the fixed-grid ones because concentrations need not be calculated for every point in the region.

Models can be described as simple or advanced based on assumptions used and the degree of sophistication with which the important variables are treated. Advanced models have been developed for such problems as photochemical pollution, dispersion in complex terrain, long-range transport, and point sources over uneven terrain. The most widely used models for predicting the impact of relatively unreactive gases, such as SO_2, released from smokestacks are based on Gaussian diffusion.

In Gaussian plume models, it is assumed that concentrations of pollutants associated with a continuously emitting plume are proportional to the emission rate and inversely proportional to wind speed. It is also assumed that, as a result of molecular diffusion pollutant concentrations in the horizontal and vertical dimensions of a plume will be normally distributed (Figure 3.12). These models assume that pollutants do not undergo significant chemical reaction or removal as they travel away from a source, and as the plume grows vertically it is evenly reflected back toward the plume centerline.

Under stable atmospheric conditions or unlimited vertical mixing ground level concentrations can be calculated from Equation 7.1.

$$C_x = \frac{Q}{\pi \sigma_y \sigma_z \bar{u}} e^{-1/2[H/\sigma_z]^2} e^{-1/2[y/\sigma_y]^2} \tag{7.1}$$

where C_x = ground level concentration at some distance x downwind ($\mu g/m^3$), Q = average emission rate (g/sec), $\bar{u}$ = mean wind speed (m/sec), H = effective stack height (m), σ_y = standard deviation of horizontal distribution of plume concentration (m), σ_z = standard deviation of vertical distribution of plume concentration (m), y = off-centerline distance (m), and e = natural log equal to 2.71828.

Equation 7.1 reduces to Equation 7.2 when y (the distance from the centerline of the plume) is equal to zero.

$$C_x = \frac{Q}{\pi \sigma_y \sigma_z \bar{u}} e^{-1/2[H/\sigma_z]^2} \tag{7.2}$$

Table 7.1   Stability Classes for Determining Dispersion Coefficients

Wind speed, 10 m (m/sec)	Day: Incoming solar radiation			Night: Thinly overcast	
	Strong	Moderate	Slight	>4/8 Cloud	<3/8 Cloud
<2	A	A–B	B	E	F
2–3	A–B	B	C	D	E
3–5	B	B–C	C	D	D
>6	C	D	D	D	D

The parameters σ_y and σ_z describe horizontal and vertical dispersion characteristics of a plume at various distances downwind of a source as a function of different atmospheric stability conditions (Table 7.1). Values for different stability classes are determined from graphs found in Figures 7.7 and 7.8.

Plume rise is typically calculated from equations developed by Briggs. These equations are used to calculate H (the effective stack height) used in Equations 7.1 and 7.2. The effective stack height is the sum of the physical stack height (h) and plume rise. For neutral and unstable stability conditions and for buoyancy

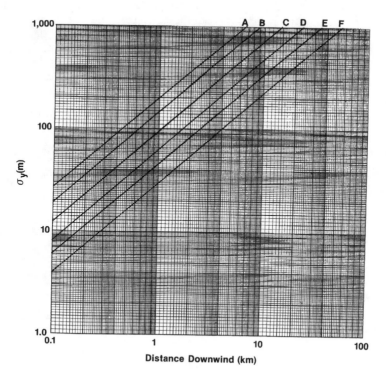

Figure 7.7   Pasquill's horizontal dispersion coefficients. (From Turner, D.B. 1969. *Workbook for Atmospheric Dispersion Estimates*. EPA Publication No. AP-26.)

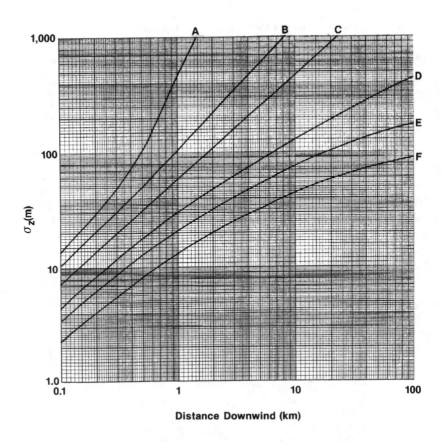

Figure 7.8   Pasquill's vertical dispersion coefficients. (From Turner, D.B. 1969. *Workbook for Atmospheric Dispersion Estimates.* EPA Publication No. AP-26.)

fluxes (F) <55 m⁴/s³, the effective plume height (m) can be estimated from Equation 7.3.

$$H = h + 21.45 \; F^{3/4}/\bar{u} \qquad (7.3)$$

when F is ≥55

$$H = h + 38.71 \; F^{3/5}/\bar{u} \qquad (7.4)$$

The buoyancy flux F (m⁴/s³) must be calculated from the following equation.

$$F = 2.45 \; v_s d^2 \; [(T_s - T)/T_s] \qquad (7.5)$$

where v_s = stack gas velocity (m/s), d = internal diameter of top of stack (m), T_s = stack gas temperature (°K), and T = ambient temperature (°K).

The horizontal distance (x_{pr}) from the stack to where the final plume rise occurs is assumed to be 3.5 x^* where x^* is the distance in kilometers where atmospheric turbulence begins to dominate entrainment. In Equation 7.3, $x_{pr} = 0.049\ F^{5/8}$, and in 7.4, $x_{pr} = 0.119\ F^{2/5}$.

Taking factors into account such as multiple eddy reflections from the ground and the bottom of the inversion layer and σ_z is evaluated as being greater than 1.6 times the mixing height (m), an equation for determining maximum ground level concentration can be derived.

$$C_{max} = \frac{2Q\sigma_z}{\pi \bar{u} e H^2 \sigma_y}$$
(7.6)

This equation is only applicable when the ratio σ_z/σ_y is constant with distance. The distance to the maximum concentration is the distance where $\sigma_z = (H/2)^{1/2}$.

Although the use of air-quality models is a subject of some controversy, there is general agreement that there are few alternatives to their use, particularly in making decisions on an action which is known in advance to pose a potential environmental problem. The debate arises as to which models should be used and the interpretation of model results. The underlying question in such debates is how well, or how accurately, does the model predict concentrations under the specific circumstances, because model accuracy may vary from 30% to a factor of 2 or more. If a model is conservative, i.e., it overpredicts ground-level concentrations, a source may be required to install costly control equipment unnecessarily. Less conservative models may underpredict concentrations, and thus violations of air-quality standards may occur. The uncertainty associated with models, however, may be no greater than that normally associated with input variables, such as wind data and source emission data. In many cases in the past, such data were usually estimated.

READINGS

1. American Conference of Governmental Industrial Hygienists. 1995. *Air Sampling for Evaluation of Atmospheric Contaminants.* 8th ed. ACGIH, Cincinnati, OH.
2. Axelrod, H.D. and J.P. Lodge, Jr. 1977. "Sampling and Calibration of Gaseous Pollutants." 145–177. In: Stern, A.C. (Ed.). *Air Pollution. Vol. III. Measuring, Monitoring and Surveillance of Air Pollution.* 3rd ed. Academic Press, New York.
3. Boubel, R.A., D.L. Fox, D.B. Turner, and A.C. Stern. 1994. *Fundamentals of Air Pollution.* 3rd ed. Academic Press, San Diego.
4. Bryan, R.J. 1977. "Ambient Air Quality Surveillance." 343–389. In: Stern, A.C. (Ed.). *Air Pollution. Vol. III. Measuring, Monitoring and Surveillance of Air Pollution.* 3rd ed. Academic Press, New York.
5. Briggs, G.A. 1975. "Plume Rise Predictions." 59–111. In: Haugen, D.A. (Ed.). *Lectures on Air Pollution and Environmental Impact Analysis.* American Meteorological Society, Boston, MA.
6. Budiansky, S. 1980. "Dispersion Modeling." *Environ. Sci. Technol.* 14:370–374.

7. Cheremisinoff, P.N. and A.C. Morresi. 1977. "Predicting Transport and Dispersion of Air Pollutants from Stacks." *Poll. Eng.* March, 26–30.

8. Couling, S. (Ed.). 1993. *Measurement of Airborne Pollutants.* Butterworth Heineman, Oxford, UK.

9. Goodin, W.R. 1981. "Advanced Air Quality Modeling." *Dames & Moore Engineering Bulletin 58.*

10. Harrison, R.M. and R.J. Young (Eds.). 1986. *Handbook of Air Pollution Analysis.* 2nd ed. Chapman & Hall, London.

11. Lodge, J.P., Jr. (Ed.). 1988. *Methods of Air Sampling and Analysis.* 3rd ed. Intersociety Committee. Lewis Publishers, Chelsea, MI.

12. Morgan, G.B. 1977. "Monitoring the Quality of Ambient Air." *Environ. Sci. Technol.* 11:352–357.

13. Noll, K.E. 1976. *Air Monitoring Survey Design.* Ann Arbor Science Publishers, Ann Arbor, MI.

14. Pasquill, F. and F.B. Smith. 1990. *Atmospheric Diffusion.* Ellis Horwood, New York.

15. Paulus, H.J. and R.W. Thron. 1977. 525–585. In: Stern, A.C. (Ed.). *Air Pollution. Vol. III. Measuring, Monitoring and Surveillance of Air Pollution.* 3rd ed. Academic Press, New York.

16. Powals, R.J., L.J. Zaner, and K.F. Sporek. 1978. *Handbook of Stack Sampling and Analysis.* Technomic Publishing, Inc., Westport, CT.

17. Turner, D.B. 1969. *Workbook for Atmospheric Dispersion Estimates.* EPA Publication No. AP-26.

18. Turner, D.B. 1979. "Atmospheric Dispersion Modeling: A Critical Review." *JAPCA* 29:502–519.

19. Turner, D.B. 1990. *Atmospheric Dispersion Estimates.* 2nd ed. CRC/Lewis Publishers, Boca Raton, FL.

20. USEPA. 1979. *Handbook, Continuous Air Pollution Source Management Systems.* EPA/625/6-79-005.

21. USEPA Air Pollution Training Institute. 1981. *Course 411: Meteorology — Instructors' Guide.* EPA/450/2-81-0013.

22. USEPA Air Pollution Training Institute. 1983. *Course 435: Atmospheric Sampling — Student Manual.* 2nd ed. EPA/450/2-80-004.

23. Winegar, E.D. and L.H. Keith. 1993. *Sampling and Analysis of Airborne Pollutants.* CRC/Lewis Publishers, Boca Raton, FL.

24. Zannetti, P. 1990. *Air Pollution Modeling: Theories, Computational Methods and Available Software.* Van Nostrand Reinhold, New York.

QUESTIONS

1. How do precision and accuracy differ?

2. In collecting a sample of ambient air, indicate what factors will determine your choice of sampling size, rate, and duration.

3. How does sampling differ from air-quality monitoring?

4. Why is equipment calibration crucial to an air-quality monitoring program?

5. In determining the concentration of a contaminant in ambient air, what factors would be a major source of error contributing to the difference between the reported value and the true value?

6. How is air-quality modeling used as a surveillance tool in air pollution control programs?

7. Distinguish between grab, static, intermittent, and continuous methods used to sample or monitor ambient pollutants.

8. What factors are important in determining an averaging time?

9. Widely used air-quality models are based on Gaussian plume behavior. What does this mean?

10. Describe the physicochemical principles by which SO_2, CO, and O_3 are analyzed in continuous monitoring instruments.

11. What is isokinetic sampling? Why is it used in source sampling for particles but not for gases?

12. What are the advantages and disadvantages of instantaneous real-time measurements?

13. What is the effect of horizontal wind speed on plume rise and ground-level concentrations?

14. What factors affect plume rise?

15. What is the effect of stability class on concentration gradients in a plume?

16. Describe how one would determine emission rates for a source in the absence of stack sampling data?

17. Under what circumstances would one model emissions from a smokestack?

18. What particle collection principles are used in a PM_{10} sampler?

8   REGULATION AND PUBLIC POLICY

Nominal regulatory efforts to control air pollution in the U.S. began in the late 19th and early 20th centuries when a number of large cities passed ordinances to control "smoke" produced by a variety of industrial activities. Though these smoke ordinances were ineffective because they lacked enforcement authority, they nevertheless served as forerunners of present-day regulatory programs.

Serious efforts to control air pollution began in California in the 1950s in response to the southern coast's increasingly worsening smog problem. As scientists began to unravel the mysteries of smog, regulatory efforts in California changed focus from controlling emissions from stationary sources to controlling emissions from automobiles. As a result, by the late 1960s, both California and the U.S. government began in earnest their significant and continuing regulatory efforts to reduce emissions from motor vehicles.

In the 1960s there was an increasing recognition at all levels of government that air pollution was a serious environmental problem. Many state and local governments began to form air pollution control agencies and develop regulatory frameworks to respond to their air pollution problems. Even these efforts (in retrospect) were relatively timid and for the most part ineffective.

With the onset of the modern environmental movement which began in earnest in 1970, the U.S. no longer accepted polluted cities and countrysides as a price of progress. A sea change occurred in American attitudes. The economic well-being which provided Americans the so-called "good life" was seen as diminishing the quality of the environment in which the fruits of economic prosperity could be enjoyed. The desire of Americans to prevent the further deterioration of the environment and to improve environmental quality gave rise to the significant regulatory efforts to control air pollution and other environmental problems which have taken place since 1970.

ALTERNATIVES TO REGULATION

Prior to modern regulatory programs to control air pollution, individuals or groups of citizens were very limited in what they could do to abate an air pollution problem which was adversely affecting them. They could, of course, ask corporate officers and others responsible for managing sources of pollution to be better neighbors. They could appeal to their sense of civic responsibility. Such appeals (if many of them were indeed made) would have for the most part gone unheeded. The abatement of pollution requires the allocation of what many economists and corporate financial officers describe as unproductive resources, that is, money which does not add to the value of the product.

In the absence of strong governmental regulatory programs, a citizen or group of citizens could file suit against a source under legal principles which existed under what is described as common law. Citizens continue to have this right to sue despite the fact that significant air pollution control regulations have been in place for almost three decades.

Under common law individuals who claim injury from emissions produced by a source can file a civil suit against the source in an effort to recover damages and to abate the problem. Such suits are called torts. Tort law involves a system of legal principles and remedies which attempt to address harm to persons and/or their property. Historically, both nuisance and trespass principles of tort law have addressed claims of harm associated with air pollution.

Nuisance

Under tort law a nuisance is an unreasonable interference with the use and enjoyment of one's property. The nuisance may result from an intentional or negligent act. It may be private (affecting some individuals and not others) or public (affecting everyone equally). Under the private nuisance concept the plaintiff, or injured party, must convince a judge or jury that the interference is unreasonable. If the suit is successful, he/she is entitled to recover damages. In order to abate the problem, however, the plaintiff must in addition seek an injunction against the source to stop its polluting actions.

To grant such an injunction, a court is required to balance the equities; that is, it must take into account the following: (1) extent of the damage, (2) cost of pollution abatement, (3) whether the pollution source was on the scene first, and (4) the economic position of the polluting source in the community. If the dispute is between a corporation with a large local work force and a single offended party, or even a group of landowners, courts invariably balance the equity in favor of the polluting source and refuse to grant an injunction. In such cases courts reason that injunctive relief will result in great economic harm to the local community if granted, and thus balance the equity in favor of the source. Although damages are awarded to the affected party or parties, injunctive relief is only rarely provided.

If all members of a community are affected equally, then no individual citizen can sue. In such cases it is necessary for public authorities to file a public nuisance suit against the pollution source which is causing "an unreasonable interference with a right common to the general public." As in private suits, courts must assess the seriousness of alleged harm relative to the social utility of the pollution source's conduct. Although public nuisance suits are provided for under common law, public authorities have been historically reluctant to file such actions.

Trespass

The trespass concept can be used whenever there is a physical invasion of a landowner's right to the exclusive possession of his property. Particulate-laden smoke could be considered to violate trespass provisions of tort law. Historically, courts have not required that the invasion be proven unreasonable or that damage be proven in awarding liability.

Reprise

The initiation of civil suits under common law to abate an air pollution problem was rarely if ever successful. In most cases, defendants settled such cases out of court by buying the plaintiff's property and/or providing a monetary award. When such cases went to trial, plaintiffs were in many cases successful in being awarded damages. They rarely were provided injunctive relief (which held the promise of abating the problem) because courts in such cases had to balance the equities involved.

REGULATORY STRATEGIES AND TACTICS

Successful abatement of air pollution and the protection of public health and welfare requires the selection of appropriate regulatory strategies and tactics to implement them. A variety of air-quality regulatory strategies have been proposed. These include cost–benefit, air-quality management, emission standard, and economic strategies. Air-quality management and emission standard strategies are the mainstay of pollution control efforts in the U.S. and other developed countries. A pollution-control strategy based on cost–benefit analysis is at best theoretical. However, a variety of economic-based approaches are being used to achieve air-quality goals and requirements.

Cost–Benefit Analysis

A cost–benefit strategy would have the support of many traditional or conservative economists. It would require the quantification of all damage costs associated with atmospheric pollutants and the costs of controlling them. Damage costs would be limited to that which is tangible, that is, the dollar value of health care and mortality, crop losses, reductions in property values, and effects on materials. Intangible values such as the loss of friends and loved ones, changes in ecosystems, and diminished aesthetic quality would not likely be included in

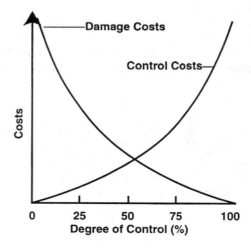

Figure 8.1 Relationship between control costs and pollutant damage costs.

damage costs. The primary focus of a cost–benefit strategy would be to select one or more pollution control options which minimizes the sum of damage and control costs.

In Figure 8.1, theoretical costs of pollution damage to public health and welfare are compared to control costs. As damage costs decrease, control costs increase. Note the steep rise in control costs as damage costs decrease toward zero. This illustrates a major problem in pollution control: a disproportionate increase in the cost of control relative to environmental benefits (decreased damage costs) achieved. In economic theory it would be desirable to select that degree of control that provides the greatest reduction of damages (or provides the greatest benefit) per unit cost. Damage and control costs would be "optimized" where the two lines cross in Figure 8.1.

Cost–benefit analysis suffers from a variety of practical and theoretical problems. First, it assumes that both damage and control costs can be easily quantified. Such quantification could be easily achieved for control costs. However, there is a considerable degree of uncertainty associated with the nature and extent of costs associated with pollutant exposures. Second, it assumes that damage costs to public health and welfare above the point of optimization are socially and politically acceptable. It would pose a public policy nightmare for regulatory officials who would have to make decisions on who or what is to be protected (more so than has been the case).

Air-Quality Management

   Air-quality management strategies are based on the widely accepted toxicological principle that pollutant exposures below threshold (lowest dose where toxic effects are observed) values are relatively safe and therefore some level of atmospheric pollution is acceptable and thus legally permissible. Pollutants identified as posing potentially significant public health risks and welfare concerns may be regulated by the promulgation of ambient air-quality standards.

An air-quality standard is a legal limit on concentrations of a regulated pollutant which can occur in the atmosphere. The promulgation of an air-quality standard for an ambient air pollutant is typically characterized by intensive reviews of the scientific evidence, recommendations on acceptable levels, and ultimate decision making by a regulatory authority.

Setting ambient air-quality standards is a difficult and arduous process. In most cases there is considerable scientific uncertainty associated with pollutant levels and their effects and the level of protection necessary. In the U.S., the Environmental Protection Agency (USEPA) is required to promulgate air-quality standards that provide an "adequate margin of safety" with special consideration for those individuals who may be the most sensitive (asthmatics, children, the elderly, the infirm, those with existing cardiovascular disease, etc.). As a result, ambient air-quality standards are much more stringent than occupational standards which are designed to protect nominally healthy working adults.

Because air-quality management strategies are based on the premise that there is a safe level of exposure (a threshold exists), they are generally not applied to contaminants which are unlikely to have threshold values. It has been widely accepted in both the scientific and regulatory communities that there is no safe level of exposure to carcinogens.

Once an air-quality standard for a pollutant or pollutant category has been promulgated, regulatory authorities at federal, state/provincial, and local levels are responsible for devising and implementing control/management practices (tactics) which will ensure that the air-quality standard is not exceeded, or if it is exceeded that the area is brought into compliance.

Air-quality management, though generally assumed to be a cost-effective approach to achieving air-quality goals, is a relatively complicated undertaking. Its complexity involves both technical and political dimensions. Regulatory authorities must select and implement tactics which in their best judgment will achieve each air-quality standard. This has proved to be very difficult particularly in regard to the O_3 standard because of the uncertainties associated with the respective roles of anthropogenic and biogenic emissions of hydrocarbons (HCs) in the atmospheric chemistry of tropospheric O_3. In a number of cases, regulatory authorities in their attempts to achieve particulate matter (PM) standards were confronted with selecting a control tactic which made both technical and economic sense but which was not politically popular. Such is the case with prohibitive bans on leaf burning in the fall. Though such a ban would be relatively cost effective, it has in some cases not been politically acceptable to the community. Prohibitive bans on leaf burning has a variety of side effects as well, most notably consuming scarce landfill space. On paper, air-quality management is a highly practical and cost effective control strategy. However, as is the case with most things in life, "the devil is in the details."

Emission Standards Strategy

In an emissions standards strategy, limits are placed on the maximum quantity (typically mass per unit time or thermal input) of a pollutant or pollutants which can be emitted from specific sources or categories of sources. Emission standards are applied to pollutants in source categories equally; that is, all members of a category must comply with the emission limit without regard to existing air quality. The same emission limits apply to sources in "clean" or "dirty" regions.

With the exception of the control of hazardous or "toxic" pollutants in the U.S., emission standards, when used as a strategy, are very much influenced by the concept of "practicability"; that is, emission standards reflect the highest degree of pollution reduction achievable taking cost of control into consideration.

The "best practicable means" or "good practice" approach is widely used for the development of emissions standards both in the U.S. and other developed countries. In Europe, the emission standards strategy based on "best practicable means" has been the traditional approach to air pollution control.

The term "practicable" indicates economic, technical, and political practicality. The "best practicable means" in practice may be the degree of control achieved by the best industrial plants in a source category, or technology which can be reasonably applied by borrowing from other industries. Implicit in this approach is that emissions standards may become more stringent (at least on new sources) as control technology improves and becomes more affordable. The "best practicable means" approach would be equivalent to emission limitation requirements used in the U.S. described as "reasonably available control technology" (RACT). Such requirements in the U.S. are generally applied to new or significantly modified existing sources under New Source Performance Standards (NSPS) to be discussed later in this chapter.

An emission standards strategy can also require higher degrees of emission reduction. For pollutants believed to pose unique health hazards (such as mercury, beryllium, benzene, radioactive isotopes, asbestos, etc.), the U.S. has required emission reductions which reflect the "state of the art," that is, the "best available control technology" (BACT). The objective of BACT standards is to achieve the highest degree of emission reduction that technology is capable of, with a more limited consideration of capital and operating costs. Emission standards or NSPS for large coal-fired steam electric generating plants promulgated in the U.S. in 1979 would also be described as requiring BACT.

Recent amendments (1990) to the Clean Air Act (CAA) require the setting of so-called maximum achievable emission reduction (MACT) standards for hazardous or "toxic" air pollutants. This concept suggests a higher degree of emission control than BACT. This may be true in the fact that emissions reductions can be achieved by a variety of control practices other than treating waste gas streams.

Emissions standards are usually specified in some numerical form. In the case of new coal-fired power plants licensed under the 1972 NSPS for SO_2, the emission limit is 1.2 lb SO_2/MBTU. In the 1979 revised standard, the emission limit was the same but the technology used to achieve the standard was specified, as was the minimum percentage of emission reduction required. As seen above, numerical standards can be based on heat input (lb/MBTU), unit of time (lb/hour), air volume (grains/SCF), or weight of process material (lb/ton). Outside the U.S. these would be prescribed in metric units.

Emission standards have the advantage of simplicity. There is no need for extensive and expensive air-quality monitoring networks, modeling to determine compliance, and complex technical and public policy decisions. All sources in the source category must meet the same requirements regardless of existing air quality. From an administrative standpoint, it is a highly attractive control strategy. As used in the U.S., it has the advantage of being applied to special pollution concerns such as new sources and pollutants which are believed to pose uniquely hazardous exposure problems.

The use of emission standards as a control strategy has, of course, significant limitations or problems. Though costs are taken into consideration in setting emission limits, sources must comply regardless of existing air quality. In areas with relatively good air quality, sources will be required to control emissions beyond that needed to protect public health and possibly welfare as well. In "dirty air regions," emission limitations based significantly on cost considerations may not be sufficiently stringent to protect public health and the environment.

Economic Strategies

A variety of economic-based approaches have been suggested as alternatives to air-quality management and emission standard strategies which dominate air pollution control throughout much of the developed world.

In an emission tax or charges strategy, emission sources would be charged or taxed using a scale related to their emission rate(s). Sources would have a choice to pay the tax or to reduce emissions sufficiently to avoid it. This strategy recognizes that sources have different costs of control, that some could control more cheaply than others, and that there would be a net savings to society as a whole if some facilities reduced emissions to a greater extent than others. Assuming that a source would want to minimize its total charges or taxes, it would in theory have an economic incentive to reduce emissions and its tax liability in a manner that best meets its needs. The charge or tax would have to be set sufficiently high to encourage sources to reduce emissions rather than pay for the "right to pollute."

Emission charges or "pollution taxes" have received only relatively limited usage. The once communist government of Czechoslovakia reportedly had a program (at least on the books) of assessing emission charges on a variety of industrial sources. In Japan, fees have been charged on SO_2 emissions from large sources in high pollution areas. France has also reportedly levied fees on combustion sources with 750 MW electrical generating capacity. In the 1970s, Norway levied taxes on oil with increases for each 0.5% of sulfur content. In the U.S., chlorofluorocarbons (CFCs) have been taxed to encourage manufacturers and users to more quickly phase them out.

Because of the significant costs involved in air pollution control, a variety of market-based alternatives have been developed and used in the U.S. The first of these was the so-called bubble policy. It allows corporations to reduce emissions from a facility as a whole from point and fugitive sources. Each facility with its imaginary bubble would be given a maximum emission limitation. Within the bubble, emissions could be increased at a point source as long as compensating reductions could be made at other locations.

This policy as well as others evolved into what can be described as emissions trading. Under it, corporations can earn emission reduction credits which could be banked for use in future facility expansions or openly traded with other companies. Constraints have to be imposed on the use of such credits to ensure that air-quality standards are not violated.

A market-oriented economic strategy has been incorporated in the 1990 CAA Amendments for the purpose of reducing SO_2 emissions 10 million tons annually from electrical utilities by the year 2000. It provides for emission allowances and the right to apply unused allowances to offset emissions at other operating units or to be traded to other sources.

Market-oriented economic strategies have been a significant innovation in air pollution control efforts in the U.S. They promise increased corporate technical and economic flexibility while at the same time meeting control requirements and achieving overall air quality goals.

Tactics

Once control strategies have been selected, regulatory agencies must develop a plan or plans on how to achieve those strategies. This is particularly the case with air-quality management. Specific actions taken to implement a control strategy can be described as tactics.

Tactics employed by regulatory agencies to reduce emissions from stationary sources have included the setting of emission standards for specific sources in a control region, restrictions on fuel use (e.g., prohibiting the use of coal or oil with sulfur content >1%), use of tall stacks, and prohibitive bans on open burning or use of certain types of combustion equipment (e.g., apartment house incinerators).

Tactics used by regulatory agencies to control emissions from motor vehicles or reduce air-quality problems associated with motor vehicle use include control requirements for blow-by gases, evaporative losses, refueling and cold starts, exhaust emission standards, inspection and maintenance, fuel-additive standards, special fuel use, use of alternative power systems, and transportation plans.

As can be seen, emission standards can be used both as an overall control strategy and as a tactic to achieve air-quality standards. Various elements of economic strategies may also be used as tactics.

Permit systems are a major component of all air pollution control programs. A permit extends a legal right to a source to emit contaminants into the atmosphere. By requiring emission sources to obtain permits, regulatory authorities can impose conditions on emission limits on the permittee or require other actions which reduce emissions to the atmosphere. Permits are important tools in obtaining information about emissions from a source and requiring compliance with emission limits. In the broad context, permitting can be viewed as a major air pollution control tactic.

FEDERAL LEGISLATIVE HISTORY

At the end of the 19th century and through the first half of the 20th century, air pollution in the U.S. was seen to be a local problem. This began to change in the 1950s as states led by California began a variety of regulatory initiatives. A federal role also began to evolve at about the same time. The legislative history of federal air pollution control efforts is described below.

1955 Clean Air Legislation

Legislation enacted in 1955 authorized the Public Health Service in the Department of Health, Education and Welfare (DHEW) to conduct air pollution research and training programs, and to provide technical assistance to state and local governments. It affirmed that state and local governments had the fundamental responsibility for air pollution control. Amendments to this law in 1960 and 1962 called for special studies relating to health effects associated with motor vehicle-related pollutants.

1963 Clean Air Act

As the quality of the nation's air continued to deteriorate, Congress responded by passing the Clean Air Act (CAA) of 1963. This legislation broadened the federal role in the air pollution abatement effort. Specifically, it provided for (1) awarding grants for program development and improvement of state and local air pollution control efforts, (2) accelerated research, training, and technical assistance, (3) federal enforcement authority to abate interstate air pollution problems, (4) federal research responsibility for automobile and sulfur oxides pollution, and (5) air-quality criteria development for the protection of public health and welfare. Although the federal role was considerably expanded, state and local governments were nevertheless primarily responsible for air pollution control.

Amendments to the CAA of 1965, the Motor Vehicle Air Pollution Control Act, authorized the Secretary of DHEW to promulgate emission standards for motor vehicles. As a consequence, federal emission standards were established for all 1968 model light-duty motor vehicles. These 1965 amendments also established the National Air Pollution Control Administration (NAPCA) within DHEW to provide regulatory leadership in the nation's efforts to control air pollution.

Air Quality Act of 1967

Although some progress was made under the CAA of 1963, control programs at all governmental levels remained relatively weak, and air pollution problems continued to worsen. At the request of the president, Congress in 1967 strengthened this country's air pollution control effort by enacting the comprehensive Air Quality Act. Many of the pollution control concepts developed in the 1967 legislation continue to provide the basic framework for our current control efforts. Of special significance was the institution of a regional approach to air pollution control and the development and implementation of air-quality standards.

The Air Quality Act of 1967 required the Secretary of DHEW to designate Air Quality Control Regions (AQCRs) and issue air-quality criteria and control technique information. After the federal agency completed these mandated requirements, individual states were to develop state air-quality standards and plan for their implementation on a fixed time schedule.

The Air Quality Act also provided for a continuation of the federal role in air pollution research and development, training programs, and grants-in-aid to state and local air pollution control programs. Although the federal government was given significant new authority to initiate lawsuits whenever interstate violations occurred, states maintained primary responsibility for enforcement.

Progress under the Air Quality Act was slow. States could not promulgate air-quality standards until NAPCA designated AQCRs and issued Air Quality Criteria and Control Technique Documents for specific pollutants. These tasks took considerable time. For example, Air Quality Criteria Documents for sulfur oxides (SO_x) and particulate matter (PM) were issued in 1969, for hydrocarbons (HCs), carbon monoxide (CO), and oxidants in 1970, and for nitrogen oxides (NO_x) in 1971. The promulgation of Control Technique Documents was carried out more expeditiously. The designation of AQCRs was an arduous process. By 1970 NAPCA had proposed 57 AQCRs but had only designated 23.

1970 Clean Air Act Amendments

In 1970 public concern for environmental quality had reached its zenith. As a consequence of dissatisfaction with progress under the Air Quality Act of 1967 and public support for strong legislation, Congress enacted tough new clean air amendments in 1970. The NAPCA was dissolved and its air pollution control functions were transferred to USEPA, an independent federal agency created by executive order of the President.

The CAA Amendments of 1970 sought to remedy many of the deficiencies of previous legislation. Most significantly, it completed the federalization of the air pollution control effort, provided significant federal enforcement authority, and placed air pollution control on a rigorous timetable.

Specific provisions of the 1970 CAA Amendments included (1) the setting of uniform National Ambient Air Quality Standards (NAAQS); (2) immediate designation of AQCRs; (3) State Implementation Plans (SIPs) to achieve NAAQS;

(4) stringent new automobile emission standards and standards for fuel additives; (5) emission standards for new or modified sources and for hazardous air pollutants; (6) the right of citizen suits; and (7) federal enforcement authority in air pollution emergencies and interstate and intrastate air pollution violations.

1977 Clean Air Act Amendments

The goal of the 1970 CAA Amendments was to achieve clean air for all regions of the country by July 1, 1975. Although significant progress for some pollutants was made, many control regions did not achieve one or more NAAQS.

Recognizing that the air-quality goals set in 1970 could not be met without major economic disruptions, Congress amended the CAA in 1977. These amendments would best be described as a "mid-course correction." Specifically, it (1) postponed compliance deadlines for national primary (health based) air-quality standards, (2) postponed, and in some instances modified, federal automobile emission standards, (3) included the concept of Prevention of Significant Deterioration (PSD) into the language of the act, (4) provided a mechanism by which USEPA could have flexibility in allowing some growth in "dirty air" regions, and (5) gave authority to USEPA to regulate stratospheric O_3-destroying chemicals. The regulation of O_3-destroying chemicals was the only new initiative in the 1977 CAA Amendments.

1990 Clean Air Act Amendments

Though it is the intent of Congress to reauthorize and update by amendment all major federal legislation on a 5-year schedule, disagreements over policies related to the control of acidic deposition contributed to an 8 year hiatus in amending the CAA. In 1990, major agreements on clean air amendments were reached by members of Congress and the president. The 1990 CAA Amendments represented a significant political achievement. Major new or expanded authorities included changes in the timetables for the achievement of air-quality standards in nonattainment areas, the regulation of emissions from motor vehicles, regulation of hazardous air pollutants, acidic deposition control, stratospheric O_3 protection, permitting requirements, and enforcement.

Federalization

In the evolution of federal legislation certainly the most significant trend was the increasing involvement of the federal government in the problem and ultimate primacy in its solution. This federalization of the air pollution control effort is not unique; it was consistent with the centralization of responsibility at the federal level for many problems that have beset American society for the past four decades. During this period, the principle of "states' rights" slowly eroded with the recognition that problems such as air pollution could only be effectively dealt with in a coordinated approach at all levels of government.

AIR POLLUTION CONTROL UNDER THE 1970, 1977, AND 1990 CLEAN AIR ACT AMENDMENTS

Prior to the enactment of the CAA Amendments in 1970, NAPCA (a predecessor agency to USEPA) and Congress considered several different control strategies. The outcome of their deliberations was an ambient air pollution control strategy that was based on establishing NAAQS with the proviso that states would develop implementation plans to achieve them. Additionally, national emission standards would be established for major new sources and for existing and new sources of hazardous air pollutants. As a result, the nation's ambient air pollution control program has been primarily based on strategies employing both air-quality management and emission standards concepts.

Air-Quality Standards

Under the 1967 Air Quality Act, states were required to set air-quality standards for AQCRs within their jurisdiction. These standards were to be developed on a fixed schedule and were to be consistent with air quality criteria developed and issued by NAPCA. Development of state air-quality standards under the 1967 legislation was slow, due in part to the slowness of NAPCA in publishing air-quality criteria and designating AQCRs. In addition, states were concerned that the adoption of stringent standards would result in the relocating of industries to other states which had weaker standards. This concern created a great deal of uncertainty, confusion and, of course, inaction.

Under the CAA Amendments of 1970, the federal government was charged with the responsibility of developing uniform NAAQS. These were to include primary standards designed to protect health and secondary standards to protect public welfare. Primary standards were to be achieved by July 1, 1975, and secondary standards in "a reasonable period of time." Implicit in this dual standard requirement was the premise that pollutant levels protective of public welfare were more stringent than those for public health, and that achievement of the primary health standard had immediate priority. In 1971, USEPA promulgated NAAQS for six classes of air pollutants. In 1978 an air-quality standard was also promulgated for lead. Additionally, the 1971 photochemical oxidant standard was revised in 1978 to an O_3 standard, with an increase in permissible levels. In 1987, the suspended PM standard was revised and redesignated as a PM_{10} standard. This revision reflected the need for a PM standard based on particle sizes (≤ 10 μm) that have the potential for entering the respiratory tract and affecting human health. National Ambient Air Quality Standards, averaging times, and recommended measurement methods are summarized in Table 8.1.

Air-Quality Criteria

The promulgation of NAAQS must be preceded by the publication of Air Quality Criteria. These are issued in document form and summarize all the relevant scientific information on the health and welfare effects of individual pollutants. The principal function of the Air Quality Criteria process is to deter-

Table 8.1 National Ambient Air-Quality Standards[a]

Pollutant	Averaging time	Primary standard	Secondary standard	Measurement method
Carbon monoxide	8 hour	10 mg/m³ (9 ppm)	Same	Nondispersive infrared spectroscopy
	1 hour	40 mg/m³ (35 ppm)	Same	
Nitrogen dioxide	Annual average	100 µg/m³ (0.05 ppm)	Same	Colorimetric using Saltzman method or equivalent
Sulfur dioxide	Annual average	80 µg/m³ (0.03 ppm)		Pararosaniline method or equivalent
	24 hour	365 µg/m³ (0.14 ppm)		
	3 hour		1300 µg/m³ (0.5 ppm)	
PM₁₀ (≤10 µm)	Annual arithmetic mean	50 µg/m³	50 µg/m³	Size-selective samplers
	24 hour	150 µg/m³	50 µg/m³	
Hydrocarbons (corrected for methane)	3 hour (6–9 am)	160 µg/m³ (0.24 ppm)	Same	Flame ionization detector using gas chromatography
Ozone	1 hour	235 µg/m³ (0.12 ppm)	Same	Chemiluminescent method or equivalent
Lead	3 month average	1.5 µg/m³	Same	Atomic absorption

[a] Standards, other than those based on the annual average, are not to be exceeded more than once a year.

mine the relationship between pollutant concentrations and health and welfare effects so that air-quality standards are supported by good scientific evidence.

Air-Quality Control Regions

The concept of a regional approach to air pollution control was first introduced in the Air Quality Act of 1967. This approach recognized that air pollution was a regional problem — one that did not respect political boundaries. Initially, the regional concept was only to be applied to interstate problems. With the enactment of the 1970 CAA Amendments, Congress mandated that both interstate and intrastate AQCRs be established for the purpose of achieving NAAQS.

State Implementation Plans

After NAAQS are promulgated, individual states are required under provisions of the CAA to develop and submit to USEPA plans for their implementation, enforcement, and maintenance. These SIPs can be approved by USEPA in whole or in part. If a portion of a SIP is disapproved, the state is obligated to revise it and seek approval for the revisions. If the plan, in whole or in part, is not acceptable, USEPA is empowered to develop a SIP for the state, which the state must then enforce.

In developing implementation plans, states must formulate policies for each air pollutant for which NAAQS have been promulgated. In this process they are required to determine and report on the nature and quantity of emissions from existing sources and require at a minimum RACT to limit emissions. They must negotiate compliance schedules with major sources and indicate what effect the emission reduction will have on air quality.

Episode Plans

State plans must also include control strategies to deal with air pollution episodes or emergencies. Episodes occur as a result of the stable atmospheric conditions associated with stagnating, migrating, or semipermanent high pressure systems. The poor dispersion conditions associated with these inversions may result in high ambient levels of toxic pollutants which pose an imminent danger to public health. States are required to develop response criteria and plans similar to those in Table 8.2. Note that each stage reflects a higher degree of pollution severity. Control actions at each stage should be sufficient to prevent the next stage from occurring. The episode plan must provide for an orderly reduction of emissions on a prearranged schedule and should therefore, in theory, protect public health.

Legal Authority

In order for a state to gain approval for its SIP, it must have adequate legal authority to (1) adopt and enforce emission regulations, (2) implement episode control plans, (3) regulate new or modified sources, (4) require stationary sources to provide emission data, and (5) gather information to determine compliance with regulations, including the right to inspect and test. Although federal law is "the law of the land," it cannot be enforced directly by state and local officials. State and local governmental regulatory agencies must enact their own enabling legislation/ordinances to enforce provisions of federal law. Although these reflect local circumstances and sentiments, they must be consistent with the major provisions of federal clean air legislation.

SIP Approval

In practice, the process of SIP development and approval has been a time consuming and frustrating experience for federal, state, and local officials. Plans had to be submitted within 9 months of primary standard promulgation. In many cases, data on emissions (emission inventory) were scant, and manpower for plan development was insufficient. Many of these hastily prepared plans were subject to numerous revision requests before federal approval could be granted. Additional revision requirements were necessitated by federal court decisions resulting from citizen suits filed against USEPA relative to requirements of SIPs. As a

Table 8.2   Suggested Criteria for Air Pollution Episodes

Alert stage

An Air Pollution Alert goes into effect if:

SO_2 reaches 0.3 ppm, 24-hour average,
  or
Particulate matter (PM) reaches 3.0 COH, 24-hour average,
  or
Combined SO_2 and PM product reaches 0.2,
  or
CO reaches 15 ppm 8-hour average,
  or
O_3 reaches 0.1 ppm, 1-hour average,
  or
NO_2 reaches 0.6 ppm, 1-hour average and 0.15 ppm, 24-hour average
  and
Meteorological conditions are such that pollutant concentrations can be expected to remain at these levels for 12 hours or more, or increase unless control actions are taken.

Warning stage

An Air Pollution Warning goes into effect if:

SO_2 reaches 0.6 ppm, 24-hour average,
  or
PM reaches 5.0 COH, 24-hour average,
  or
Combined SO_2 and PM product reaches 0.8,
  or
CO reaches 30 ppm, 8-hour average,
  or
O_3 reaches 0.4 ppm, 1-hour average,
  or
NO_2 reaches 1.2 ppm, 1-hour average and 0.3 ppm, 24-hour average
  and
Meteorological conditions are such that pollutant concentrations can be expected to remain at the above levels for 12 hours or more, or increase unless control actions are taken.

Emergency stage

An Air Pollution Emergency goes into effect if:

SO_2 reaches 0.8 ppm, 24-hour average,
  or
PM reaches 7.0 COH, 24-hour average,
  or
Combined SO_2 and PM product reaches 1.2,
  or
CO reaches 40 ppm 8-hour average,
  or
O_3 reaches 0.6 ppm, 1-hour average,
NO_2 reaches 1.6 ppm, 1-hour average and 0.4 ppm, 24-hour average
  and
Meteorological conditions are such that this condition can be expected to continue for 12 hours or more.

From USEPA. 1971. GPO Publication No. 1972 0-452-7291.

result of these decisions, states in the early 1970s had to revise their implementation plans to provide for maintenance of air quality once standards were
achieved, and for the prevention of significant deterioration (PSD) of air quality
in regions in which air quality was better than the standards. Since then, SIPs
have had to be revised in response to changes in standards and regulatory policies
initiated by USEPA or mandated by the 1977 and 1990 CAA Amendments.

Nonattainment of Air-Quality Standards

As the 1975 statutory deadline for the attainment of NAAQS was approaching, it was evident that many AQCRs would not be in compliance with one or
more standards. Congress recognized that an extension of the deadlines was
necessary. Although the CAA was scheduled for reauthorization in 1975, Congress could not agree on proposed changes. As a result, the law was not amended
until 1977. Under these amendments, attainment deadlines for SO_2 and PM were
extended to 1982; those for auto-related pollutants were extended to 1987.

Because many AQCRs were not in compliance with one or more air-quality
standards, these were classified by USEPA as nonattainment areas. This nonattainment status had considerable significance, because under the 1970 CAA
Amendments, the introduction of new major sources in a dirty air region would
be expected to aggravate the threat to public health and would therefore be
prohibited. The prospect of limitations on economic development where economic growth would naturally occur created considerable political and economic
concerns. In response to these concerns, USEPA administratively developed a
"compromise" emission offset policy which would allow growth as long as new
sources did not produce a net increase in emissions. Under this offset policy, a
new industrial facility would have to persuade existing sources to achieve sufficient additional emission reductions in order to produce no net increases in
emissions. In some cases the new industrial source might even find it desirable
to provide the capital for an existing source to sufficiently reduce its emissions
so that an offset could be achieved. In cases where there was to be an expansion
of an existing facility, sufficient emission reductions were made in existing facilities to qualify under the emissions offset policy.

The 1977 CAA Amendments mandated that the USEPA offset policy be
incorporated into the SIP process. In an alternative approach proposed in these
amendments, a major new source could be developed in a nonattainment region
if it would attain the "lowest achievable emission rate" (LAER). To be eligible
under this concept, an industrial source having other facilities within a state must
be in compliance with all relevant emission reduction programs for these facilities.
In practice LAER has meant the use of BACT.

The problem of nonattainment of primary air-quality standards was addressed
in the 1990 CAA Amendments. Deadlines were again postponed, and USEPA
was given increased flexibility in setting deadlines for the attainment of standards
in nonattainment areas. In general, the attainment date for a nonattainment area
was one that could be achieved as expeditiously as practicable but no later than
5 years after designation. This could be further extended for no more than 10

Table 8.3   Classification and Attainment Dates for
             1989 Nonattainment Areas Under 1990
             CAA Amendments

Area class	Design value (ppmv)	Attainment date
Ozone		
Marginal	0.121–0.138	3 years after enactment
Moderate	0.138–0.160	6 years after enactment
Serious	0.160–0.180	9 years after enactment
Severe	0.180–0.280	15 years after enactment
Extreme	≥0.280	20 years after enactment
Carbon monoxide		
Moderate	9.1–16.4	December 31, 1995
Serious	≥16.5	December 31, 2000
PM_{10}		
Moderate		December 31, 1994
Serious		December 31, 2001

years if USEPA deemed it appropriate considering the severity of the nonattainment problem and availability and feasibility of control measures.

For O_3, CO, and PM_{10}, attainment dates were to be determined on the basis of appropriate classifications. These classifications and applicable deadlines are summarized in Table 8.3. As can be seen, the more severe the problem and the more difficult it is to control, the more time a state has/had to bring it into compliance with primary air quality standards.

Prevention of Significant Deterioration

Although the concept of PSD in areas where air quality was better than the standards was not specifically mentioned in the CAA Amendments of 1970, environmental groups maintained that nondegradation of air in clean air areas was implicit from the wording of the preamble of the Act "to protect and enhance" and reflected Congressional and federal agency intent documented in Congressional hearings before its passage. In a suit before the Federal District Court of the District of Columbia, the Sierra Club contended that in regions where air was of higher quality than that required by NAAQS, it should not be permitted to deteriorate to the level of the standards. The Federal District Court agreed, and it was upheld unanimously by the Court of Appeals for the District of Columbia. It was appealed to the U.S. Supreme Court where the justices split equally. As a result, the principle of PSD was upheld, and USEPA was compelled to develop regulations to protect air quality in clean air areas.

In 1975, USEPA developed a classification system which would allow some economic development in clean air areas and still protect air quality from significant deterioration. To eliminate any legal uncertainty, the classification system and provisions for its implementation were included in the 1977 CAA Amendments. Clean air areas were divided into three classes. Very little deterioration would be allowed in Class I regions. Congress mandated that all International

Table 8.4   Allowable Increments for PM_{10}, SO_2, and NO_2 in Prevention of Significant Deterioration Regulated Regions

Pollutant	Standard	Class	Allowable increment ($\mu g/m^3$)
PM_{10}	Annual mean	I	4
		II	17
		III	34
	24-hour maximum	I	8
		II	30
		III	60
SO_2	Annual mean	I	2
		II	20
		III	40
	24-hour maximum	I	5
		II	91
		III	182
	3-hour maximum	I	25
		II	512
		III	700
NO_2	Annual mean	I	2.5
		II	25
		III	50

Parks, National Wilderness Areas exceeding 5000 acres, National Monuments exceeding 5000 acres, and National Parks exceeding 6000 acres would be automatically designated as Class I. These included 155 areas in 36 states and one International Park. USEPA was required to develop and implement regulations which would protect visibility in these pristine air areas. Class II areas would allow moderate air-quality deterioration, and Class III areas would allow air-quality deterioration up to the secondary standard. Maximum allowable increases above baseline values for each of the three PSD classes are summarized in Table 8.4.

New Source Performance Standards

In addition to establishing a program of air pollution control by means of air-quality standards, the CAA Amendments required that new or significantly modified existing sources comply with NSPS. Emissions standards established under NSPS were to be applied to all new sources in designated source categories uniformly, regardless of whether they were being constructed in clean or dirty air regions. These emission standards were to reflect the degree of emission limitation achievable through the application of the best system of emissions reduction, taking cost into consideration. The purpose of these emission limitations was to prevent the occurrence of new pollution problems by requiring the installation of control measures during construction, when they would be least expensive.

New Source Performance Standards are applied to specific source categories, the definition of which is the responsibility of the USEPA administrator. A

Table 8.5   Pollutants Regulated Under NSPS in Selected Source Categories

Sources	Pollutants
Fossil fuel-fired steam generators	Particulate matter, opacity, SO_2, NO_x
Coal preparation plants	Particulate matter, opacity
Incinerators	Particulate matter, opacity
Primary lead, zinc, and copper smelters	Particulate matter, SO_2, opacity
Secondary lead smelters	Particulate matter, opacity
Secondary bronze and brass smelters	Particulate matter, opacity
Primary aluminum reduction plants	Fluorides
Iron and steel plants	Particulate matter, opacity
Ferroalloy production facilities	Particulate matter, opacity, CO
Steel plants — electric arc furnaces	Particulate matter, opacity
Sulfuric acid plants	SO_2, acid mist, opacity
Nitric acid plants	NO_x
Portland cement plants	Particulate matter, opacity
Asphalt concrete plants	Particulate matter, opacity
Sewage treatment plants	Particulate matter, opacity
Petroleum refineries	Particulate matter, SO_x, CO, reduced sulfur
Kraft pulp mills	Reduced sulfur, particulate matter, opacity
Grain elevators	Particulate matter, opacity
Lime manufacturing plants	Particulate matter
Phosphate fertilizer industry	Fluorides

sampling of source categories and pollutants regulated under NSPS is indicated in Table 8.5. The application of NSPS to pollutant types is quite flexible. In addition to criteria pollutants, NSPS have required emissions limits on fluoride from phosphate and aluminum plants, acid mist from sulfuric acid plants, and total reduced sulfur from paper mills. New Source Performance Standards can be applied to new sources for pollutants not regulated under NAAQS or hazardous pollutants standards. Although the formulation and promulgation of NSPS for any category of sources and pollutants is a federal responsibility, USEPA can delegate enforcement authority to a state with an acceptable SIP.

Hazardous Air Pollutants

Under the 1970 CAA Amendments, Congress authorized USEPA to regulate pollutants to which ambient exposures were deemed to be more hazardous than those regulated under NAAQS. As a result of this authority, USEPA promulgated national emissions standards (NESHAPs) for mercury, beryllium, and asbestos in the early 1970s. By 1990, NESHAPs had only been promulgated for seven pollutants/pollutant categories. These also included vinyl chloride, inorganic arsenic, benzene, and radioactive isotopes. In addition to those indicated above, coke oven emissions had been listed, but no specific regulations had been finalized by 1990.

Over 20 years, USEPA designated and regulated only seven hazardous air pollutants. Many states had gone their own way in regulating what they called "air toxics." State air pollution control agencies by 1990 had established emission standards for approximately 800 "toxic" chemicals.

The relatively slow pace of designating and regulating hazardous air pollutants reflected policy uncertainties associated with this provision of the 1970 CAA Amendments. USEPA interpreted NESHAP provisions in the 1970 CAA Amendments as requiring the setting of emission limits which would provide an ample margin of safety. Because many hazardous pollutants were carcinogenic, USEPA had at various times interpreted the statutory language of the 1970 CAA Amendments as requiring an emission standard of zero for carcinogenic substances. It was therefore reluctant to regulate emissions of economically important substances which were potentially carcinogenic in humans because such regulation could have required a total ban on production.

The regulation of hazardous air pollutants received special attention under the 1990 CAA Amendments. In the 1990 CAA Amendments, Congress listed 189 hazardous air pollutants for which USEPA was to identify source categories and promulgate emission standards by the year 2000. USEPA was also authorized to add substances to this list if there was adequate data to show that emissions, ambient concentrations, bioaccumulation, or deposition could reasonably be anticipated to cause adverse human health or environmental effects.

One unique feature of Title III of the 1990 CAA Amendments is that hazards were not limited to human health. By reference to adverse environmental effects, Congress indicated its intent to protect wildlife, aquatic life, and other natural resources including populations of endangered or threatened species or the degradation of environmental quality over broad areas.

Under Title III USEPA was required to publish a list of categories and subcategories of major sources (>10 ton/year single pollutant, >25 ton/year aggregate of hazardous air pollutants) and area sources for pollutants listed and promulgate emission standards for each in accordance with a congressionally mandated schedule. In promulgating technology-based (rather than health based) emission standards, USEPA was given the flexibility of distinguishing among classes, types, and sizes of sources. USEPA was to require the maximum degree of reduction in emissions, taking into account cost of achieving such emission reductions and any nonair-quality health and environmental impacts, as well as energy requirements. Maximum achievable emission reduction could be effected by (1) process changes, substitution of materials, or other modifications, (2) enclosing systems or processes, (3) collecting or treating pollutants released from a process, stack, storage, or fugitive source, or (4) design, equipment, work practice, or operational standards. Maximum achievable control technology (MACT) is the term commonly used to describe these emission reduction requirements for hazardous air pollutants.

The maximum degree of emission reduction deemed achievable for new sources has to be at least as stringent as emission control that is achieved by the best controlled similar source. Within certain limits, emission standards for new sources may be less stringent than for existing sources. For major existing sources, MACT requirements were to be based on the average rate achieved by the best performing 12% of existing sources or average emission of the best performing five existing sources in the respective category.

Under the 1990 CAA Amendments, USEPA is required to establish MACT standards for listed hazardous pollutants according to the following schedule: (1) ≥40 categories and subcategories not later than 2 years after passage of amendments; (2) coke oven emissions by the end of 1992; (3) 25% of listed categories and subcategories not later than 4 years after the passage of amendments; (4) additional 25% of listed categories or subcategories no later than 7 years; and (5) emission standards for all categories/subcategories no later than 10 years. Priorities for promulgating standards are to be based on known or anticipated adverse effects, quantity and location of emissions, and efficiency of grouping categories.

Stratospheric O_3 Depletion

The 1977 CAA Amendments provided USEPA clear statutory authority to regulate substances, practices, processes, and activities that could be reasonably anticipated to damage the O_3 layer. Using this authority USEPA banned the use of CFCs as aerosol propellants in 1978. As a result of this ban, U.S. emissions of CFCs declined by 35 to 40% of their pre-ban levels. The 1977 amendments also provided USEPA the authority needed to meet the emission reduction targets of the 1987 Montreal Protocol and subsequent amendments to it.

The 1990 CAA Amendments repealed the O_3 protection section of the 1977 CAA Amendments. The CAA was amended by adding Title VI: Stratospheric O_3 Protection. These amendments updated USEPA authority and added specific requirements which reflected the more recent needs of protecting the O_3 layer and policies to achieve such protection.

Major provisions of Title VI include a listing of Class I and Class II substances, monitoring and reporting requirements, phaseout of the production of Class I and Class II substances, accelerated phaseout authority, a national recycling and emission reduction program, servicing of motor vehicle air conditioners, regulating nonessential products containing CFCs, labeling requirements for Class I and II substances, a policy for the development of safe alternatives, transfers among parties to the Montreal Protocol, and international cooperation.

The classification of chemicals into Class I and II substances reflects differences in O_3 depletion potential and a recognition that Class II substances which are used as substitutes for CFCs can also destroy O_3, albeit this potential is at least an order of magnitude less. The production of Class I substances including CFCs, three types of bromine-containing halons, and carbon tetrachloride were to be banned by the year 2000, and methyl chloroform by the year 2002, with some exceptions for essential uses of methyl chloroform for medical devices, aviation safety, and national security. Production of Class II compounds which include hydrochlorofluorocarbons (HCFCs) was to be banned in the year 2030 with a phaseout beginning in the year 2015.

Title VI recognizes that regulatory bans on production are not sufficient to deal with the problem of O_3-depleting substances. Chlorofluorocarbons, for example, in refrigerators and air conditioners can be captured during servicing and recycled. Capture and recycling would also be appropriate for the disposal of appliances using CFCs.

In addition to stratospheric O_3 protection, Title VI contains provisions that require USEPA to prepare reports on methane (CH_4) emissions from a variety of domestic sources, global CH_4 emissions, preventing increases in CH_4 concentrations, and measures to limit growth in atmospheric CH_4 levels. These requirements reflect concerns about the global warming potential of methane rather than any potential threat to the O_3 layer.

Acidic Deposition

For years, Congressional efforts to amend the CAA were stymied by contentious debate over the control of acidic deposition. In 1990 Congress put that debate to rest and voted to enact the CAA Amendments with specific authority and language addressing the control of acidic deposition.

The purpose of Title IV Acid Deposition was to minimize adverse effects of acidic deposition by reducing annual emissions of SO_x and NO_x by 10 and 2 million tons, respectively, from 1980 emission levels. This goal was to be achieved in two phases.

Mindful of potentially significant economic costs and dislocations that may have occurred as a result of traditional "command and control" regulatory approaches, Congress developed an alternative approach which they believed would be both more flexible and cost effective. Emission reductions needed to meet the goals of the acidic deposition program would be achieved by establishing a program of emission allowances. An allowance under the Act was equal to one ton of SO_x.

Emission reduction requirements for SO_x using the allowance concept applied to large fossil fuel-fired electric boilers specifically identified in the Act. These requirements were to be achieved using a phased approach, with phase 1 commencing in 1995 and phase 2 in the year 2000. Once phase 2 emission reduction requirements were to take effect, it was anticipated that total annual emissions in the 48 contiguous states and the District of Columbia would not exceed 8.9 million tons of SO_x annually.

In the allowance program, each unit of each affected source is allocated allowances which are determined from total emissions and emission reduction requirements to achieve phase 1 and phase 2 goals. Each source is then required to develop a plan which describes a schedule and methods for achieving compliance. Sources are free to choose among emission reduction options which best suit their needs and capabilities. Congress encouraged affected sources to use energy conservation, renewable and clean alternative technologies, and pollution prevention in their emission reduction efforts.

Allowances are allocated annually for each operating unit of an affected source. It is unlawful for the source operator to emit SO_2 in excess of the number of allowances held for that unit for that year. If the affected source exceeds its allowances, it is liable for the payment of an excess emissions penalty for a given year with the penalty calculated as a product of tons in excess of allowance multiplied by \$2000 (adjusted for the Consumer Price Index). The source is also liable to offset the excess emissions by an equal tonnage in the next year or period allowed by USEPA.

Unused allowances at one unit or source can be used to offset emissions at other units or sources by a utility operator. Emission allowances can also be traded among sources. The number of allowances which can be traded in spot and advance auctions are prescribed in the Act. Unused allowances in a given year can be banked for use in future years.

Title IV of the 1990 CAA Amendments repealed the percentage reduction requirements for SO_2 prescribed for coal-fired power plants under the 1977 CAA Amendments. Congress mandated USEPA to promulgate a NSPS for coal-fired power plants which required the best system of emission reduction taking into account the costs of achieving the reduction and any nonair-quality health and environmental impact and energy requirements.

Motor Vehicle Emission Control

The pioneering atmospheric chemistry studies of Dr. Arie Hagen-Smit provided the fundamental understanding of the role of automobile emissions in the development of photochemical smog. As was seen in Tables 2.1 to 2.3, the automobile is the major anthropogenic source of three of the six primary pollutants listed. In addition to being the single largest source of CO, HCs, and lead, the automobile is the second largest source of NO_x. Indirectly, it is also the major factor in the production of photochemical oxidants such as O_3. It is no wonder then that automobile emissions have been subject to considerable regulation.

Automobile air pollution has been of greater significance in southern California than any other place in the world. This has been primarily due to a culture heavily dependent on automobile usage, abundant sunshine, topographical barriers to air movement, and frequent elevated inversions during the summer months. As a consequence, California has been the leader in both identifying and controlling automobile-related pollution. California motor vehicle emission requirements have historically preceded those required nationally.

Vehicle Emissions. Emissions from light-duty motor vehicles include exhaust and blow-by gases and volatile HC losses from the carburetor and fuel tank. Exhaust gases account for approximately 92% of all automobile emissions. Blow-by gases which slip by the piston rings into the crankcase during the combustion cycle accounted for 4% of total emissions in uncontrolled vehicles and fuel evaporative losses another 4%. In addition, significant emissions of HCs may also occur when the vehicle is being refueled.

The control of motor vehicle emissions has required regulations to reduce crankcase emissions, exhaust gases, and evaporative losses. Fuel composition standards which permitted the use of certain technologies such as catalytic systems have also been necessary. At the time of this writing, regulations for controlling HC emissions from vehicle refueling have been promulgated in California and have been proposed for national adoption.

Table 8.6 **Regulatory History of Control Requirements on New Light-Duty Motor Vehicles**

Year/model year	Control actions
1961	California — Crankcase emission controls.
1963	U.S. — Crankcase emission controls.
1966	California — Exhaust emission controls for CO and NMHCs.
1968	U.S. — Exhaust emission controls for CO and NMHCs.
1970	California — Evaporative control systems. New, more stringent emission standards for CO and NMHCs. Required reduction of NO_x emissions for the first time.
1970	Congress passes Clean Air Act Amendments. Mandates stringent new motor vehicle exhaust emission standards for CO and NMHCs to be achieved by 1975 model year. Mandates stringent NO_x emission standard to be achieved by 1976 model year. Provides for fuel composition standards.
	U.S. — USEPA imposes more stringent CO and NMHC standards.
1971	U.S. — Evaporative control systems.
1972	California — Emission standards for NO_x are tightened.
	U.S. — USEPA imposes tighter standards for CO and NMHCs. First federal emission standard for NO_x.
1974	California — Stringent emission standards mandated for 1975 model cars are postponed. California imposes interim standards for all new cars sold in the state.
	U.S. — Statutory emission standards for CO and NMHCs and NO_x mandated under 1970 Clean Air legislation are postponed. USEPA imposes interim standards for CO and NMHCs.
1977	California tightens NMHC and NO_x emission standards.
	U.S. — USEPA imposes interim emission standards for NO_x mandated by Congress in 1974 legislation.
	Clean Air Act Amendments passed. Interim standards for CO and NMHCs continued through 1979 model year. 1975 statutory emission requirements for NMHCs required in 1980, CO in 1981. NO_x standard relaxed with 1981 deadline.
1990	Clean Air Act Amendments passed. Significant additional reduction in NMHCs required. Emission limits for CO and NO_x under 1970 CAA amendments to be met in the period 1993 to 1995.

Note: NMHC, nonmethane hydrocarbons; CO, carbon monoxide.

Regulatory History. The regulatory history of control requirements mandated for both California cars and those sold nationwide is summarized in Table 8.6. Because of the relatively simple and low-cost technology (Chapter 9) required, crankcase emissions were effectively controlled in the early 1960s. Significant control of evaporative emissions (1970 to 1971) was also attained by the application of relatively simple and low-cost technology. The control of exhaust gases, which account for the largest percentage of emissions, has been far more difficult. Exhaust gas control has had a tortuous path of regulatory requirements, technology development, cost, and public acceptance.

Exhaust Gas Emission Standards. Emission standards for exhaust gases were first required on 1966 model light-duty vehicles sold in California and in 1968 models sold nationwide. Emission standards were applied only to nonmethane hydrocarbons (NMHCs) and CO, as these were more easily controlled than NO_x.

A history of exhaust gas emission reduction requirements is summarized in Table 8.7. All emission standards are given in grams per mile (g/mi). Note,

Table 8.7 History of Light-Duty Motor Vehicle Exhaust Emission Standards

Model year	Authority	HCs (g/mi)	CO (g/mi)	NO_x (g/mi)	Evap. (g/test)	PM (g/mi)	HCHO (g/mi)
Pre-control		10.6	84.0	4.1	47		
1966	California	6.3	51.0				
1968	U.S.	6.3	51.0				
1970	California	4.1	34.0				
	U.S.	4.1	34.0		6		
1971	California	4.1	34.0	4.0	6		
1972	California	2.9	34.0	3.0	2		
	U.S.	3.0	28.0				
1973	U.S.	3.0	28.0	3.0			
1975	California	0.9	9.0	2.0	2		
	U.S.	1.5	15.0	3.1	2		
1977	California	0.41	9.0	1.5	2		
	U.S.	1.5	15.0	2.0	2		
1978	California	0.41	9.0	1.5	6[a]		
	U.S.	1.5	15.0	2.0	6[a]		
1980	California	0.39[b]	9.0	1.0	2		
	U.S.	0.41	7.0	2.0	6		
1981	California	0.39	7.0	0.7	2		
	U.S.	0.41	3.4	1.0	2		
1983	California	0.39	7.0	0.4	2		
1984	California	0.39	7.0	0.4	2	0.60	
1985	California	0.39	7.0	0.4	2	0.40	
1986	California	0.39	7.0	0.4	2	0.20	
1989	California	0.39	7.0	0.4	2	0.08	
1993	California	0.25 (0.31)[c]	3.4 (4.2)	0.4 (0.6)	2	0.08	
1994	California	0.125 (0.156)[d]	3.4 (4.2)	0.4 (0.6)	2	0.08	0.015 (0.018)
	U.S.	0.25 (0.30)[c]	3.4 (4.2)	0.4 (0.6)	2		
1997	California	0.075 (0.09)[e]	3.4 (4.2)	0.2 (0.3)	2	0.08	0.015 (0.018)
		0.04 (0.055)[f]	1.7 (2.1)	0.2 (0.3)	2	0.08	0.008 (0.011)
2003	U.S.	0.125[g]	1.7	0.2			

[a] Testing procedure change.
[b] Nonmethane HC standard or 0.41 g/mi total HCs.
[c] Standards are for nonmethane organic gas for 5 years/50,000 mi and 10 years/100,000 mi, respectively.
[d] Transitional low emission vehicles.
[e] Low emission vehicles.
[f] Ultra low emission vehicles.
[g] Standards are for 10 years/100,000 mi.

From Calvert, J.G., et al. 1993 *Science* 261:37–45. With permission.

however, that methods of measuring emissions, as well as of expressing their concentrations, have changed several times since emission limits were initially mandated.

Emission Concentrations.
The first emission standards were based on concentration, for example, 275 ppmv NMHCs and 1.5% CO. To take into account the lower mass emission rate for smaller cars, California and U.S. standards for 1968 to 1969 were based on engine size. To simplify the problem of having different emission standards for various motor sizes, the 1970 standards for both California and the U.S. specified emissions on a mass basis (g/mi).

Emission Test Cycles.
Exhaust emission standards must reflect to some degree actual emissions under normal driving conditions. Emissions may, however, vary considerably as automobiles are operated under many different driving conditions, e.g., long trips, short trips, stop-and-go urban traffic, and superhighway travel. In addition, individual drivers may operate their vehicles differently under similar driving conditions. Because it is impossible to duplicate all of these driving factors in specifying emission requirements, a standard emission assessment procedure was developed to represent traffic that is responsible for emissions in areas where the most severe motor vehicle-related air pollution occurs.

The certification of automobile engines for compliance with emission standards requires that a large number of emission tests be conducted under identical conditions. As a result, both California and U.S. test procedures utilize a dynamometer test cycle that simulates early morning driving conditions in downtown Los Angeles.

A seven-mode dynamometer cycle was used to certify vehicles from 1966 to 1972. To reflect changed traffic conditions (the seven-mode test was based on 1956 Los Angeles traffic conditions), a new test procedure was developed to certify 1972 model cars. The dynamometer cycle (CVS-1) used in 1972 federal and California test procedures simulated a 7.5-mile, 23-minute, urban trip rather than a number of driving modes. This procedure was revised for 1975 model vehicles to take hot and cold motor vehicle operation into account. This procedure is referred to as CVS-2.

Dynamometer tests are conducted under laboratory conditions. They are lengthy, complicated, and expensive. As a result, certification requirements for compliance with emission standards do not include testing of vehicles before they are sold. Certification does, however, require testing of engine systems that will be installed in vehicles for sale.

Inspection and Maintenance.
Once motor vehicles are sold, regulation of emissions from in-service vehicles becomes a state and local function. Emission control devices need to be properly maintained to ensure that vehicle emission standards are met during continued vehicle operation. Motor vehicle inspection programs are therefore necessary to ensure that emission standards for in-service vehicles are not exceeded, and if they are, to require appropriate adjustment or

repair. These inspection programs are particularly necessary where NAAQS for automobile-related primary pollutants such as CO are being violated.

1970, 1977, and 1990 Light-Duty Motor Vehicle Emission Control Requirements.

The 1970 CAA Amendments required that emissions of NMHCs and CO for 1975 and NO_x for 1976 model light-duty motor vehicles be reduced by 90% of their 1970 levels. Because an approximate 60% reduction in emission of NMHCs and CO had already been achieved, total emissions reductions from a precontrol base would actually be on the order of 95 to 96%. Under provisions of the 1970 CAA, these emission standards were to be extended by 1 year if USEPA found, in consultation with the National Academy of Sciences (NAS), that technology was not available to meet the statutory requirements. Congress, anticipating that compliance with these emission standards would require the use of catalysts which would be "poisoned" by lead fuel additives, authorized USEPA to regulate fuel content.

In the early 1970s, considerable controversy arose concerning the availability of control technology to meet the statutory deadline. A study of the problem by a NAS panel was inconclusive. The administrator of USEPA, however, concluded that the technology was indeed available, and a 1-year extension of the deadline was not necessary. Motor vehicle manufacturers argued that the technology was not available and sued USEPA in federal court for a 1-year extension. The federal courts decided in favor of the motor vehicle manufacturing industry, and USEPA extended the deadline in accordance with the court's order.

Congress in 1974 passed the Energy Supply and Coordination Act in response to the energy problems caused by the Arab oil embargo in 1973/1974. It amended the 1970 CAA to extend the motor vehicle emission standards deadline from 1975 to 1977 model cars for NMHCs and CO and relaxed the NO_x standard for 1977 model vehicles from 0.4 to 2.0 g/mi. In lieu of the 1975 statutory requirements for NMHCs and CO, Congress authorized USEPA to set interim emission standards which would reflect the greatest degree of emission reduction achievable through the use of existing technology, taking into account the costs of applying this technology within the period of time available to auto manufacturers. USEPA then imposed interim emissions standards for 1975, 1976, and 1977 model vehicles.

Recognizing the difficulties in achieving the emission standards mandated in 1970, Congress began in 1975 to both reauthorize and amend the CAA. Controversy associated with the PSD issue resulted in a Senate filibuster. As a result, provisions for further extensions and some relaxation of auto emission standards which had strong Congressional support were not enacted. Automobile manufacturers, anticipating that Congress would extend the interim standards, began to build 1978 model cars which could not meet the statutory requirements. Had not Congress finally amended the law in late summer of 1977, the automobile manufacturers would not have been legally able to sell 1978 model light-duty motor vehicles. Congressional agreement came at the eleventh hour; the 1977 amendments were enacted, postponing the statutory deadline for emission standards.

The 1977 amendments extended the original emission deadlines, set for 1975, to 1980 for NMHCs and 1981 for CO. The NO_x standard was relaxed from 0.4

to 1.0 g/mi, with the deadline extended to 1981. Under these amendments USEPA could extend the 1981 deadlines for 2 years if technology was not available. Both the CO and NO_x deadlines were again postponed. After nearly two decades of postponements and extensions, Congress under the 1990 CAA Amendments set the final course of achieving the tough motor vehicle emission requirements it set under the 1970 CAA Amendments. The original emission standards for CO (3.4 g/mi) and NO_x (0.4 g/mi) and a tougher NMHC standard (0.25 g/mi) were to be achieved on all new light-duty cars and trucks by 1996 model vehicles with a phase-in period beginning with 1994 model vehicles. These standards were based on vehicle operating lives of 5 and 10 years or 50,000 and 100,000 miles. For an operating life of 10 years or 100,000 miles, the standards are NMHC, 0.3 g/mi; CO, 4.2 g/mi; and NO_x, 0.6 g/mi. In addition, Congress mandated that manufacturers of light-duty motor vehicles up to 6000 lb meet a PM standard of 0.09 g/mi based on a 5-year, 50,000-mile operating life and 0.10 g/mi based on a 10-year, 100,000-mile operating life. These PM standards, like the other motor vehicle emission standards, were to be phased in over 3 years starting with 1994 model vehicles.

Recognizing that motor vehicles operating under cold weather conditions posed unique CO emission control concerns, Congress in the 1990 CAA Amendments mandated cold-start standards with a reference temperature of 20°F (−6.7°C). Under phase I of these requirements, light-duty motor vehicles were to achieve an emission rate not to exceed 10 g/mi. This cold-start standard was to be phased in from 1994 to 1996. Under phase II of these requirements, USEPA was required to assess the need for further reductions and their achievability. Assuming that the need exists and technology is available, a cold-start emission standard for CO of 3.4 g/mi would be required on model year 2002 light-duty vehicles.

Though evaporative emissions were regulated by California for 1970 model motor vehicles and by the U.S. for 1971 models, evaporative emissions of NMHC continue to be a problem. As a result, the 1990 CAA Amendments mandated that USEPA promulgate regulations for evaporative emissions of NMHCs from all gasoline-fueled motor vehicles during operation and over 2 or more days of nonuse for O_3-prone summertime conditions. These regulations were to require the greatest degree of emission reduction achievable by methods reasonably expected to be available with consideration of fuel volatility, cost, energy, and safety factors.

Emission Requirements for Trucks and Other Vehicles. Because of differences in weight, vehicle use and population, and engine types, trucks (except those that are less than 3750 lb) by necessity are regulated differently than light-duty cars. Emission standards for larger trucks (up to 6000 lb) and heavy-duty trucks are less stringent than those for light-duty motor vehicles. Under the 1990 CAA Amendments, for example, light-duty trucks with a gross vehicle weight rating of 3750 lb (1705 kg) to 6000 lb (2727 kg) are required to meet emission limits of 0.32 g/mi NMHC, 4.4 g/mi CO, and 0.7 g/mi NO_x for a 5-year/50,000-mile operating life and 0.40 g/mi NMHC, 5.5 g/mi CO, and 0.97 g/mi NO_x for a 10 year/100,000-mile operating life. Light-duty trucks with a gross vehicle weight rating greater than 6000 lb (2727 kg) have slightly higher emission reduction requirements for vehicle operating lives of 11 years/120,000 mi.

Emissions of NMHCs, CO, NO_x, and PM from heavy-duty trucks have been regulated since 1983. In the 1990 CAA Amendments, USEPA is given authority to revise emission standards for heavy-duty trucks promulgated prior to 1990. These revisions are to include an emission standard for NO_x from gasoline and diesel-fueled heavy-duty trucks which will not exceed 4 grams per brake horse-power. USEPA had previously been given authority to regulate emissions from motorcycles in the 1977 CAA Amendments.

Throughout the history of federal involvement in regulating emissions from motor vehicles, California alone has been permitted to regulate emissions more stringently than the federal government. This "California exemption" recognized California's leadership role in motor vehicle emissions control and the unique motor vehicle-related air pollution problems it experienced. In the 1990 amendments, other states were allowed to adopt California emission standards and requirements.

Fuel Additives. Anticipating that catalytic converters were going to be used on 1975 and later model cars, Congress gave USEPA the authority to regulate fuel additives. Lead additives which had been universally used to increase octane ratings of gasolines were anticipated to coat catalytic materials, rendering them ineffective. Using this authority, starting with 1975 model cars, USEPA began a phase-down in the use of lead in gasoline. The 1990 CAA Amendments prohibited misfueling, that is, the introduction of leaded gasoline into a vehicle labeled unleaded gasoline only, the use of gasoline or diesel fuel with a sulfur concentration >0.05% by weight, the sale and use of high volatility (Reid Vapor Pressure >9 PSI) gasoline during the high O_3 season, and the sale and highway use of lead additives by 1996.

Specific provisions of the 1990 CAA Amendments addressed the issues of reformulated gasoline and oxygenated fuels. USEPA was to promulgate regulations establishing requirements for reformulated gasoline to be used in motor vehicles in specified O_3 nonattainment areas. These regulations were to require the greatest reduction in emissions of O_3-forming volatile organic compounds (VOCs) during the O_3 season and emissions of toxic air pollutants during the entire year. Regulations affecting reformulated gasoline had to meet some general and more stringent formula and performance requirements. The performance standard for reformulated gasoline required a 15% reduction in aggregate emissions of O_3-forming VOCs from baseline values with a 25% potential reduction by the year 2000 and an aggregate emission reduction of toxic air pollutants of 15% with a 25% reduction effective in the year 2000. Toxic pollutants included in these regulations were benzene, 1,3-butadiene, polycyclic organic matter (POM or PAH), acetaldehyde, and formaldehyde.

In CO nonattainment areas, oxygenated gasoline containing no less than 2.7% O_2 by weight was required to be sold in CO prone seasons. Systems dispensing oxygenated fuels are required to be labeled.

Other Motor Vehicle Control Requirements. A major new requirement of the 1990 CAA Amendments is the use of onboard vapor recovery systems to control NMHC emissions during vehicle refueling. USEPA is mandated to promulgate standards which require that new light-duty vehicles be equipped with vapor recovery systems after standards are promulgated. These requirements are to be phased in over a three-year period. The minimum evaporative emission capture recovery is to be 95%.

The 1990 CAA Amendments address the use of clean-fuel vehicles. USEPA was mandated under the Act to promulgate standards for clean-fuel vehicles. So-called clean fuels include any fuel, reformulated gasoline, diesel, natural gas, liquefied petroleum gas, and hydrogen, or power source (including electricity) used in vehicles that comply with clean-fuel vehicle standards.

Emission standards for light-duty cars and trucks are specified for phase 1 beginning in model year 1996 and phase 2 beginning in model year 2001. These include both 50,000- and 100,000-mi standards for nonmethane organic gas (NMOG), CO, NO_x, PM, and formaldehyde. Phase 2 standards are more stringent than phase 1; the 50,000 standards are more stringent than the 100,000-mi standards. In certain O_3 nonattainment areas, clean-fuel vehicles and alternative fuels are to be phased in for fleet vehicle use over the period of 1998 to the year 2000.

Inspection and Maintenance. The regulation of emissions from in-service vehicles is a state and local responsibility. Over the decades there has been considerable evidence to indicate that poorly maintained, malfunctioning, and tampered-with emission control systems contribute significantly to air quality problems. There is evidence to indicate that in some urban areas as few as 10% of in-service vehicles may be responsible for approximately 50+% of total emissions from motor vehicles. This confirms the need for inspection programs to identify vehicles with poorly operating emission control systems so that proper repair and maintenance can be conducted to achieve the emission reduction for which they are designed. Inspection and maintenance is particularly necessary in AQCRs where one or more motor vehicle related NAAQS is not being attained.

Inspection and maintenance programs have a significant potential in improving air quality. Such programs have been difficult to implement because many vehicle owners have seen them as intrusive and burdensome. Though many profess to want clean air, they often are skeptical that they are significant contributors to the problem. Nevertheless, inspection and maintenance programs are conducted by regulatory fiat in many nonattainment areas for CO and O_3.

In the 1990 CAA Amendments, Congress mandated that USEPA promulgate regulations that would require manufacturers to install diagnostics systems on light-duty vehicles which would detect emission-related systems deterioration or malfunction, alert the vehicle owner/operator to the likely need for the maintenance and repair of emission systems, and storage and recovery of information on emission systems operation. States with inspection and maintenance programs are then required to provide for the inspection and evaluation of information provided by onboard diagnostic systems.

Tampering Prohibitions. One major problem that motor vehicle emission control programs have had to contend with has been the willful disabling of emission control systems by a variety of individuals. Prohibitions on tampering with emission control systems were included in both the 1970 and 1977 CAA Amendments. In both of these cases, prohibitions were placed on motor vehicle dealers, repair shops, etc. In the 1990 amendments, this prohibition was also placed on individuals who manufacture products which are designed to disable or interfere with the proper functioning of emission control systems. The 1990 amendments prohibited anyone from tampering with motor vehicle emission control systems.

Enforcement

The absence of enforcement authority at all levels of government before 1970 was the major reason air pollution programs were ineffective. The 1970 and subsequent CAA Amendments were a significant departure from previous legislative efforts. These laws provided considerable federal enforcement authority.

Under the 1970 CAA Amendments, the administrator of USEPA had the authority to take enforcement actions when provisions of SIPs, NSPS, NESHAPs, and requirements related to inspection, monitoring and right-to-entry were being violated. In subsequent amendments these authorities were extended to other provisions such as stratospheric O_3 depletion, acidic deposition, etc.

Under current statutory language, USEPA can, in response to violations of SIPs and other major provisions of the CAA, issue administrative orders requiring compliance. It can also issue administrative orders against violators assessing penalties of up to $25,000 per day of violation.

Should violators choose not to respond to these administrative orders, USEPA has the authority to commence civil suits in federal district courts. In such actions, USEPA may seek a permanent or temporary injunction or civil penalties. In cases where violations are willful, USEPA can request the Attorney General to commence criminal actions which may result in penalties such as fines and/or imprisonment. In addition, Congress in 1990 provided USEPA authority to pay awards (or bounties) to individuals who furnished information leading to a criminal conviction or an administrative or judicial civil penalty for violations under various 1990 CAA titles.

Although USEPA has significant enforcement authority, in practice, most enforcement occurs at state and local levels. Federal enforcement authority typically is used when state and local agencies are reluctant to enforce provisions of their own SIPs and other programs (e.g., NSPS) for which state and local agencies have been granted authority. Federal enforcement authority, of course, is applied to violations of regulatory requirements for which only USEPA has authority. Most notable of these are regulatory requirements for motor vehicles, O_3-destroying chemicals, and acidic deposition.

Permit Requirements

The most effective way to achieve compliance with emission reduction requirements is to use a permit system. Permit requirements put the burden of proof of compliance on the emission source. Permits may be required for existing sources to discharge any pollutants into the atmosphere and to install and operate a pollution control device. For new or significantly modified existing sources, permits may be required before construction commences. Legally, a source cannot operate, that is, emit pollutants, without a permit. To be granted a permit, a source must comply with all existing emission reduction regulations.

In many cases it is not possible for an existing source to immediately comply with the conditions of the permit. This may be due to time requirements for designing, ordering, and installing an emission control device. For the source to operate in violation of emission regulations, a legal exception or variance to the regulations may be granted for a specified period, e.g., 6 months. These variances provide a reasonable time for a source to install control equipment necessary to attain compliance with emission regulations.

The role of permitting in achieving air-quality goals was given significant status and attention in the 1990 CAA Amendments. Title V directed that USEPA promulgate regulations which would establish minimum elements of air pollution permit programs. Among these were (1) requirements for permit applications including a standard application form and criteria for determining the completeness of an application, (2) monitoring and reporting requirements, (3) requirements for permittees to be assessed an annual fee, (4) requirements for personnel and funding to administer the program, and (5) requirements for adequate authority to issue permits and assure compliance.

Pollution Prevention

In 1990, Congress passed the Pollution Prevention Act (PPA). Under the PPA of 1990 Congress expanded the scope of federal pollution policy to cover all releases of pollutants to all media. The focus of the Act is to set a new policy for waste reduction (interpreted broadly). The guiding principle is that source reduction is fundamentally different and more desirable than waste management and pollution control. It focuses on source reduction as the primary means of reducing the production of wastes. A source reduction strategy deemphasizes command-and-control regulation and emphasizes voluntary cross-media pollution prevention activities. The Act mandates that USEPA promote governmental involvement to overcome institutional barriers by providing grants to state technical assistance programs under a new federal pollution prevention program. Pollution prevention is given a new emphasis in the 1990 CAA Amendments under both Title I Air Pollution Prevention and Control and Title III Toxic

Substances. The amendments required that USEPA conduct an engineering research and development program to develop, evaluate, and demonstrate non-regulatory technologies for air pollution prevention. In Title III, USEPA, in establishing emission standards for hazardous air pollutants, is required to consider cross-media impacts, substitution of materials, or other modifications to reduce the volume of hazardous air pollutants, the potential for process changes, and eliminating emissions entirely. Other pollution-related provisions include requirements for the sale of cleaner-burning reformulated gasoline in cities not in compliance with O_3 and CO air-quality standards, and the phaseout of CFCs and halons.

Citizen Suits

One major concept embodied in air pollution legislation since 1970 is the right of citizens to bring suit against both private and governmental entities to enforce air pollution requirements. Before this, citizens did not automatically have legal standing to seek court review of federal agency decisions. The basic premise of citizens' suits is to provide a mechanism to make the federal enforcement agency (USEPA) accountable when its actions are deemed to conflict with the laws it is to enforce. These provisions have been used in a number of instances by environmental groups such as the Natural Resources Defense Council, the Environmental Defense Fund, and the Sierra Club when they determined that USEPA was not complying with its statutory obligations in implementing provisions of clean air legislation. Successful citizen suits have required USEPA to promulgate regulations for air-quality maintenance, indirect source review, and PSD, and an air-quality standard for lead.

One principal effect of citizen suits has been to compel USEPA to implement provisions or concepts embodied in clean air legislation when it was reluctant to do so. The reasons for such reluctance are many, including practical and economic considerations, availability of manpower, administrative priorities and, of course, bureaucratic inertia. An attendant result of citizen suits is the legal clarification of the language of clean air legislation.

Although citizen suits have for the most part been used to compel action by USEPA, citizen legal action can also be taken against a pollution source to force compliance with emission reduction requirements. As such, citizen suits are a valuable adjunct to federal enforcement efforts.

Citizens were given additional legal rights under the 1990 CAA Amendments. Citizens are now permitted to file lawsuits for the purpose of seeking civil penalties against parties in violation of the CAA. Such penalties would be placed in a special fund to finance compliance and enforcement activities.

STATE AND LOCAL AIR POLLUTION CONTROL FUNCTIONS

Prior to the 1970 CAA Amendments, all states and many municipalities had some kind of air pollution legislation or ordinance. These state and local programs varied markedly in scope, enforcement authority, administration, and public support. All control programs were limited in their ability to deal adequately with the complex problems of achieving control objectives.

The passage of the 1970 CAA Amendments, with their attendant requirements for federal approval of SIPs, imposed considerable homogeneity on state and local programs. As a result, many state and local control programs strongly resemble each other. Despite this homogeneity, differences still exist since these must in general reflect unique state and local control problems.

Although the major emphasis of this chapter has been federal air pollution control legislation, in reality most control of stationary sources is carried out at the state and local level. Federal legislation is basically an umbrella providing direction for effective pollution control at these levels. This direction is provided in the USEPA regulations promulgation process, the SIP approval process, grants-in-aid to state and local agencies, and backup federal enforcement authority should state and local agencies fail to perform their responsibilities.

The federal posture in working with state and local agencies is a combination of carrot and stick. State and local programs which meet federal requirements receive monies to maintain their air pollution programs. In the past these federal monies provided as much as 50% of control agencies' budgets. Noncompliance with federal requirements can result in the loss of such federal support.

Most state and local programs operate under the philosophy that local authorities best know how to deal with state and local problems, that is, what control tactics should be used and how they are to be applied. In theory, of course, this is true. It is also true that political pressure to hamper or slow down pollution control efforts is the most intense at these levels. Because of this, a federal presence is necessary to enforce, if necessary, the implementation of air pollution control regulations should state or local agencies choose not to effectively deal with their own problems. In most instances, state and local agencies prefer not to invite "federal interference."

Boards and Agencies

Regulation of air pollution is carried out by state and local authorities through the "police powers" delegated to the states under the U.S. Constitution. The utilization of these police powers requires the enactment of state legislation and local ordinances to provide a legal framework for air pollution control. Because legislation/ordinances cannot deal with or anticipate all of the individual circumstances of source control, they must provide for the establishment of an administrative agency to carry out the provisions embodied in them. These agencies may have quasilegislative powers to adopt rules and regulations and to recommend appropriate legal action. In some instances these powers are vested in separate pollution control boards or commissions. Both functions may also be carried out by a single agency.

Air pollution control agencies have the day-to-day executive responsibilities of planning to achieve NAAQS, source data collection, air-quality monitoring, development of emission standards, development of source compliance schedules and permit review, surveillance of sources to determine compliance status, and recommendations for enforcement action in cases of noncompliance. As previously discussed, states are required to submit implementation plans for USEPA approval which outlines in detail how NAAQS are going to be met in each AQCR under their jurisdiction. States can delegate responsibility to local agencies to develop local control plans. These local plans must be compatible with the SIP because they will be incorporated into it.

Once it has been determined that ambient levels of a specific pollutant exceed ambient air-quality standards, it is necessary to develop and promulgate emission standards for sources in the AQCR. These standards may be based on computer models which use emission source data and dispersion characteristics of the atmosphere to predict ambient concentrations. Once the amount of emission reduction is determined from these models, compliance schedules are negotiated with industrial sources in the region.

Effectiveness of State and Local Programs

The effectiveness of state and local air pollution control programs varies considerably from state to state and from one municipality to another. Many factors determine the effectiveness of air pollution control efforts, including (1) a good legal framework for regulation promulgation and enforcement, (2) the availability of economic resources, (3) sufficient trained manpower, (4) the commitment of pollution control officials in enforcing control requirements, and (5) the cooperation of public officials and public acceptance of needed pollution control measures. Even though each of these is important, the most vital factor is the acceptance of pollution control requirements by private citizens, public officials, and, of course, sources. Without this support, air pollution control efforts will fail.

PUBLIC POLICY ISSUES

The 1970, 1977, and 1990 CAA Amendments have provided the U.S. with a framework for ambient pollution control. Implementation of these statutes is the responsibility of USEPA, a regulatory agency that answers to the president. As an administrative agency, USEPA has the authority to formulate policy relative to different air-quality issues and to promulgate regulations to implement policy decisions.

The regulation of air pollution is not a static kind of responsibility. Regulatory programs must continually be evaluated relative to their effectiveness, the identification of new problems, new data and understanding of specific problems, requests from the regulated community and from public interest groups, and legal decisions. Although the regulatory framework does not change without congressional action, air pollution control policies are continuously evolving to reflect current realities. Policy considerations may involve existing regulatory authority, and/or they may require new authority which Congress must provide.

Changes in policy or the proposed implementation of new regulatory policies generally result in some measure of controversy. Changes or proposed changes may have significant economic implications for those regulated. On the other hand, public interest groups who strongly favor air pollution-related regulation may not accept that regulatory policy is sufficiently stringent.

The regulatory aspect of air pollution control involves a continual involvement in the formulation, public debate, and ultimately, promulgation and implementation of policy decisions. These are reflected in such issues as the revision and attainment of air-quality standards, the regulation of toxic pollutants, motor vehicle emission control, PSD, acidic deposition, stratospheric O_3 depletion, and global warming. At this writing, these are some of the major public policy issues that USEPA must address. For illustrative purposes, public policy issues associated with air-quality standards, stratospheric O_3 depletion, and global warming are described below.

Air-Quality Standards

In the 1970 CAA Amendments, Congress mandated that the goal of the U.S. in protecting and enhancing the quality of the nation's air was to be achieved by the setting of NAAQS for air pollutants that posed a major threat to public health and welfare. Such standards were to be achieved by July 1, 1975. Although considerable progress has been made, there are still numerous AQCRs in the country that have failed to achieve air-quality standards promulgated by USEPA. At this writing, 20+ years after the 1975 deadline, the nation has been unable to achieve the clean air goals that were set in 1970.

Many cities and counties still have been unable to achieve one or more of the seven air-quality standards. The estimated number of people living in such areas where ambient concentrations are excessive is illustrated in Figure 8.2. Of note is O_3, with an estimated 50+ million people living in areas which do not comply with the air-quality standard for O_3.

Recent epidemiological studies discussed in Chapter 5 indicate that the current PM_{10} standard may not be sufficiently stringent to safeguard public health. As a result, it is apparent that some revision will be required. This will require a reevaluation of the present air-quality standard, recommendations for revision, public comment, and ultimately possible promulgation of a revised standard.

Because of the economic and political realities involved, revision of the PM_{10} standard and promulgation of a new standard is expected to involve significant controversy.

Public policy initiatives involving air-quality standards involve a number of facets. USEPA maintains a continuing program of updating air-quality criteria documents and reviewing the health and monitoring data to determine whether one or more air-quality standards should be revised. Such reviews have been recently conducted to determine the need for short-term standards for NO_2 and SO_2, a multiple hour standard for O_3, and for a respirable particle standard.

For purposes of illustration, let's look at the public policy issues involved in changing the TSP standard to a PM_{10} standard. The former was promulgated in

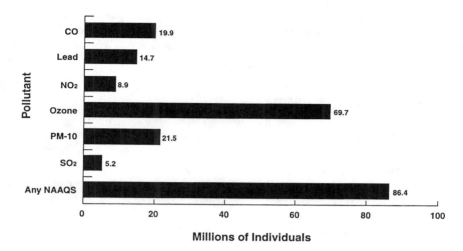

Figure 8.2 Estimates of U.S. population living in areas in which exceedances of NAAQS occurred in 1991. (From USEPA. 1992. EPA/400-R-92-013.)

1971 and was in effect until 1987. It was based on sample collections by filtration using high-volume samplers. Particles were collected over a broad size range including those particles (>10 μm) that have no apparent health significance (they cannot penetrate the upper respiratory system). Because of their large mass, particles greater than 10 μm contributed disproportionately to the concentrations of individual samples. State and local air pollution agencies that had the burden of bringing AQCRs in their jurisdiction into compliance with the TSP standard were unhappy with it, as were those industries that had to comply with very stringent PM emission limitations. The standard was designed to protect public health, but the significance of individual PM samples relative to human health was more uncertain because the sampling method gave undue emphasis to larger, nonhealth-affecting particles.

The revision of the TSP standard to reflect a need for size discrimination should have, in theory, been a relatively simple matter, but it was not. Health effects studies and dose–response health effects evaluations were based on TSP measurements or conversions to TSP values from other types of measurements. No historical health effects database existed for particles less than 10 μm. The regulatory dilemma thus became one of setting a health effects standard applied only to those particles of health significance when the available database was severely limited. Though TSP measurements were a relatively crude means of relating particle levels to health effects, there was at least a historical database available.

After approximately a ten-year period of review, debate, and proposals, USEPA finally promulgated a revision of the TSP standard, now a PM_{10} standard. Values of the PM_{10} standard are indicated in Table 8.1. Note that these values are considerably lower than the 24-hour (260 μg/m³) and annual (75 μg/m³) TSP standards. Measurements are to be conducted using sampling devices that collect only those particles less than or equal to 10 μm mass median diameter. These include fractionating devices and modified high-volume samplers with size-selective inlets (Figure 7.4). At this writing, USEPA has initiated proposed rulemaking which would require that AQCRs implement regulatory actions to achieve a more stringent $PM_{2.5}$ standard.

Achieving the O_3 Standard

In 1971 USEPA promulgated an ambient air-quality standard for photochemical oxidants which was in reality a de facto standard for O_3, the predominant photochemical oxidant in urban areas. This standard was revised in 1979 to its present status as an O_3 standard with an allowable daily maximum 1-hour concentration of 0.12 ppmv (235 $\mu g/m^3$) not to be exceeded more than once per year.

Elevated urban and suburban O_3 levels have proved to be the most intractable air-quality problem faced by regulatory authorities in the U.S. Exceedances of the O_3 standard continue to occur in a large number of AQCRs in the country. Lists of nonattainment areas for 1990 and 1991 include counties with populations exceeding 50+ million people (Figure 8.2).

The goal of the 1970 CAA Amendments was to achieve all NAAQS for all criteria pollutants in every AQCR in the country by 1975. As indicated above, this and the revised standard for O_3 have yet to be achieved in many AQCRs. This has required repeated statutory and regulatory extensions of compliance deadlines. It has also required increasingly stringent O_3 precursor emission reduction requirements.

Because O_3 is a secondary pollutant produced as a result of photochemical reactions involving HCs and NO_x, a control strategy has to be based on reducing one or both of the precursors responsible for producing elevated O_3 levels. Since 1971, USEPA's O_3 control strategy has been based on the premise that in most polluted urban atmospheres HC/NO_x ratios are less than ten where HC control is more effective than NO_x control. Such a strategy was convenient because the control of HC emissions to the atmosphere was deemed to be technically an easier task than controlling NO_x emissions.

After 25 years, there is little evidence to indicate that USEPA's O_3 control strategy has been effective. Indeed, it appears to have been a failure. If it has been a failure, it is reasonable to ask why USEPA continues to stay with it. USEPA officials would rightly point out that the complexities associated with the O_3 problem make it very difficult to both assess the present control strategy's effectiveness and to develop a new strategy which would be both relatively easy to implement and be effective as well.

Assessing trends in O_3 levels is a difficult task. Ozone levels are very strongly associated with elevated temperatures. The summers of 1983, 1988, and 1995 were relatively hot, with numerous exceedances of the O_3 standard in many AQCRs in the U.S. Such annual fluctuations in O_3 levels associated with variable meteorological conditions make it very difficult to determine long-term trends and their nature in a given area.

Control strategies have relied largely on unverified estimates in HC emission reductions. A variety of evidence exists to indicate that emission inventories for

HCs significantly underestimated anthropogenic HC emissions. As a result, actual emissions from mobile sources may have been two to four times greater than emission inventories ascribed to them. These underestimations have several significant implications. First, it is apparent that older emission control technology was limited in its effectiveness. More importantly, a significant upward correction in HC emission inventories could result in high HC/NO_x ratios and indicate a need for fundamental changes in strategy to control O_3 in many geographic areas, that is, control strategies based on NO_x emission reductions.

In addition to the problems with anthropogenic emissions described above, the combination of both biogenic HCs and anthropogenic NO_x can have significant effects on photochemical O_3 formation both in urban and rural areas. Measurements of isoprenes and other HCs from vegetation, as well as estimates of total biogenic emissions, suggest that biogenic HCs contribute to elevated O_3 levels in some urban areas and other areas affected by anthropogenic emissions of NO_x. In rural areas biogenic HC concentrations often exceed anthropogenic HC concentrations. As anthropogenic HCs are reduced as a result of emission controls, biogenic HCs which serve as a background for HC concentrations become a larger and more significant fraction of total HCs. Based on an analysis conducted by the Committee on Tropospheric Ozone Formation and Measurement of the National Research Council (NRC), O_3 levels above the air-quality standard can be expected in many urban cores and surrounding areas on hot summer days even if anthropogenic HCs were totally eliminated. Based on improved air-quality models and improved knowledge of ambient concentrations of HCs and NO_x, the NRC committee concluded that an O_3 control strategy based on NO_x emission reduction will be necessary in addition to, or instead of, the control of HCs in many areas of the country.

In areas where biogenic HC emissions contribute relatively little to O_3 chemistry, anthropogenic HC emissions are relatively high, and HC/NO_x ratios are less than ten, control strategies based on reducing HC emissions appear to be appropriate. Such conditions are commonly the case in urban areas with the worst O_3 problems, notably cities on the southern coast of California.

The most widely used HC control strategy has been to reduce the total mass (excluding methane) of HC emissions. The appropriateness of the approach has been challenged as our understanding of O_3 chemistry has matured. Because of differences in O_3 reactivity, each of the hundreds of HCs emitted to the atmosphere may have a different effect on O_3 levels. Indeed, HC compounds may have O_3 reactivities which may differ by an order of magnitude or more. Though they may be somewhat more complicated and difficult to administer, control strategies based on O_3 reactivities are likely to be both more effective and less costly. Predicted annual costs of mass-based and reactivity-based strategies as a function of controllable O_3 are illustrated in Figure 8.3. A reactivity-based control strategy is currently being used in California and may be a desirable strategy for other areas of the country as well.

Achieving the 1979 O_3 standard is but one of several major public policy issues which both regulators and the nation must address. As a part of a settlement of a lawsuit brought by the American Lung Association against USEPA for its

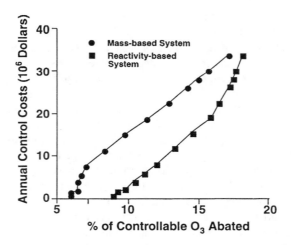

Figure 8.3 O_3 control costs predicted for mass-based vs. reactivity-based NMHC control programs. (From Russel, A. et al. 1995. *Science* 269:491–495. With permission.)

failure to assess and possibly revise the O_3 standard every 5 years as required by the CAA, USEPA agreed to review the standard but determined in 1993 that no revision was necessary. The American Lung Association sued USEPA again in 1994 citing recent studies that pulmonary health effects occurred at levels below 0.12 ppmv over an 8-hour period. USEPA, in response, announced that it would review the standard and possibly revise it by 1997. In 1995 USEPA staff recommended an 8-hour standard of 0.07 to 0.09 ppmv. As of this writing, USEPA has initiated rulemaking to require AQCRs to implement regulatory actions to achieve an 8-hour O_3 standard.

Stratospheric O_3 Depletion

After the CFC-O_3 depletion theory was proposed in 1974, three western countries, the U.S., Canada, and Sweden, implemented regulations to limit the nonessential uses of CFCs. In the 1977 CAA Amendments, Congress granted USEPA the authority to regulate O_3-destroying chemicals. Using this authority, the use of CFCs as aerosol propellants was banned in 1978. This resulted in a total reduction of U.S. CFC emissions of approximately 40%. This ban, as well as those by Sweden and Canada, had the effect of offsetting the increasing emissions of CFCs from other countries until 1983. Other countries took the economically safe position that O_3 depletion by CFCs and other halogenated HCs was an unverified theory.

Scientific and public policy recognition that the Antarctic O_3 hole was real and that Cl radicals from CFCs and other halogenated HCs were primarily responsible resulted in an international agreement among many nations to freeze CFC production with subsequent significant reductions in production and emissions. This accord, known as the Montreal Protocol, was completed in 1987. A few years later it became clear that O_3 depletion was an even more serious problem than previously thought. In a follow-up 1990 London conference, the Montreal Protocol was amended to require (1) a phaseout of CFC production levels by the year 2000 and an 85% reduction by 1997, (2) a phaseout of halons by the year 2000, (3) an 85% reduction (based on 1989 production levels) in the production of carbon tetrachloride (CCl_4) by 1995 and a complete phaseout by the year 2000, and (4) a freeze on the production of methyl chloroform (CH_3CCl_3) by 1993 with a reduction of 30% by 1995, 70% by the year 2000, and complete cessation of production by the year 2005. A variety of countries (Australia, Canada, the European Community, New Zealand) committed themselves to a complete phase-out of CFCs by 1997.

From 1974 to 1987, manufacturers were not surprisingly very skeptical of the CFC–O_3 depletion theory. Responding to the aerosol propellant ban in the U.S., they began to develop and evaluate alternatives, an effort that was largely abandoned during the conservative Reagan administration.

As we move to achieve the goals of the Montreal Protocol (as amended), the challenge is to develop replacements for CFCs and other O_3-destroying chemicals. Not surprisingly, replacement chemicals are inferior, and substitutions cannot easily be made in some cases without significant equipment modifications. Replacements include various hydrofluorocarbons (HFCs) and hydrochlorofluorocarbons (HCFCs). HFC-134a (CF_3CFH_2) can be easily used as a refrigerant and contains no chlorine. HFCs cannot be used in existing refrigerators and air conditioners without replacing the compressor.

The HCFCs can be oxidized in the troposphere by OH, and although they contain Cl, they are not expected to damage the O_3 layer. Though they may not adversely affect the O_3 layer, CFC substitutes such as the HCFCs are nevertheless greenhouse gases, albeit HFC-134a is only one fourth as potent a greenhouse gas and is only one fourth as effective as a thermal absorber as some CFCs. As a consequence, amendments to the Montreal Protocol call for the voluntary phase-out of many CFC substitutes between 2020 and 2040.

There is broad consensus among atmospheric scientists and public policy makers in most countries that the contamination of the atmosphere by CFCs and other halogenated HCs is responsible for the Antarctic O_3 hole and decreasing global trends in stratospheric O_3 levels. There are, of course, dissenters. These dissenters do not include the industries affected who have come to believe that the phenomenon is real and that significant actions are warranted. The dissent comes from a few scientists, political commentators, and public policy advocates associated with so-called conservative think tanks. This "revisionist school" has proposed a variety of alternative theories including the unusual meteorology of the Antarctic, chlorine associated with seawater, chlorine injection into the stratosphere by volcanos, prior existence of an O_3 hole in the 1950s before significant CFC production, etc. None of these alternative theories have withstood scientific scrutiny, and their proponents have failed to provide more than speculation to substantiate their claims. Nevertheless, they persist and attempt to explain away the O_3 depletion problem no matter what practicing atmospheric scientists have demonstrated and developed into an overwhelming scientific consensus. It is the nature of science to challenge the established paradigm. The revisionist school unfortunately has more to do with political beliefs than with the science of the atmosphere.

Global Warming

Though there is increasing scientific evidence to support a causal relationship between observed increases in global temperature (surface air, sea water, and ground surface) and increases in greenhouse gases associated with human activities, there is nevertheless considerable uncertainty about such a relationship both in the scientific community and the larger socioeconomic-political community in the world who must make public policy decisions on an environmental issue that poses potentially enormous environmental consequences and which may require draconian control measures on a global scale.

Atmospheric scientists have been voicing concern about the threat to earth's climate system posed by increasing greenhouse gases for decades. Despite the uncertainties involved, the United Nations Environment Program (UNEP) conference scientists in 1985 urged that nations begin to determine policy options for limiting greenhouse gases and/or adapting to various global warming projections. At a 48 nation conference in Toronto, Canada in 1988, UNEP urged the world's nations to reduce their CO_2 emissions by 20% by the year 2005 and warned that ultimately CO_2 emissions may have to be reduced by 50%. At the world climate conference in 1990, recommendations to reduce emissions of CO_2 and other greenhouse gases were presented to 137 heads of state. Various countries in Europe, Australia, and New Zealand proposed a United Nations convention requiring nations to stabilize greenhouse gas emissions and reduce them by 20% by the year 2000 and no later than 2010. Such a convention was opposed by large fossil fuel-consuming nations including the U.S. In 1990 the European Community agreed to stabilize CO_2 emissions at the 1990 level by the year 2000. Significant CO_2 emission reductions were also proposed at the UN Environment and Development Conference in Rio de Janeiro, Brazil in 1992. The U.S., responsible for 20% of the world's CO_2 emissions, refused to sign the Rio agreement both because of scientific uncertainties about global warming and greenhouse gases and the potentially enormous adverse effects that a commitment to reduce greenhouse gases would have on the American economy, the most sensitive of political issues.

As with the stratospheric O_3 depletion theory in the 1970s and early 1980s, nations are reluctant to act without convincing scientific proof. Of course, once the Antarctic O_3 hole was discovered and its cause determined, nations acted to phaseout CFCs and other halogenated HCs which posed a threat to the O_3 layer.

As was the case when the CFC-O_3 depletion theory was proposed, and for global warming which may be forced by increases in greenhouse gases, nations must make the hard decision to act now on major potential global environmental threats or wait for further scientific proof. Though more scientific information is desirable, delay may further increase the severity of the problem.

Several policy options have been proposed to deal with global warming concerns: adaptation, mitigation, and limitation.

Adaptation. In the first case, we would seek to maintain the status quo presumably for short-term economic reasons and hope for the best. We would accept the inevitability of global warming (if we would accept it at all) and adapt to whatever changes would be necessary as has been the case throughout the history of human civilization. Infrastructure changes would be made as needed in response to rising sea levels and changes in precipitation. Land-use changes would be made in response to climate changes which would affect crop production. This approach would be appealing to economic and business interests where short-term concerns are primary determinants of decision making.

Mitigation. In the mitigation approach, we would to a considerable degree accept the inevitability of continued CO_2 and trace gas emissions to the atmosphere. We would, however, attempt to mitigate their effects on global warming. One more novel and in fact extreme approach would be to spread dust in the stratosphere as a counterbalance to decreasing atmospheric emissivity. The dust would in theory reflect sunlight back to space and thereby cool the planet. However, such a mitigation effort would be filled with enormous uncertainties and could even result in a different type of climatic peril for the planet.

Another novel but less extreme approach would require a program of massive global reforestation. The newly planted trees would absorb CO_2 from the atmosphere and store it as woody biomass. Reforestation would be attractive because it would have other benefits as well, including slowing soil erosion, improving watersheds, providing timber, and sheltering a wide diversity of life forms. Because the amount of carbon released into the atmosphere each year by the burning of fossil fuels is approximately 5 billion tons, an estimated 7 million km^2 of trees would have to be planted to absorb 5 billion tons of carbon per year. This is an area approximately the size of Australia. If we were to attempt to stabilize (and not reverse) CO_2-related global warming, we would only have to plant enough trees to absorb 3 billion tons of carbon, since this is the actual amount accumulating in the atmosphere each year. Such a reforestation program would require that global deforestation currently proceeding at a rapid pace be curtailed. Additionally, using forests to store carbon is only a temporary solution; sooner or later, carbon stored in trees must be released.

Global reforestation would be a massive undertaking. It would conflict with the development aspirations spurred by enormous population growth in many of the nations of the world. It would require a form of global planning and cooperation never before known.

Limitation. The third approach, and the one likely to receive the most attention in future public policy initiatives, is limitation. Limitation in this case would involve the implementation of policies and practices that would reduce emissions of heat-absorbing gases such as CO_2, CFCs, and CH_4. The phaseout of CFCs to protect the O_3 layer is an initial first step to reduce emissions of greenhouse gases.

A limitative approach that could have a significant effect in slowing the pace of global warming would be to sharply curtail global deforestation, particularly the tropical forests. This is likely to have multiple benefits. Although there are considerable uncertainties, the clearing and burning of forests in the tropics is itself contributing several billion tons of carbon to the atmosphere annually. Reduced forest clearing should be expected to result in similar reductions of CO_2 emissions. Land not cleared would then serve as a sink for atmospheric CO_2. There is also evidence to indicate that clearing of the tropical forests is one of the major factors in increased atmospheric levels of CH_4. Reduced tropical forest clearing would, of course, be expected to result in reduced emissions of CH_4 to the atmosphere.

Because CO_2 appears to be the major greenhouse gas of anthropogenic origin, limitation initiatives have to be focused on reducing CO_2 emissions. One of the best ways to slow global warming would be to implement energy policies favoring the use of alternatives to fossil fuels, to use fossil fuels that emit less CO_2 to the atmosphere, and to use energy more efficiently. Alternatives to fossil fuel use would include solar energy and nuclear power. Though considerable advancements have been made in solar energy technologies, solar energy in the developed part of the world has heretofore failed to be an economically attractive alternative to fossil fuels. Nuclear power, once an attractive energy alternative, has failed to receive needed acceptance because of unresolved safety and waste disposal problems.

Fossil fuels differ in their emission of CO_2 per unit of energy. Emission of CO_2 to the atmosphere could be reduced by shifting from coal, a high CO_2-emitting fuel, to natural gas. On a long-term basis, however, it appears that increased production and consumption of coal is likely to occur, because the earth has enormous reserves of coal that, unlike natural gas, are unlikely to be depleted any time soon. Depletion forecasts for natural gas indicate that reserves are only adequate for another 30 years or so.

Significant reduction in CO_2 emissions from industrialized countries could be achieved by reductions in per capita consumption of energy. This would require using energy more efficiently. To achieve a more energy-efficient world, economic disincentives would have to be implemented. Several policy measures have been suggested. These include consumption taxes on commercial and private energy use proportional to the carbon content of the fuel consumed and environmental surcharges incorporated into the production price of energy supplies.

Carbon dioxide could be removed from the flue gases of major emission sources such as fossil fuel-fired power plants by flue gas decarbonization techniques. Such techniques have to be developed and proven effective. Like flue gas desulfurization, decarbonization systems would be expected to be very costly to build and operate.

The problem of global warming and its slowing is different from other global environmental problems that we face. It is more complex both scientifically and politically than acid rain or stratospheric O_3 depletion. In both of these latter cases, the industrialized world has for the most part the ability to resolve these problems. This is much less so with global warming. Slowing of global warming will require much more extensive and complicated global cooperation.

READINGS

1. Beaton, S.B., G.A. Bishop, Y. Zhang, L.L. Ashbaugh, D.R. Lawson, and D.H.Stedman. 1995. "On-Road Vehicle Emissions: Regulations, Costs, and Benefits." *Science* 268:991–993.

2. Bedrock, N.R. 1993. "Enforcing the CAA." *Environmental Protection*, November, 10–14.

3. Breslin, K. 1995. "The Impact of Ozone." *Environmental Health Perspectives* 103:660–664.

4. Burke, R.L. 1992. *Permitting for Clean Air — A Guide to Permitting Under Title V of the CAA Amendments of 1990.* Air and Waste Management Association, Pittsburgh, PA.

5. Calvert, J.G., J.B. Heywood, R.F. Sawyer, and J.H. Seinfeld. 1993. "Achieving Acceptable Air Quality: Some Reflections on Controlling Vehicle Emissions." *Science* 261:37–45.

6. Campbell, W.A. and M.S. Health, Jr. 1977. "Air Pollution Legislation and Regulations." 355–377. In: A.C. Stern (Ed.). *Air Pollution. Vol. IV. Air Quality Management.* 3rd ed. Academic Press, New York.

7. Cannon, J.A. 1986. "The Regulation of Toxic Air Pollutants: A Critical Review." *JAPCA* 36:562–573.

8. Chaemeides, W.L., R.W. Lindsay, J.L. Richardson, and C.S. King. 1988. "The Role of Biogenic Hydrocarbons in Urban Photochemical Smog: Atlanta as a Case Study." *Science* 241:1473–1474.

9. Clairborne, M.L. 1992. "The New Air Toxics Program." *Natural Resources & Environment*, Fall, 21–23 and 54–55.

10. Committee for Economic Development. 1993. *What Price Clean Air? A Market Approach to Energy and Environmental Policy.* pp. 96. Committee for Economic Development, Washington, DC.

11. deNevers, N.H., R.E. Neligan, and H.H. Stater. 1977. "Air Quality Management, Pollution Control Strategies, Modeling, Evaluation." 4–38. In: A.C. Stern (Ed.). *Air Pollution. Vol. V. Air Quality Management.* 3rd ed. Academic Press, New York.

12. General Accounting Office. 1986. *Air Quality Standards — EPA's Standard Setting Process Should be More Timely and Better Planned.* GAO/RCED-87-23.

13. Gibson, J.A. 1992. "The Roads Less Traveled? Motoring and the Clean Air Act Amendments of 1990." *Natural Resources & Environment*, Fall, 13–16 and 50–51.

14. Henz, D.J. 1984. "Air Pollution Standards and Regulations." 919–969. In: S. Calvert and H. Englund (Eds.). *Handbook of Air Pollution Technology.* Wiley, New York.

15. Lave, L.B., C.T. Hendrickson, and F.C. McMichael. 1995. "Environmental Implications of Electric Cars." *Science* 268:993–995.

16. Lee, B. 1991. "Highlights of the Clean Air Act Amendments of 1990." *JAWMA* 41:48–55.

17. Lindsay, R.W., J.L. Richardson, and W.L. Chaemeides. 1989. "Ozone Trends in Atlanta, GA: Have Emission Controls Been Effective?" *JAPCA* 39:40–43.

18. Lloyd, A.C., J.M. Lents, C. Green, and P. Nemeth. 1989. "Air Quality Management in Los Angeles: Perspectives on Past and Future Emission Control Strategies." *JAPCA* 39:696–703.
19. Maga, J.A. 1977. "Emission Standards for Mobile Sources." 506–567. In: A.C. Stern (Ed.). *Air Pollution. Vol. V. Air Quality Management.* 3rd ed. Academic Press, New York.
20. Mintzer, I.M. 1987. "A Matter of Degrees: The Potential for Controlling the Greenhouse Effect." *World Resources Institute Research Report #5.*
21. National Research Council. 1981. *On Prevention of Significant Deterioration of Air Quality.* National Academy Press, Washington, DC.
22. National Research Council. 1989. *Ozone Depletion, Greenhouse Gases and Climatic Change.* National Academy Press, Washington, DC.
23. National Research Council. 1991. *Rethinking the Ozone Problem in Urban and Regional Air Pollution.* National Academy Press, Washington, DC.
24. Newell, V.A. 1977. "Air Quality Standards." 445–503. In: A.C. Stern (Ed.). *Air Pollution. Vol. V. Air Quality Management.* 3rd ed. Academic Press, New York.
25. Office of Technology Assessment. 1984. *Acid Rain and Transported Air Pollutants: Implications for Public Policy.* OTA-O-205.
26. Office of Technology Assessment. 1989. *Catching Our Breaths: Next Steps for Reducing Urban Ozone.* OTA-O-412.
27. Padgett, J., and H. Richmond. 1983. "The Process of Establishing and Revising Ambient Air Quality Standards." *JAPCA* 33:13–16.
28. P.L. 91-604. Clean Air Act as Amended. 1970. 42 U.S.C. 1857 et seq.
29. P.L. 95-95. Clean Air Act as Amended. 1977. 42 U.S.C. 1857 et seq.
30. P.L. 101-549. Clean Air Act as Amended. 1990. 42 U.S.C. 1857 et seq.
31. Roberts, J.J. 1984. "Air Quality Management." 969–1030. In: S. Calvert and H. Englund (Eds.). *Handbook of Air Pollution Technology.* Wiley, New York.
32. Rowland, F.S. 1989. "Chlorofluorocarbons and the Depletion of Stratospheric Ozone." *American Scientist* 77:36–45.
33. Russel, A., J. Milford, M.S. Bergin, S. McBride, L. McNair, Y. Yank, W.R. Stockwell, and B. Croess. 1995. "Urban Ozone Control and Atmospheric Reactivity of Organic Gases." *Science* 269:491–495.
34. Schneider, S.H. 1989. "The Greenhouse Effect: Science and Policy." *Science* 243:771–780.
35. Stenvaag, J.M. 1991. *Clean Air Act Amendments: Law and Practices.* Wiley, New York.
36. Stern, A.C. 1977. "Prevention of Significant Deterioration: A Critical Review." *JAPCA* 27:440–453.
37. Stern, A.C. (Ed.). 1977. *Air Pollution. Vol V. Air Quality Management.* Academic Press, New York.
38. Stern, A.C. 1982. "History of Air Pollution Legislation in the United States." *JAPCA* 32:44–61.
39. Stewart, R. 1992. "Stratospheric Ozone Protection: Changes over Two Decades of Regulation." *Natural Resources & Environment*, Fall, 24–27 and 53–54.
40. Tietenberg, T.H. 1985. *Emissions Trading — An Exercise in Reforming Pollution Policy.* 222. Resources for the Future, Washington, DC.
41. USEPA. 1971. *Air Pollution Episodes: A Citizen's Handbook.* GPO Publication No. 1972-0-452-729.
42. USEPA. 1990. *The CAA Amendments of 1990.* Summary materials, November.

43. USEPA. 1992. *Report of the Office of Air and Radiation to Administrator William K. Reilly.* "Implementing the 1990 Clean Air Act: The First Two Years." EPA/400-R-92-013.

44. Watson, A.Y., R.R. Bates, and D. Kennedy (Eds.). 1988. *Air Pollution, the Automobile, and Public Health.* Health Effects Institute. National Academy Press, Washington, DC.

45. Wolff, G.T. et al. 1988. "The Scientific and Technical Issues Facing Post-1987 Ozone Control Strategies: A Conference Summary." *JAPCA* 38:895–900.

QUESTIONS

1. What is the significance of the following in achieving air-quality standards (1) air-quality control regions, (2) state implementation plans, and (3) emission standards?

2. Both New Source Performance Standards (NSPS) and National Emissions Standards for Hazardous Air Pollutants (NESHAPs) employ the "Emission Standard" concept of controlling pollution. In practice, what are the differences in the application of NSPS and NESHAPs?

3. Despite enactment of clean air rules and regulations, some property owners may still wish to sue a source which is alleged to be depositing particulate matter on their property. What would they reasonably expect to gain from filing a civil suit, and what legal principles might they use?

4. Describe the principle of the allowance system in achieving the objectives of acidic deposition control. What are advantages and disadvantages of this approach?

5. What options does an industry have if it wishes to build a new facility or expand an existing one in an Air Quality Control Region in which an ambient air-quality standard for a pollutant it emits has not been attained?

6. How are Prevention of Significant Deterioration regulatory requirements different from most other clean air requirements?

7. What are specific advantages and disadvantages of air pollution control programs based on air-quality standards? Emissions standards?

8. What tactics are used to achieve air-quality standards?

9. What enforcement tools does USEPA have under the 1970, 1977, and 1990 CAA Amendments?

10. The deadline for achieving air-quality standards in all Air Quality Control Regions (AQCRs) in the United States was July 1, 1975. It is many years thereafter, and many AQCRs have yet to attain these standards. Why should this be the case? Use SO_2 and O_3 as a focus for your response.

11. In achieving the objectives of clean air legislation, what are the specific functions of the federal government and state and local air pollution control agencies?

12. Describe factors which have made achieving the O_3 standard so difficult.

13. What are the responsibilities of federal and state agencies relative to air-quality standards, NSPSs, NESHAPs, regulation of motor vehicle emissions, and regulation of O_3 destroying chemicals?

14. Why are permits so important in air pollution control programs?

15. Describe policy options that might be applied to the potential problem of global warming associated with anthropogenic emissions of greenhouse gases?

16. Describe briefly the regulatory authority USEPA has in dealing with the problem of stratospheric O_3 depletion.

17. Describe briefly the regulatory history of motor vehicle emission control efforts.

18. How does regulation of "toxic" air pollutants under the 1990 CAA Amendments differ from previous NESHAP requirements?

19. Describe differences between air pollution control strategies and tactics.

20. Why are inspection and maintenance programs needed in some AQCRs?

21. How do reasonably available and best available control technologies differ? What are their respective applications?

22. Under Prevention of Significant Deterioration (PSD) requirements, what areas receive the highest degree of protection and why? What is the relationship of PSD to the secondary standard for SO_2?

9   MOTOR VEHICLE EMISSIONS CONTROL

THE INTERNAL COMBUSTION ENGINE

The regulation of emissions from light-duty motor vehicles has been described in Chapter 8. As indicated in that discussion, emission standards have become increasingly more stringent. Continued changes in regulatory requirements have necessitated the development and improvement of emission control technologies. Indeed, motor vehicle emissions standards mandated under the 1970 Amendments to the CAA were intended to "force" the development of new control technology to achieve standards.

The development of effective motor vehicle controls requires a thorough understanding of emission production processes in the gasoline-powered, spark-ignition, internal combustion engine (ICE). The four-cycle operation of an ICE is illustrated in Figure 9.1.

In operation gasoline is pumped from a fuel tank to a carburetor where it is mixed with air. The cycle begins as the air-fuel mixture is drawn or injected into a combustion chamber (cylinder). The subsequent upward movement of the piston compresses the air-fuel mixture to the design compression ratio, where it is spark-ignited. On ignition, the explosion of the gasoline forces the piston downward, providing power to the drive shaft which propels the vehicle. The hot combustion gases are vented through the exhaust port into the exhaust manifold during the upward movement of the piston. The cycle is continuously repeated in each cylinder.

Motor vehicle emissions are generated in several different ways and locations during engine/vehicle operation (Figure 9.2). The most important sources are, of course, those produced in combustion and vented through the exhaust pipe. These exhaust gases consist mainly of unburned HCs, CO, and NO_x and account for approximately 90–92% of all motor vehicle emissions. Some products of combustion are not vented through the exhaust system, as they slip by the piston rings and the cylinder walls. These "blowby" gases consist mainly of unburned HCs

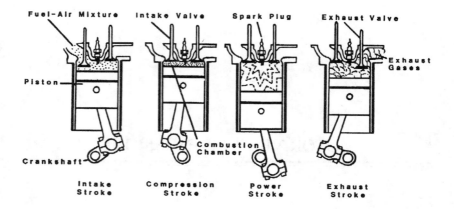

Figure 9.1 Internal combustion engine combustion cycle.

that accumulate in the crankcase. In uncontrolled vehicles they would normally be vented to the atmosphere through a crankcase exhaust port. A third source of emissions is the volatilization of HCs through the carburetor and fuel tank vents. Carburetor emissions are pronounced during the "hot soak" period immediately following vehicle operation.

Combustion Chemistry

Gasoline basically consists of a mixture of non-methane HC compounds that, when ignited in the presence of sufficient O_2, produces H_2O, CO_2, and energy. Equation 9.1 summarizes the relevant chemical reactions, assuming that ambient air is the source of O_2 to support combustion and that gasoline has an average molecular formula roughly equivalent to C_8H_{18}.

$$C_8H_{18} + 12.5\ O_2 + 47\ N_2 \rightarrow 8\ CO_2 + 47\ N_2 + 9\ H_2O \tag{9.1}$$

On the basis of these reactions, 1735 g of air are required for every 114 g of gasoline for complete combustion. In theory, complete combustion will occur at an air-fuel (A/F) ratio of 15.1:1, or the stoichiometric ratio. The actual ratio in

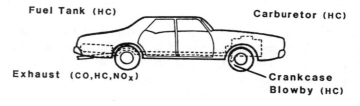

Figure 9.2 Emissions and emission sources associated with a light-duty motor vehicle.

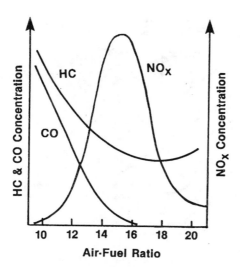

Figure 9.3 Relationship between air/fuel ratios and emissions of CO, NMHCs, and NO_x.

an operating vehicle may vary considerably. Prior to emission control requirements, ratios of 12.5 to 14:1 were common. These rich A/F ratios provide maximum power output. Because of insufficient O_2, rich A/F mixtures generate large quantities of CO and HCs (Figure 9.3). Note that minimum CO emissions are produced near the stoichiometric ratio.

Carbon monoxide is produced by incomplete combustion, the degree of which is determined by the A/F ratio. Although A/F ratios also affect emissions of unburned HCs (Figure 9.3), the primary source of HC emissions is the quenching of the combustion reaction as the flame front of burning fuel approaches the relatively cool walls of the cylinder. This quenching permits a film of unburned or partially burned gasoline a few thousandths of an inch (25 to 50 μm) wide to remain near the cylinder walls during the combustion phase of engine operation. This unburned fuel is expelled in the exhaust with other combustion gases.

To simplify the discussion of HC combustion, Equation 9.1 has been written so as to exclude chemical reactions involving nitrogen. Indeed, because of the high combustion chamber temperatures (>1350°C), O_2 and N_2 may combine to form NO_x. The influence of A/F ratios on NO_x produced is also illustrated in Figure 9.3. Note that emissions of NO_x are highest near the stoichiometric ratio, where combustion is most efficient.

Crankcase Blowby Gases

Crankcase emissions can be controlled by a relatively simple and inexpensive technology called positive crankcase ventilation (Figure 9.4). In this control technique, blowby gases are recirculated to the combustion chamber for reburning.

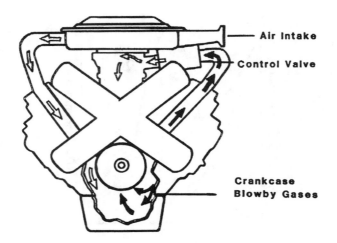

Figure 9.4   Positive crankcase ventilation of blowby gases.

Evaporative Emissions

Evaporative emissions include diurnal, hot soak, and running losses. Diurnal emissions occur as the fuel tank of a parked vehicle takes in air as it cools at night and exhales air and gasoline vapors as it heats up during the day. Hot soak emissions occur after the engine is shut off as heat from the engine heats the fuel system. Running losses occur when gasoline vapors are forced from the fuel tank while the vehicle is being operated and the fuel tank becomes hot. Evaporative losses can also occur during refueling.

Evaporative emission control charcoal canisters sorb gasoline vapors (Figure 9.5). Sorbed vapors are purged from charcoal into the engine during normal driving. In California, vapor control systems at service stations reduce potential refueling vapor losses.

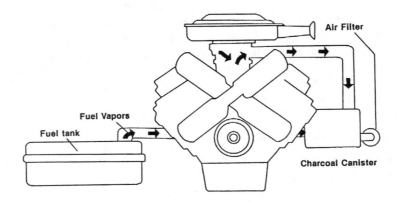

Figure 9.5   Evaporative emissions control system.

Exhaust Gases

Unlike blowby gases and evaporative emissions, which are controlled by relatively simple and inexpensive technologies, exhaust gas emissions have proven to be more difficult to control. The complexity of the exhaust gas problem is increased by the fact that control measures employed to reduce emissions may significantly affect vehicle operating characteristics to which consumers are particularly sensitive. These include vehicle power, driveability, and fuel economy. Adding further to the dilemma is the paradox of CO/HC and NO_x control measures. In general, measures employed to control CO/HC increase NO_x; conversely, NO_x control measures may increase CO and HC.

Non-Methane Hydrocarbon and Carbon Monoxide Control — Pre-1975 Model Vehicles

Since HCs and CO are products of incomplete combustion, control measures that increase combustion efficiency will reduce emissions of both but not to the same extent.

When NMHC and CO exhaust emission requirements for 1966 California and 1968 U.S. light-duty vehicles were first imposed, automobile manufacturers attempted to meet these standards by using two basically different control systems. One of these was an air injection or thermactor system in which air was pumped into the exhaust manifold. This air-rich environment in the exhaust manifold allowed combustion to continue. The second and more widely employed approach was that of engine modification, or the improved combustion system. The most important of the modifications was the use of lean (above the stoichiometric ratio) A/F mixtures. Lean air-fuel ratios increase combustion efficiency and reduce emissions of CO and NMHCs. As the A/F ratio is increased from 11 to 16:1, CO in exhaust gases decreases from 7.5 to 0.2%. Note, however, that at very lean A/F ratios (Figure 9.3) the concentration of NMHCs begins to increase. In such very lean A/F mixtures, fuel particles may be so far apart that the flame may not advance over the intervening space. The flame may become extinguished (misfiring) and unburned HCs are exhausted.

The control of emissions by manipulation of A/F ratios would be more effective if driving conditions were always the same. Driving conditions, of course, vary considerably. A driving cycle may include starting, acceleration, cruising, deceleration, and idling in various combinations. No two vehicles are exactly alike. Emissions will affect the A/F mixture changes needed to meet changing driving conditions. Air-fuel mixtures that favor complete combustion usually are too lean to prevent the engine from stalling during idling or during a cold start. Engines utilizing lean A/F mixtures for emission control therefore require a much higher idle speed to prevent stalling.

Ignition timing also has an important effect on NMHC emissions. The effect of spark timing on NMHC and NO_x emissions is illustrated in Figure 9.6.

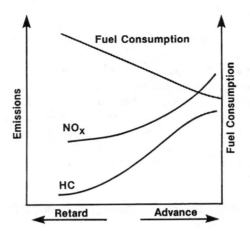

Figure 9.6 The effect of spark timing on NMHC and NO$_x$ emissions and fuel composition. (From Starkman, E.S. 1975. *Environ. Sci. Tech.* 9:820–824. With permission.)

Non-Methane Hydrocarbon and Carbon Monoxide Control — Post-1974 Model Vehicles

The 1970 Amendments to the CAA mandated that NMHC and CO emissions for 1975 model vehicles, and NO$_x$ emissions from 1976 vehicles, be reduced by 90% of their 1970 levels. When these amendments were enacted, technology to meet such stringent standards did not exist. After considerable litigation, political wrangling, and Congressional action, deadlines mandated by the 1970 CAA Amendments were postponed several times. In lieu of this, EPA imposed interim emission standards effective for 1975 model vehicles. Achievement of these relatively stringent interim standards necessitated the utilization of oxidation catalysts as well as other auxiliary measures.

In the early history of emission control, NMHCs and CO were reduced by a variety of engine modifications designed to improve combustion. In contrast, catalytic converters are after devices, reducing emissions by combusting exhaust gases after they leave the combustion chamber. They are mounted before the muffler close to the engine manifold (Figure 9.7).

The term catalyst means helper — in this case it helps O$_2$ to combine with CO and the unburned HCs to produce CO$_2$ and H$_2$O. Although the catalytic

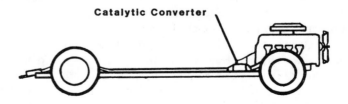

Figure 9.7 Placement of catalytic converter on vehicle assembly.

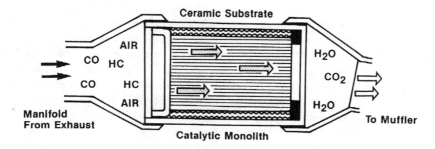

Figure 9.8 Catalytic converter — monolithic design.

materials participate in combustion reactions, they do not themselves undergo chemical conversion. In catalyzing combustion, the catalytic converter acts as an incinerator. However, this incineration occurs at much lower temperatures (350–400°C) than is required for thermal incineration (700–750°C) and consequently does not require a supplemental energy source for efficient combustion. As combustion takes place in the converter, temperatures may rise to over 500°C.

Two basic geometric configurations have been used in catalyst design: the monolith and the pellet. Both designs use the noble metals platinum and palladium as catalytic materials. They may be used individually or in a platinum-palladium blend. Although these are more expensive than other catalytic materials, they have the advantage of not being "poisoned" by sulfur compounds in gasoline. Because of their catalytic efficiency, less of these metals is required per vehicle.

The monolithic design consists of a cylinder, 3–6 in. in diameter, which has an internal ceramic honeycomb structure (Figure 9.8) whose surface is coated with the highly catalytic material. In the pellet design (Figure 9.9) small pellets are impregnated with catalytic materials. In both designs there are thousands of passages which allow the exhaust gases to flow freely through the converter with relatively low back pressure. The high surface area — in the thousands of square feet — allows the catalytic material to come into direct contact with the exhaust gases, oxidizing them to CO_2 and H_2O vapor.

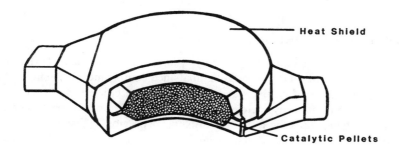

Figure 9.9 Catalytic converter — pellet design.

For catalysts to function efficiently, it is essential that they not be contaminated with chemicals that may "poison" them, that is, render them inactive. For this reason, motor vehicles with catalytic systems cannot use fuels that contain lead or phosphorus. Anticipating that auto manufacturers would employ catalyst technology for 1975-model vehicles, USEPA required petroleum companies to make available for distribution unleaded, phosphorus-free fuels. To prevent customer use of leaded fuels in catalytic systems, USEPA also required auto manufacturers to employ small-diameter fuel tank filler necks, and fuel distributors to use smaller-diameter gasoline pump nozzles.

If leaded gasoline were to be used in a catalytic system either accidentally or on purpose, the catalytic system would be rendered inactive since the lead coats the catalyst and makes it inaccessible to the exhaust gases. A few fill-ups with leaded gasoline should not destroy the catalytic system; they would, however, reduce the system's effectiveness for a period of time. Upon returning to the exclusive use of unleaded gasoline, catalyst performance should return to normal.

In addition to the converter itself, such systems also utilize a quick heat intake manifold, electronic ignition, improved carburetion and, in California, an air pump.

Catalytic converters are least effective during cold-start and engine warm-up. A rich air-fuel mixture must be used during cold starts to compensate for the low volatility of cold fuel. This rich fuel mixture is not completely combusted and high NMHC and CO emissions result. To reduce cold-start emissions, auto manufacturers have employed leaner air-fuel mixtures, preheating of air and fuel, improving carburetor air-fuel mixture control, and shorter choking periods.

In the quick heat intake system, air is drawn over the exhaust manifold where it is warmed to simulate summer ambient temperatures. The heating of the intake air improves fuel vaporization. The high volatility of the fuel allows for leaner air-fuel engine operation and a shorter choking period during warmup.

Catalyst-equipped vehicles manufactured for sale in California are also equipped with air pumps or thermactor systems. In these vehicles, air is pumped into the exhaust ports to allow combustion to continue in the exhaust manifold, reducing the amount of NMHCs and CO that the converter must oxidize. As a consequence, better control of NMHCs and CO is realized. An air pump system is illustrated in Figure 9.10.

Control of NO$_x$ Emissions

Nitrogen oxides emission control has lagged behind that of NMHCs and CO. The control of NO$_x$ poses a dilemma since control measures employed to decrease emissions of NMHCs and CO increase the efficiency of combustion. As combustion efficiency increases, higher combustion chamber temperatures result, which enhances the formation of NO$_x$. As previously discussed, combustion efficiency can be improved by using lean air-fuel mixtures. Nitrogen oxides, on the other hand, can be significantly reduced by utilizing rich air-fuel mixtures (Figure 9.3). These rich mixtures result in lower combustion chamber tempera-

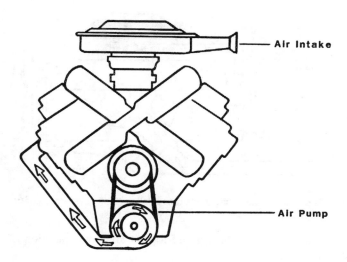

Figure 9.10 Air pump system used to reduce exhaust gas emissions.

tures and decreased emissions of NO_x, but NMHC and CO emissions are increased significantly.

Historically, control measures for NO_x have employed techniques that reduce combustion chamber temperatures without significantly increasing emissions of NMHCs and CO. These have included the retardation of spark timing, decreased compression ratios, and exhaust gas recirculation.

Spark Timing. Spark ignition normally takes place 15–20° before the piston reaches top dead center. Retardation of the spark so that ignition occurs late in the compression stroke reduces the time available for the production of higher temperatures and, as a consequence, results in decreased production of NO_x. As the spark is retarded, the surface area-to-volume ratio of the combustion chamber is also decreased. Since there is less wall surface available for quenching of the flame front, NMHC emissions are also decreased. Note the effect of spark timing on both NMHC and NO_x emissions in Figure 9.6. Also note the effect of spark retardation on fuel economy.

Compression Ratio. Before ignition, the air-fuel mixture is compressed by the upward motion of the piston. At maximum compression the air volume of the combustion chamber is only a fraction of that existing before compression was initiated. The compression ratio is the relationship between the initial and final combustion chamber volumes.

Lowering compression ratios produces lower combustion chamber temperatures. Vehicles produced before the early 1970s had relatively high compression ratios, approximately 8.5 to 11.1:1. Modern day compression ratios have been lowered to about 8.1 to 8.5:1 to reduce NO_x emissions and to facilitate the use

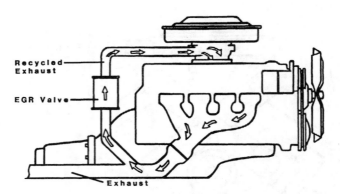

Figure 9.11 Application of an exhaust gas recirculation (EGR) system to control emissions of NO_x.

of lower-octane unleaded fuels required for the effective operation of catalyst-equipped vehicles.

Exhaust Gas Recirculation. One of the most effective ways of reducing NO_x emissions has been the use of exhaust gas recirculation (EGR). In this control method, NO_x emissions are reduced by recirculating controlled amounts of exhaust gas back to the intake manifold where it mixes with air and fuel entering the combustion cycle. These recirculating gases are relatively inert (containing very little combustible material). Carbon dioxide and H_2O vapor in this recirculated exhaust gas serve as heat sinks, reducing peak combustion temperatures. An exhaust gas recirculation system is illustrated in Figure 9.11.

Future Control Technologies

The achievement of stringent CO, NMHC, and NO_x emission standards originally mandated in the 1970 CAA Amendments will require the development of more sophisticated control technology. Future emission control systems may utilize dual catalytic or three-way catalytic systems. In the dual catalytic system the HC-CO oxidation catalyst is located downstream of a NO_x reduction catalyst (Figure 9.12). The two catalysts may be in single or separate containers. The oxidation catalyst would utilize platinum and/or palladium, or base metals promoted with these noble metals. The NO_x reduction catalysts would use platinum, palladium, ruthenium, base metals, or base metals promoted with the above noble metals.

EMISSION CONTROL PROBLEMS

The development of motor vehicle emission control technologies has resulted in problems for both auto manufacturers and consumers. Some control technologies have been associated with undesirable performance characteristics, including decreased power, driveability, and fuel economy.

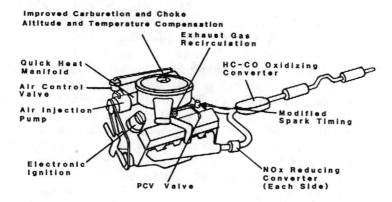

Figure 9.12 Dual catalytic converter system. (From NRC. 1973. Report of Committee on Motor Vehicle Emissions. National Academy Press, Washington, DC. With permission.)

Power and Driveability

The most noticeable side effects of control measures have been the loss of vehicle power and decreased driveability. This decrease in performance has been attributed to the use of lean A/F mixtures, retarded spark timing, exhaust gas recirculation, lower compression ratios, and lower-octane unleaded fuels.

The relationship between A/F ratios and engine performance is shown in Figure 9.13. Note that the A/F ratios to the rich side of the stoichiometric ratio (15.1:1) provide the highest power output. These ratios, of course, also result in the highest emissions of NMHCs and CO. In pre-catalyst controlled vehicles, lean A/F mixtures were the primary means by which CO and NMHC emissions were controlled. Vehicles operated with lean A/F ratios have a tendency to stall

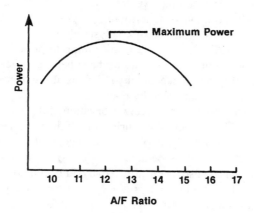

Figure 9.13 Effect of air/fuel ratios on power performance of vehicles. (From Wildeman, T.R. 1974. *J. Chem. Ed.* 51:290–294. With permission.)

during idling or when the engine is cold. To minimize such problems, auto manufacturers have adjusted carburetion so that the engine operated faster during idling and cold starts. If the engine maintains this faster idling condition longer than necessary for warm-up, it becomes hot, resulting in what consumers refer to as dieseling. The higher engine temperatures result in autoignition of the A/F mixture. As a consequence, the engine may continue to operate after a motorist shuts off the ignition. This effect produces malodors similar to those of diesel engine-equipped vehicles.

In pre-emission control motor vehicle operation, ignition took place at about 15 to 20° before the piston reached top dead center. Such spark timing provided excellent performance. Retardation of the spark to decrease NO_x and NMHC emissions decreased both power and vehicle driveability.

The use of exhaust gas recirculation to cool combustion chamber temperatures for NO_x control also results in decreased vehicle power and driveability. Under full load the combination of exhaust gas recirculation and lean A/F mixtures may reduce engine power by 25%.

High compression ratios (10 to 11:1) and high-octane fuels maximize power output. Reduced compression ratios used for emission control have significantly reduced vehicle power.

Fuel Economy

Since exhaust emission controls were first required on vehicles sold in California (1966) and nationwide (1968), diminished fuel economy had been an undesirable trade-off of most exhaust emission control systems. By 1974, the average fuel penalty resulting from control measures was about 16%. Decreased fuel economy was associated with using leaner A/F mixtures, retardation of spark timing, lower compression ratios, and exhaust gas recirculation.

Catalytic control systems have been used to control NMHC and CO exhaust emissions on most American-made vehicles since 1975. These systems, for the most part, reversed the trend of deteriorating fuel economy that characterized emission control systems. The use of catalysts allowed auto manufacturers to substantially restore A/F mixtures and spark timing to pre-control days. Consequently, both fuel economy and driveability have been significantly improved.

Power, driveability, and fuel economy losses resulting from emission control measures were not readily accepted by many motorists. The promotion of high-performance vehicles was a major feature of automobile manufacturer competition and motor vehicle advertising during the 1950s and 1960s. The performance consciousness that this promotion aroused conflicted with the cold reality of diminished performance associated with emission control requirements. Motorist sensitivity had also been affected by fuel penalties associated with control measures. These fuel penalties exacerbated the already significant problems of deteriorating fuel economy that had resulted from ever-increasing motor vehicle weight, and the use of automatic transmissions and air conditioning. Although benefits of increased vehicle size, automatic transmission, and air conditioning appeared to outweigh the associated fuel economy penalties, economy losses attributed to emission control measures received no such acceptance. Motorist perceptions of decreased performance and fuel economy resulted in widespread tampering with control devices. In many cases, this tampering significantly decreased the efficiency of emission control measures, thus defeating their purpose.

Catalytic Converters

Although catalytic systems have been able to meet the 1975 interim standards for CO and NMHCs, they are not without problems. One of the first of these surfaced in 1972 when Ford Motor Company scientists reported that catalysts could oxidize sulfur to sulfuric acid. Their limited tests on prototype devices indicated that as much as 85% of fuel sulfur could be oxidized.

Although the sulfur content of gasoline is low, 0.024% or so, early estimates indicated that catalyst-equipped cars would produce 9–50 times higher sulfuric acid emissions than those not equipped with catalysts. This posed a potential new public health problem, as these respirable, ultrafine (0.01 to 0.1 μm) sulfuric acid particles would be emitted directly into the human breathing zone. USEPA scientists employed a computer model to predict what the impact of these emissions would be as each new model year increased the proportion of catalyst-equipped cars on the nation's highways. From this model, USEPA predicted that under certain circumstances concentrations of sulfuric acid would become sufficiently great to endanger public health. Consequently, USEPA took initial steps to require petroleum companies to remove the small amount of sulfur in gasoline and for auto manufacturers to phase out the use of catalytic systems.

In response to the above concerns, General Motors, in conjunction with USEPA researchers, conducted a large-scale experiment at the GM proving ground. They ran 350 catalyst-equipped cars on a four-lane track on schedules that duplicated heavy traffic on a freeway.

This study conclusively demonstrated that catalytic systems did not oxidize as much sulfur as was initially predicted. Actual sulfur oxidation was only 12 to 24%. In addition, the computer model used by USEPA tended to overestimate pollutant concentrations by 2–3 times under certain circumstances and by a factor of 20 times under others. In a USEPA-sponsored Los Angeles catalyst study, similar results were also observed. As a result, USEPA concluded that no regulatory action was necessary and the catalyst sulfuric acid concern was laid to rest.

This, however, was not the only problem with which makers of catalytic systems, USEPA regulators, and consumers have had to contend. Under USEPA emission control mandates, catalytic systems were required to have a 50,000-mile durability. After 50,000 miles the cost and burden of replacing the catalyst was placed on the vehicle owner. Interestingly, although many model years of vehicles equipped with catalysts have been produced by automakers, no state has adopted a regulatory mechanism to compel vehicle owners to replace catalysts once 50,000 miles of vehicle operation have occurred.

Consumer acceptance of catalysts had been significantly less than over-whelming. Although fuel economy was much better than with pre-catalyst control technology, catalysts required the use of higher priced unleaded fuel.

Consumer sensitivity may also have been affected by malodors associated with some catalytic systems. Although these are oxidative catalysts, under some operating conditions a reducing environment may result, producing H_2S, the rotten egg odor that can be perceived at concentrations as low as 0.5 ppbv.

AUTOMOTIVE FUELS

Gasolines

Gasolines are complex mixtures of paraffinic, olefinic, and aromatic HC compounds. Their exact composition may vary from refinery to refinery and from one geographical area to another, as gasolines are refined or formulated to provide optimum engine performance under a variety of driving and climatic conditions. The various constituents of gasolines differ as to ignition temperatures and other combustion characteristics.

Engine performance is significantly affected by fuel volatility, antiknock quality, and deposit-forming tendencies. Volatility can be defined as the ability of gasoline to vaporize and mix with the air under a variety of engine operating and ambient temperature conditions. Volatility is an important factor in the for-mulation of winter-grade and summer-grade fuels. Winter-grade fuels have a higher percentage of very volatile HCs. The volatile HC content will be even higher in fuels formulated for very cold climates.

Octane Rating

Antiknock quality is the ability of gasoline to burn smoothly and evenly during the combustion process. When combustion is too rapid, a sharp metallic noise called knock is produced. Components of gasoline that resist this tendency to ignite or burn at very rapid rates have a high octane quality or high octane number. The octane rating is an arbitrary number, with the HC compound iso-octane being equal to 100. Fuels formulated with high octane values have a reduced tendency to knock.

The octane rating can be increased by a variety of fuel additives. Over much of the history of light-duty motor vehicle use, lead alkyls (ethyl and tetramethyl lead) have been the additives of choice to boost octane ratings. Lead contents of gasolines have in the past averaged approximately 0.01% by volume or 2 g/gal. On combustion, lead alkyl additives form lead oxide, which tends to remain within the engine, fouling engine components and reducing performance. Chlorides and bromides were added to gasoline to scavenge lead as lead halides, which were more readily exhausted.

Octane ratings in unleaded gasolines are boosted by the addition of aromatic HCs. In order to produce unleaded fuels comparable in octane rating to regular and premium-grade leaded gasolines, the aromatic content must be raised by approximately 50%. This requires that low-octane fractions of crude oil be converted to high-octane aromatic HCs such as benzene, toluene, ethylbenzene, and xylene (BTEX). This additional processing produces less fuel for a given quantity of crude oil when compared to leaded gasolines. Because of the increased energy usage and additional refining steps, unleaded fuels have been more expensive to produce.

Gasoline Composition and Emissions

Fuel composition is an important determinant of emissions from motor vehicles. Gasolines with a high content of very volatile HCs will result in higher emissions from the vehicle itself as well as during refueling. Volatile components are less likely to be sorbed and retained on evaporative emission control systems. Fuel volatility had increased significantly from 1975 to the early 1990s. The use of lead in gasolines in some countries contributes to elevated atmospheric lead levels and human exposures.

Alternative Fuels

A variety of fuels have been evaluated or are being used as lower emission alternatives to conventional gasoline blends. These include reformulated gasoline, alcohol-gasoline blends, ethanol, methanol, liquefied natural gas (LNG), and liquefied petroleum gas.

Such fuels have the potential to improve air quality, particularly in urban areas. The extent of such improvement is uncertain and may vary depending on the type of fuel used and on climatic location. Alternative fuels may improve air quality by reducing mass emission rates from motor vehicles, or reducing their O_3-forming potential or reactivity.

Reformulated Gasolines and Oxygenated Additives. The composition of what has been described as reformulated gasoline has been altered to reduce the reactivity of exhaust products and to lower emissions of NMHCs, CO, and NO_x. Reformulated gasoline can be used in conventional ICEs with no modification of the automotive propulsion system. It is produced by modifying the refining process and by adding oxygenated compounds. It has lower olefinic and aromatic HC content and a lower vapor pressure. Oxygenated additives enhance octane rating which is lowered when the aromatic and olefinic HC content is reduced. They improve the efficiency of combustion and reduce CO emissions.

Commonly used oxygenated HC additives include methyl-t-butyl ether (MTBE), ethyl-t-butyl ether (ETBE), methanol, and ethanol. Gasoline producers tend to use ethers, although ethanol has been widely used in the midwestern U.S. Alcohols can increase the vapor pressure of gasoline, increasing evaporative emissions. The use of oxygenated compounds can increase emissions of formaldehyde, a potent mucous membrane irritant. The contents of oxygenates may vary from 2–3 to 15%. Reformulated gasolines have been introduced into limited fuel markets, those that have severe smog or CO problems.

Oxygenated compounds such as MTBE, ETBE, tertiary amyl methyl ether (TAME), and ethanol have been added to gasoline used during the winter months in 39 cities to achieve emission reductions necessary to meet the CO air quality standard. The typical oxygenate content is 15%. The introduction of MTBE into gasoline in 1992 and its use to reduce CO emissions during winter months was followed by scattered health complaints (headaches, dizziness, nausea) from residents in some areas, most notably Fairbanks and Anchorage, Alaska. As a consequence, a number of epidemiologic studies have been conducted to determine whether the use of MTBE in gasoline may have adverse health effects on exposed populations. Such studies have failed to show a significant relationship between MTBE use in gasoline in targeted cities and health complaints. Nevertheless, MTBE has become very controversial in some areas of the U.S.

MTBE is the most widely used oxygenated compound in the national reformulated gasoline program which went into effect in 1995. USEPA has estimated that the use of reformulated gasoline in nine targeted metropolitan areas will reduce smog-producing emissions by 300,000 tons annually. It has further estimated that reformulated gasoline will reduce annual per car emissions of CO by 13%, NMHCs by 15%, NO_x by 3%, and "toxic" air pollutants by 24%.

In addition to the MTBE controversy which began with the oxyfuels program in 1992, there is concern about potential consumer resistance to the use of much high-priced reformulated gasoline.

Alcohol Fuels. There has been considerable interest in the use of methanol, ethanol, and alcohol-gasoline blends as motor vehicle fuels. Methanol is a colorless, toxic liquid with a low vapor pressure and high heat of vaporization. Both properties can lower emissions under warm conditions, but tend to make pure methanol-operated vehicles difficult to start when cold, thus resulting in increased emissions. Additives such as unleaded gasoline (5–15% by volume) are blended into methanol to increase vapor pressure and improve starting under colder conditions. The most common blend containing 85% methanol by volume is described as M85.

Flexibly-fueled and dedicated methanol-fueled vehicles have been developed. The former are designed to be operated on any combination of gasoline and M85 fuel in order to ease the transition when methanol is not yet widely available.

The largest emission component (on a mass basis) from methanol-fueled vehicles is methanol itself, followed by formaldehyde (HCHO), and gasoline-like NMHC components. Of greatest concern is HCHO. Even after catalytic reduction, M85 vehicles may emit 30–40 mg HCHO/mile, about 3–6 times more than conventional vehicles. Formaldehyde is a potent mucous membrane irritant and has high photochemical reactivity.

Limited studies of flexibly fueled vehicles indicate that NO_x may be slightly reduced. However, dedicated methanol-fueled vehicles would have to use a higher compression ratio and thus result in increased NO_x emissions. Evaporative emissions would be less reactive, and the mass emission rate would be expected to be lower.

The use of ethanol would be similar to methanol. Flexibly-fueled vehicles can run on ethanol as well as methanol. Air quality benefits associated with ethanol are less than those of methanol. Ethanol is also more reactive, producing 15% more O_3 on a carbon-atom basis. Its use also leads to the formation of peroxyacyl nitrate (PAN). Less information is available on the emission characteristics of well-controlled ethanol vehicles. Limited tests have shown high emissions. The use of ethanol as a gasoline additive, while simultaneously allowing an increase in vapor pressure, would not have any air quality benefits. Increased evaporative emissions and increased emissions of ethanol and acetaldehyde would increase atmospheric levels of PAN.

Compressed and Liquefied Gases. Both natural gas (which is primarily methane, >90%) and propane can be used to power motor vehicles. Natural gas is compressed and stored at pressures of 4500 PSI or liquefied for use as an automotive fuel. Because liquefied natural gas must be cryogenically cooled and stored, its use has been limited compared to compressed natural gas (CNG).

Emissions from natural gas vehicles (NGV) primarily consist of methane (CH_4). Because of its low photochemical reactivity, it has the potential for reducing O_3 formation. This benefit can be reduced by impurities such as ethane and propane which may produce up to 25 times more O_3 than CH_4. The presence of olefins would also reduce the benefits of using CNG. Assuring a low olefin content is vital. Though combustion by-products such as aldehydes are quite reactive, mass emission rates are likely to be small.

Carbon monoxide emissions from the lean-burn operating conditions of NGV are very low, 90% less than gasoline-powered vehicles. Because of lean-burning conditions used, CNG vehicles should produce lower engine-out (released prior to catalytic converter) emissions of NMHCs, NO_x, and CO, with higher fuel economy. Evaporative emissions from CNG vehicles would be very small and only slightly reactive.

Liquefied petroleum gas (LPG) consists primarily of propane, a petroleum refining by-product. It has many of the attributes of CNG but does have several disadvantages. These include limited supply and higher exhaust reactivity than CNG. It has a higher energy output per unit volume, thus requiring smaller tanks.

In the U.S., CNG is used in fleet vehicles by some corporations and institutions. It is not available to consumers at service stations. It is available at "petrol" stations in Australia.

ALTERNATIVES TO THE INTERNAL COMBUSTION ENGINE

The ICE has been the dominant propulsion system used in light-duty motor vehicles for a century. It is reliable, economical, and gives excellent performance. Because of pollutant emissions associated with the ICE, a number of alternative propulsion systems have been evaluated by motor vehicle manufacturers and others. These have included the stratified charge, diesel, gas turbine, wankel, and rankine cycle engines and electric-powered vehicles.

Most have been shown to have significant limitations both from technical and economic standpoints. Though diesel engines are widely used in light-duty motor vehicles in Europe, diesel model vehicles manufactured in the U.S. performed poorly and were not acceptable to American consumers. Diesel engines have the capacity of easily achieving standards for NMHCs and CO, but produce significant emissions of NO_x, PM, and odor-causing substances.

Electric Vehicles

Because of stringent air pollution regulatory requirements, it is probable that significant use of electric-powered vehicles will occur in southern California in the next decade. Because of their limited driving range (circa 50–75 miles), electric vehicle use will be primarily limited to fleet operations.

In addition to their short driving range, electric vehicles provide only modest acceleration and require the use of expensive batteries. The development and use of electric vehicles to achieve air quality goals has long been advocated by critics of the automotive industry. In addition to "zero emissions," electric vehicles use energy more efficiently than ICEs; in the former case, 100%; in the latter, 14%. Electricity, of course, must be produced from other energy sources such as fossil fuels and nuclear power. The conversion efficiency from these sources is 30–35%. Though more efficient, electric vehicles are not really pollution-free, since in the case of fossil fuels, significant pollutant emissions do occur. Where power generation is distant and local problems with motor vehicle emissions are severe, the use of electric vehicles may prove to be a desirable approach to meeting air quality goals.

READINGS

1. Chang, T.Y. 1991. "Alternative Transportation Fuels and Air Quality." *Environ. Sci. Technol.* 25:1190.
2. Dekiep, E. and D.J. Patterson. 1984. "Emission Control and Internal Combustion Engines." 484-512. In: S. Calvert and H. Englund (Eds.). *Handbook of Air Pollution Technology.* John Wiley & Sons, New York.
3. Harmon, R. 1992. "Alternative Vehicle-Propulsion Systems." *Mech. Eng.* 105:67–74.
4. Malin, H.M., Jr. 1973. "An Automotive Engine That May Be Cleaner." *Environ. Sci. Technol.* 7:688–689.
5. Medlen, J. 1995. "MTBE: The Headache of Cleaner Air." *Environ. Health Perspectives* 103:666–670.
6. National Research Council. 1973. *Report by the Committee on Motor Vehicle Emissions.* National Academy Press, Washington, DC.
7. National Research Council. 1991. *Rethinking the Ozone Problem in Urban and Regional Air Pollution.* National Academy Press, Washington, DC.
8. Olsen, D.R. 1977. "The Control of Motor Vehicle Emissions." 596-652. In: A.C. Stern (Ed.). *Air Pollution. Vol. IV. Engineering Control of Air Pollution.* 3rd ed. Academic Press, New York.

9. Starkman, E.S. 1975. "Emission Control and Fuel Economy." *Environ. Sci. Technol.* 9:820–824.

10. USEPA. 1974. *Progress in the Implementation of Motor Vehicle Emission Standards Through June 1974.* EPA/230/3-74-103.

11. Watson, A.Y., R.R. Bates, and D. Kennedy (Eds.). 1988. *Air Pollution, the Automobile and Public Health. Health Effects Institute.* National Academy Press, Washington, DC.

12. Wildeman, T.R. 1974. "The Automobile and Air Pollution." *J. Chem. Ed.* 51:290–294.

QUESTIONS

1. What specific pollutants are controlled by the following technologies: (a) positive crankcase ventilation, (b) charcoal filtration, (c) spark retardation, (d) exhaust gas recirculation, (e) A/F ratio manipulation?

2. What are the advantages and disadvantages of catalytic systems used for controlling emissions of NMHCs and CO in light-duty motor vehicles?

3. What motor vehicle performance variables have been adversely affected by the application of emission control technologies?

4. What operating principles and technologies are used to control NO_x emissions?

5. Diesel engines produce very low NMHC and CO emissions. Why are they not more widely used in light-duty vehicles?

6. What is the effect of increasing air/fuel ratios on emissions of CO, NMHCs, NO_x, and on vehicle power output?

7. How are motor vehicle compression ratios related to vehicle fuel octane requirements, emissions of NO_x, and power output?

8. Why was lead added to gasoline? Why is benzene now added to unleaded gasoline?

9. Why are oxygenated compounds added to gasoline in some regions of the U.S.?

10. What effect (if any) would the altitude of Denver, Colorado have on motor vehicle emissions and the effectiveness of control devices?

11. Describe the air quality benefits associated with methanol-fueled vehicles.

12. What engine/operating factors are responsible for emissions of (a) CO and NMHCs, (b) NO_x, (c) blowby gases, (d) evaporative losses, and (e) sulfuric acid?

13. Comparing the use of methanol and ethanol fuels, what are some of the questions that have been raised concerning the use of both fuels?

14. Exhaust emissions are at their highest level under cold-start conditions. What principles/technologies have been used to reduce these emissions?

15. Formaldehyde emissions appear to be increased in vehicles operated on methanol blends and gasoline blended with other oxygenates. What environmental concerns are associated with these emissions?

16. Air pumps are used on light-duty vehicles sold in California. Why are they used, and how do they reduce emissions?

17. What principle is used to reduce NO_x emissions by exhaust gas recirculation?

18. Characterize reformulated gasoline and air quality benefits derived from its use.

19. What are the advantages and disadvantages of using electric cars for purposes of achieving air quality standards?

10

CONTROL OF EMISSIONS FROM STATIONARY SOURCES

Stationary sources as seen in Tables 2.1 and 2.2 are a significant source of emissions to the atmosphere. These emissions are quite varied in their quantity and characteristics, depending on the industrial and commercial activities involved. Pollutants may be vented to the atmosphere through specially designed stacks and exhaust ventilation systems or escape as fugitive emissions.

Emissions from stationary sources can be controlled or their effects reduced by the application of a number of control practices. These may include the use of tall stacks, changes in fuel use, implementation of pollution prevention programs, fugitive emissions containment, and "end of the pipe" control devices which remove pollutants from waste gas streams before they are emitted into the ambient environment through stacks.

TALL STACKS

Smokestacks do not, of course, reduce emissions from a source. Rather, they reflect efforts by sources to reduce potential pollution problems by elevating emissions above the ground where they may be more effectively dispersed. Historically, when public concern was limited and control devices were not available, a sufficiently sized smokestack was a relatively effective way to reduce ground-level concentrations and human exposure. The effectiveness of a smokestack in dispersing pollutants depends on stack height, plume exit velocity and temperature, atmospheric conditions such as wind speed and stability, topography, and proximity to other sources (see Chapters 3 and 7).

Utility and smelter operations have used tall stacks (200 to 400 m) to reduce ground-level concentrations and thus comply with air-quality goals or standards for more than three decades. Such stacks have been necessitated by the high sulfur dioxide (SO_2) emission potentials of such sources. The higher the stack, the higher the plume height, wind speed, and, of course, the greater the air volume between the plume and the ground available for dilution. The effect of tall stacks is to disperse pollutants over a wider area.

FUEL-USE CHANGES

Significant reductions in emissions from combustion processes can be achieved by changes in fuel use. Indeed much of the early progress in emissions reductions in the U.S. in the 1970s after the passage of the 1970 Clear Air Act (CAA) Amendments were achieved by switching to cleaner-burning fuels such as natural gas and low-sulfur oil and coal, in contrast to using high-sulfur fuel oil and coal. The cleanest burning fuel is, of course, natural gas.

These fuel-use changes have not been problem-free. Supplies of low-sulfur fuels such as oil have always been limited, and therefore, they are more expensive. Low-sulfur oil must be imported, and thus it has increased our reliance on sometimes unreliable foreign suppliers. The use of low-sulfur western coals has decreased markets for high-sulfur eastern coals.

Significant improvements in air quality have been achieved in Chicago and the surrounding area as a result of utility decisions to use nuclear fuels rather than coal to generate electricity. More than 40% of the electricity used in that region is produced in nuclear power plants. Because of the accident at Three Mile Island in the late 1970s, increased construction costs, and opposition to nuclear power, no new nuclear power plants have been constructed in the past decade.

FUGITIVE EMISSIONS

Fugitive emissions represent a special case of pollution control from stationary sources. Significant fugitive emissions of vapor and particulate-phase materials occur within industrial and commercial facilities and from outdoor materials handling activities. For much of the last three decades, fugitive emissions have for the most part gone unregulated. This reflected a basic lack of information on the nature and extent of such emissions and a regulatory preoccupation with stack emissions which were amenable to reduction using existing control technology.

Fugitive emissions are increasingly being controlled. Such control reflects both economic considerations and the fact that fugitive emissions may themselves be a significant problem. Under the bubble concept which treats a single facility as a source, it may be more economical to control fugitive emissions than to apply more expensive control technology to treat stack gases. A case in point may be a steel mill where considerable fugitive emissions occur from process or road surface dusts which become entrained on disturbance by vehicle movement. Such dusts can be (and are) controlled by periodic wetting of roadway surfaces. Fugitive emission control can be achieved by implementing a variety of operating practices, materials handling, and process changes, as well as technology.

Considerable potential exists to reduce fugitive emissions by equipment and process changes, equipment maintenance, and containment strategies. Examples of the last case include the use of negative pressure within a facility to minimize the escape of hazardous materials such as lead from secondary lead smelters, or to construct and operate negative pressure enclosures. Such enclosures are required under occupational safety and health rules for asbestos and lead abatement projects. They nevertheless have the benefit of reducing emissions to the outdoor environment. Under NESHAP asbestos abatement requirements, friable asbestos materials must be wetted prior to and during removal operations to prevent "visible emissions."

Fugitive emissions will be receiving considerable attention under Title III, air toxics provisions of the 1990 CAA Amendments where maximum achievable control technology (MACT) standards will apply to 189 hazardous pollutants.

POLLUTION PREVENTION

As indicated in Chapter 8, pollution prevention as a waste control strategy was codified in the Pollution Prevention Act of 1990 and given new attention in achieving air-quality goals under the 1990 CAA Amendments. Though pollution prevention has its origin in efforts to reduce solid waste disposal to the land, its application has been broadened to include other media including air.

The goal of all pollution prevention activities is to reduce waste generation at the source. The concept is applied both narrowly (focusing primarily on industrial processes) and broadly (focusing on such things as energy conservation, emissions from finished products, etc.).

At the level of an industrial or commercial facility, pollution prevention may include the substitution of process chemicals (materials which are less hazardous to human health and/or the environment than those presently being used), changes in process equipment, changes in plant operating practices, enhanced maintenance of plant equipment which may emit pollutants, changes in plant processes, etc.

Substitution

Examples of substitution include the use of N-methyl pyrrolidone (NMP) as an alternative for methylene chloride-based paint strippers, the use of reformulated gasoline, the use of hydrochlorofluorocarbons (HCFC) and hydrofluorocarbons (HFCs) for chlorofluorocarbons, and the use of isocyanate adhesives for high solvent-emitting adhesives. NMP is less toxic and less volatile than methylene chloride. The use of reformulated gasoline to achieve air-quality goals was discussed in Chapter 9, and the use of HFCs and HCFCs to protect the O_3 layer was discussed in Chapter 8.

Process Equipment Changes

A variety of process equipment changes can be used to reduce pollutant emissions. These may include the use of completely enclosed vats in place of open vats where solvent emissions may occur and, in the case of coke ovens, retrofitting or replacing leaky coke oven doors with state-of-the-art seals, use of flexible doors, and use of automatic spotting devices to reduce damage to doors and jambs.

Plant Operating Practices

Excess production of pollutants or emissions may occur as a result of poor equipment and plant operation. This is particularly true in operating boilers and other combustion equipment. Good boiler and incinerator operation requires the use of an adequate supply of combustion air. Insufficient combustion air results in incomplete combustion and production of particulate matter (PM) and a variety of gas-phase substances. Good operating practices are particularly important in achieving emission reductions that pollution control equipment is designed for.

Maintenance

The performance of equipment in terms of producing products and releasing contaminants to the environment may be comprised by inadequate maintenance. This is true for process and pollution control equipment. It is particularly important to maintain combustion equipment in good operating condition. Maintenance is also important in reducing leakage of solvents and other chemicals from vats, valves, and transmission lines. Maintenance is also important in reducing spill-related emissions. In coke oven operation, it is important to maintain adequate seals around coke oven doors. Such seals require considerable maintenance attention.

Process Changes

Emissions in many cases are related to processes used in product manufacturing. Emissions of solvent vapors, for example, can be reduced in painting operations by dry powder painting. This technique sprays specially formulated thermoplastic/thermosetting, heat-fusible powders on metallic surfaces which are subsequently heat cured. Dry powder painting can also be done electrostatically. In either case, significant reductions in hydrocarbon (HC) emissions can be achieved.

Energy Conservation

Reduction in energy use by the application of a variety of energy conservation measures can result in significant emission reductions at various levels. All fuels when combusted or partially combusted produce byproducts which pollute the atmosphere. Any measure which reduces energy consumption associated with fossil fuel use will result in both a reduction in fuel use and associated emissions.

Energy conservation measures may include manufacture, sale and use of fuel-efficient motor vehicles, development and use of energy-efficient combustion and heat-recovery systems in both industrial and domestic environments, use of mass-transit and/or car pooling, construction and retrofitting of buildings to reduce energy loss, manufacture of energy-efficient appliances and equipment, use of secondary (as compared to primary) materials in product manufacturing, recycling and reuse, etc.

Examples of energy conservation practices are varied, and significant reductions in energy usage have occurred in the U.S. since the energy crises experienced in the 1970s. Most notable has been improvement in motor vehicle fuel economy and energy use reductions in manufacturing facilities. Programs with less impact but of some note are USEPA's Energy Star Computer and Green Lights programs.

The Energy Star Program is a voluntary partnership between USEPA and computer manufacturers. Computers identified with the Energy Star logo are designed to power down automatically when not in use. The Green Lights Program encourages voluntary reductions in energy use through the installation of more efficient lighting in buildings.

GAS CLEANING TECHNOLOGY

Compliance with emission limits required under state implementation plans will in many instances necessitate the application of control equipment. Because of the complexity and expense involved in gas cleaning, the selection of appropriate equipment must be based on a careful engineering evaluation of the specific nature of the emission problem, including a physical or chemical characterization of pollutants and a determination of the gas discharge rate and process conditions. Control equipment selection requires a consideration of equipment performance characteristics, including ability to achieve mandated emission limitations. Equipment selection must also consider capital, operating, and maintenance costs.

Equipment Performance

Gas cleaning equipment may be designed to remove PM, individual gases, or both PM and gases. Equipment design will be determined by specific control requirements. The most important characteristic of any control technique and equipment design is performance. Performance can be quantified by determining collection efficiency, which is defined as a percentage of influent dust and/or gas collected or removed from the waste gas stream. The collection efficiency of control equipment can be calculated from Equation 10.1.

$$E = 100 \ (1 - B/A) \qquad (10.1)$$

where A is the concentration of the pollutant entering the control system, and B is the emission level achieved or emission reduction required. For example, if the influent to a dust collector were 600 lbs of dust per hour and effluent gas emissions were 30 lbs of dust per hour, the calculated collection efficiency required would be 95% (Equation 10.2).

$$E = 100 \ (1 - 30/600) = 95\% \qquad\qquad (10.2)$$

Emission limitations generally do not specify collection efficiency of control equipment. Efficiency or performance must be calculated from emission data and applicable emission standards. For example, if process emissions (A) are 600 lb/hour and the allowable emissions (B) are 60 lb/hour, the collection efficiency necessary for compliance will be 90%. Most industrial sources would elect to meet the emission limitation incurring the least allocation of economic resources.

Particle Collection Efficiency

The major determinants of particle collection efficiency are the collection principle and particle size. Some collection techniques are inherently more efficient than others. One or several of the following collecting principles may be utilized for a specific control application: gravitational settling, inertial impaction, filtration, and electrostatic attraction. The overall efficiencies and collection techniques for commonly used particle/dust collectors are summarized in Table 10.1. Collection techniques and devices will be described more fully in the following sections.

Variation in collection efficiency associated with particle size is illustrated for a hypothetical control device in Figure 10.1. Note that the fractional collection efficiency decreases sharply for small particles, whereas large particles over a broad range are collected at efficiencies approaching 100%. Fractional collection efficiencies for commonly used particle and dust collectors are indicated in Figure 10.2.

Although fractional collection efficiency is important in equipment design, performance based on overall collection efficiency is the principal determinant of compliance with emission limits. As shown in Figures 10.1 and 10.2, the efficiency of dust collectors decreases markedly in the submicrometer range. Paradoxically, it is these small uncollected particles which have the greatest human health significance.

Table 10.1   Collection Efficiencies of Particle and Dust Control Devices

Equipment type	Overall collection efficiency, %
Cyclone	
Medium efficiency	68–85
Multitube collector	70–90
Electrostatic precipitator	70–99.5+
Fabric filter	99.5+
Wet scrubber	
Spray tower	90
Baffle plate	95
Venturi	99.5+

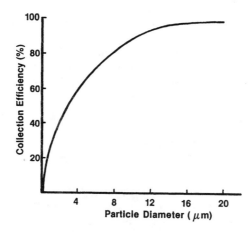

Figure 10.1 Generalized fractional collection efficiency of a particle/dust control device. (From Wark, K. and C.F. Warner. 1981. Harper & Row Publishers, New York. With permission.)

As we have seen in Table 10.1, average overall collection efficiencies depend on the collection technique or principle employed. Under actual operating conditions, performance may vary considerably from the average values of Table 10.1. This variation may be due to other factors, including gas flow rate, dust loading, and particle properties such as specific gravity, shape, viscosity, and resistivity.

Gas Collection Efficiency

The collection efficiency of equipment designed to control gas-phase substances also depends on the mechanism of collection or treatment. Commonly

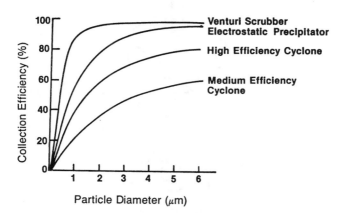

Figure 10.2 Fractional collection efficiencies of particle/dust control devices. (From Wark, K. and C.F. Warner. 1981. Harper & Row Publishers, New York. With permission.)

used control methods include incineration, absorption, and adsorption. In general, the collection efficiency of control devices employing each of these control principles will be influenced by such factors as gas flow rate and pollutant concentration. Other factors limiting efficiency may be specific to the treatment mechanism and/or equipment design. These will be discussed more fully in later sections on gas-phase contaminant control.

PARTICLE COLLECTION SYSTEMS

Cyclone Collectors

Particles can be removed from a waste gas stream by the induction of cyclonic air flow. In a common cyclone, waste gas enters a tangential inlet near the top of a cylindrical body (Figure 10.3). This creates a main vortex which spirals downward between the walls of the cyclone and the centrally placed discharge outlet. Air flow continues downward until it reaches the bottom of the cyclone where the vortex changes its direction of flow, forming an inner vortex traveling upward to the gas outlet. The centrifugal forces induced by the main vortex cause the inertial impaction of particles on the walls of the cyclone. The concentrated particle or dust layer forming on the walls swirls downward to a hopper, where dust is removed.

Cyclones may be classified as conventional or high efficiency. The latter have body diameters up to 9 in (22.9 cm) which allow them to achieve greater separating forces. Fractional efficiency ratings of conventional and high-efficiency cyclone collectors are summarized in Table 10.2.

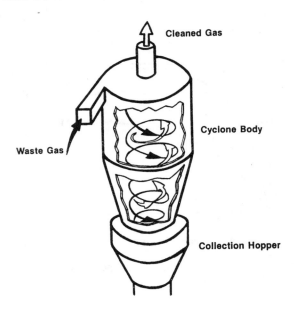

Figure 10.3 Schematic diagram of a common cyclone particle/dust collector.

Table 10.2 Fractional Collection Efficiencies of Cyclone Particle/Dust Control Devices

Particle size range, μm	Efficiency range — wt. % collected	
	Conventional	High-efficiency
<5	<50	50–80
5–20	50–50	80–95
15–40	80–95	95–99
>40	95–99	95–99

From Wark, K. and C.F. Warner. 1981. Harper & Row Publishers, New York. With permission.

The collection efficiency of cyclonic collectors may be determined by particle size or density, the velocity of the incoming gas stream, dust loadings, and equipment design parameters such as the ratio of the body to gas outlet diameter and the cyclone body or cone length. In general, collection efficiency increases with increased particle size and density, dust loading, and collector size.

Multiple Tube Collectors

Large common cyclones (84 in, 2.13 m diameter) may have gas flow capacities of up to 40,000 actual ft³/min (ACFM), 18.9 m³/sec. For higher gas flows or when higher efficiency is required, it is common to use arrays of cyclone tubes. The multiple cyclones, or tubular collectors, vary in diameter from 0.5 to 24 in (1.27 to 61.0 cm), with flow capacities of up to 4500 ACFM (2.12 m³/sec). They are usually mounted in parallel, with as many as 600 tubes in one housing. Dust collected from each tube is discharged to one of several dust hoppers. A multiple cyclone collection system is illustrated in Figure 10.4.

Applications. Cyclones have one of the lowest efficiencies and capital costs of commercially available dust collectors. They are often used to control relatively large particles which are produced in the processing of wood, grain, and mineral

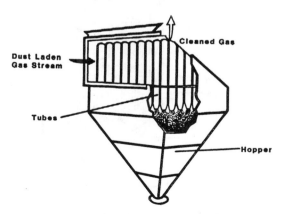

Figure 10.4 A multiple cyclone particle/dust collector.

rock materials. They are also used in series as precleaners for more efficient collectors such as fabric filters, wet scrubbers, and electrostatic precipitators.

Filtration

In filtration, solid particles are physically collected on fibers on the inside of bags (fabric filters) or on filter beds. Filtration is one of the most reliable and efficient particle/dust collecting techniques, exceeding 99.5+% collection efficiencies in some systems.

Fabric Filters. Fabric filters are employed to control particle/dust emissions from a variety of industrial sources including cement kilns, foundries, oil-fired boilers, carbon black plants, and electric and oxygen furnaces for steelmaking operations. They are also used to collect fly ash from electric generating plants burning low-sulfur coal. Fabric filters are commonly used to control particle emissions where dust loading is high, particle sizes are small, and high collection efficiencies are required. Fabric filters are designed to handle gas volumes in the range of 10,000 to 50,000 ACFM (4.72 to 23.58 m³/sec).

The fabric filter collection system consists of multiple tubular collecting bags suspended inside a housing. A single housing, called a baghouse (Figure 10.5), may contain several hundred to several thousand bag filters. Bags are made from a variety of fabrics. Fabric choice depends on temperature, moisture, and chemical composition of the gas, as well as the physical and chemical nature of the particles to be collected. Temperature and chemical ratings of commonly used fabric materials are summarized in Table 10.3. In the typical baghouse unit (Figure 10.5), waste gas enters from the side and flows downward toward a hopper where

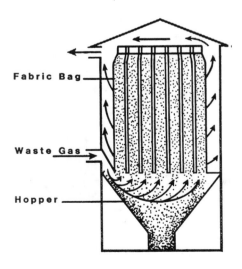

Figure 10.5 Fabric filter (baghouse) particle/dust collector.

Table 10.3 Use Ratings for Baghouse Fabrics

Fabric	Recommended maximum temperature, °C	Chemical resistance	
		Acid	Alkali
Cotton	81	Poor	Fair
Wool	103	Good	Poor
Nylon	103	Good	Poor
Dacron	134	Excellent	Good
Glass	285	Excellent	Excellent

From Tomany, J.P. 1975. American Elsevier Publishing Co., Inc., New York. With permission.

the flow is reversed upward into the array of bags. As the gas changes direction, large particles are removed by inertial separation and collected in the hopper. As gas passes through the tubular bags, dust is collected on the inside of the bag surface, and the filtered gas is discharged to the atmosphere.

Fabric filtration is similar to the process employed in a home vacuum cleaner except that positive, rather than negative, air pressure causes dirty air to pass into the collection bag. Like a home vacuum cleaner, collected dust must also be periodically removed. Bags may be cleaned by mechanical shaking, which is the most commonly used bag-cleaning method. As can be seen in Figure 10.5, bags are connected to a supporting frame at the top. The supporting frames are shaken periodically by mechanical means to dislodge collected dust particles.

Dust collection on bag filters involves more than the collection of particles on single fibers, because open spaces between fibers are often much larger than particles to be collected. Efficient collection of small particles is accomplished by the formation of a particle mat across the fibers of the filter. Collection efficiencies are lowest when filters are first installed and immediately after cleaning. Maximum efficiency occurs when mat buildup has been completed.

Although fabric filters provide high overall collection efficiencies (99+%) and are the most promising technology for controlling submicrometer particles, they do have limitations. These include high capital costs, flammability hazards for some dusts, high space requirements, flue gas temperatures limited to 285°C, and sensitivity to gas moisture content.

Because of the abrasiveness of dust and sensitivity to chemicals, bag wear is a major maintenance concern. Average bag life is about 18 months, although many industries change them once a year.

Electrostatic Precipitation

Electrostatic precipitators remove solid and/or liquid particles from effluent gases by imparting a charge to entrained particles which then are attracted to positively charged collection plates. Specifically, a high DC voltage is applied to a dual-electrode system which produces a corona between the discharge electrode and the collection plates or electrodes (Figure 10.6). The corona produces negatively charged ions. Particles entrained in the gas stream become negatively

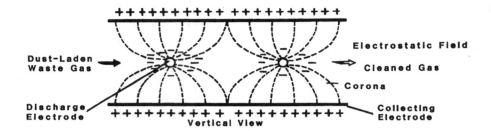

Figure 10.6 Discharge and collection electrodes of an electrostatic precipitator.

charged. These negatively charged particles migrate toward the positively charged plates where they are deposited and held by electrical, mechanical, and molecular forces. Solid particles are usually removed by periodic rapping of collecting electrodes. Collected dust falls into a hopper, from which it is manually removed. A typical plate-type electrostatic precipitator is illustrated in Figure 10.7.

The electrostatic precipitator was first developed in 1907 to collect mist from sulfuric acid plants. Historical applications have been the collection of metal oxides and dusts from a variety of metallurgical operations, including ferrous and nonferrous processes. It is commonly used in steelmaking to collect particles from blast, basic O_2 and open-hearth furnaces, sintering plants, and coke ovens. Presently, the largest single application of electrostatic precipitation is the collection of flyash from coal-fired electric power plants.

Collection efficiency is primarily influenced by particle retention time in the electrical field and particle resistivity. Retention time is determined by gas path length (distance gas travels across the electrical field) and gas velocity. Highest collection efficiencies are achieved in precipitators employing long gas path lengths and relatively low gas velocity.

The resistivity of particles to accepting an electrical charge is one of the most important factors affecting collection efficiency. When dust resistivity is high as in the case of flyash from low-sulfur western coals, the charge is not neutralized

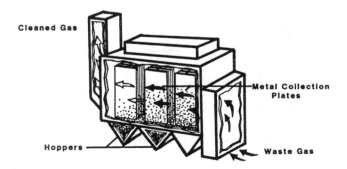

Figure 10.7 Schematic diagram of an electrostatic precipitator.

at the collection plate; as a result, an electrical potential builds up on collected dusts. As the potential increases, the incoming particles do not receive a maximum charge, and the collection efficiency is decreased. If this condition continues, a corona discharge may occur at the collection plate, causing the collector to malfunction. Paradoxically, low-sulfur coals, which are widely used to comply with SO_2 emission standards in many coal-fired utility boilers, produce flyash of high resistivity. High-sulfur coals produce flyash with low resistivity, and thus collection efficiency is high.

Resistivity is a direct function of coal sulfur content with low-sulfur coals producing flyash with the highest resistivity. Control methods employed for high-resistivity particles/dusts require a high level of technical sophistication and, of course, cost. As a result of the resistivity problem, fabric filtration is used as an alternative to the electrostatic precipitator for utility boilers using low-sulfur coal.

On the other end of the spectrum are dusts with low resistivity, such as carbon blacks. In such cases collection efficiency is markedly decreased because carbon particles are highly conductive and lose charges too readily. Because the collected particle loses its negative charge, it may reenter the waste stream, where it becomes negatively charged and is again attracted to the collection plate. The process is repeated until particles reach the precipitator outlet where they are discharged to the atmosphere. Because of this phenomenon, low resistivity also results in excessive dust reentrainment.

Electrostatic precipitation is a widely used particle/dust collection method because it has high collection efficiency for all particle sizes, including submicrometer particles. Collection efficiencies in excess of 99% can be achieved even on very large gas flows. Electrostatic precipitators have low operating and power requirements because energy is only necessary to act on particles collected and not the entire waste stream. Consequently, energy costs are low compared to other particle/dust collection systems.

Although electrostatic precipitation has advantages that make it well suited to many particle/dust collection applications, it also has the disadvantages of high capital costs and space requirements.

Wet Scrubbers

Particle/dust collection by scrubbing systems requires the introduction of a liquid into the effluent stream, with the subsequent transfer of particles to the scrubbing liquid. This transfer is accomplished primarily by conditioning which increases particle size and secondarily by the entrapment of particles on a liquid film. Both of these transfer mechanisms are used in commercially available wet scrubber systems.

Scrubber designs vary considerably from one manufacturer to another. However, all scrubbers have two basic components: (1) a section where liquid-gas contact occurs, and (2) a deentrainment section where wetted particles are removed. In scrubbers particles are brought into contact with liquid to form a particle–liquid agglomerate. The contact process is achieved by forcing a collision

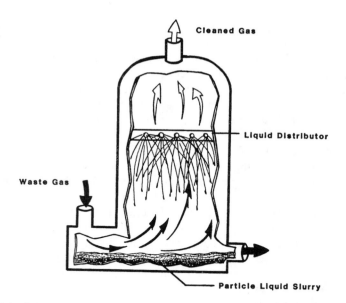

Figure 10.8 Schematic diagram of an open spray tower particle/dust scrubber.

between the liquid and particles. Collisions may be promoted by gravity, impingement, and mechanical impaction. When contact is made, particles increase significantly in mass and size. The resultant particle–gas agglomerates are removed by inertial devices. Commonly used deentrainment mechanisms include impaction on extended baffles and centrifugal separation. For simple dust collection systems, the contact liquid is water.

Open Spray Tower. In the open spray tower (Figure 10.8) a scrubbing liquid is sprayed downward through low-pressure nozzles. The dust-laden waste gas enters from the side and moves downward toward the liquid pool at the bottom of the tower. Large particles are removed by impingement on the liquid surface. The waste air then changes direction and moves countercurrently to the flow of the scrubbing liquid and is discharged at the top. Dust particles are captured by the falling droplets. The liquid-particle agglomerate is collected in the liquid pool at the bottom of the spray tower. Spray towers are limited to low gas flows to prevent liquid drop entrainment in the scrubbed gas stream. This scrubber attains relatively low efficiencies (80 to 90%) and is usually employed as a precleaner to remove particles larger than 5 μm.

Venturi Scrubber. A venturi scrubber is used where high collection efficiency is desired. In the system illustrated in Figure 10.9, effluent gases enter the venturi section where they flow against the wetted cone and throat. As waste gas enters the annular orifice of the venturi, its velocity increases. This high velocity results in a shearing action which atomizes the scrubbing liquid into many fine droplets. The high differential velocity between the gas stream and the liquid promotes

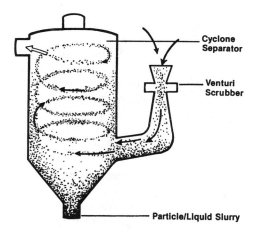

Figure 10.9 Schematic diagram of a venturi scrubber particle/dust collection system.

the impaction of dust particles on the droplets. As waste gas leaves the venturi section, it decelerates, and impaction and agglomeration of particles and liquid continue to occur. Particle-liquid droplets are removed from the waste stream by centrifugal forces in a cyclonic deentrainment section.

Wet scrubbers can provide high collection efficiencies in a variety of applications. Because they are effective in controlling fine particles, they are almost exclusively used for such applications. Venturi scrubbers, for example, are used to collect particles from basic oxygen steelmaking processes.

Although capital costs are relatively low, operating and maintenance costs are often very high. These high operating costs are primarily the result of the energy requirements needed to provide high collection efficiencies for fine particles. Additional operating costs are associated with the disposal of collected dusts and waste liquid. High maintenance requirements often result from the erosion and corrosion of scrubber surfaces produced by abrasion and the action of corrosive acids.

CONTROL OF GAS-PHASE EMISSIONS

Public and regulatory concern for the control of particulate emissions has, for the most part, predated concern over gas-phase pollutants by several decades. Consequently, the performance of control technology for many gas-phase pollutants does not approach that routinely achieved by particle/dust collecting devices. The lower performance of equipment designed to remove gases cannot, however, be singularly attributed to the historical lag in regulatory requirements and development of technology. In dust collection, individual techniques can be utilized for a variety of particle/dust collection problems and sources. The major consideration in selecting such equipment is performance required. The application of control technology for gas-phase pollutants cannot be as generically applied. Gas-cleaning equipment must be developed or designed to control specific gas-phase pollutants or pollutant categories which vary considerably in their chemical and physical properties. The control of gas-phase pollutants is an inherently more complex technological problem. It is also an inherently more expensive one, as both capital and operating costs for control equipment are often high.

Major physical and/or chemical principles used in removing gas-phase contaminants from effluent gas streams include combustion, adsorption, and absorption. These principles and their applications are described below.

Combustion

Combustion involves a series of complex chemical reactions in which O_2 is combined with organic molecules with a resultant release of heat, light, and simple oxidation products such as CO_2 and H_2O. Although combustion releases considerable quantities of heat energy, it requires a flame and/or high temperatures for its initiation.

Flammable Limits. Combustion of gaseous molecules can occur only when the concentration of combustible (flammable) substances in air is within a given range. This range is determined by gas–O_2 levels called the upper and lower explosion or flammability limits. For a mixture of natural gas and air, the lower (LEL) and upper (UEL) explosion limits, respectively, are 5.3 and 15 volume percent. Within these limits, a flame will be self-propagating; beyond these, a flame will be extinguished. Flammability levels vary, as they are affected by the temperature and pressure of the combustion mixture, the presence of contaminants, and the geometry of the combustion chamber.

Incineration. The application of combustion processes for control of effluent gases is commonly referred to as incineration or afterburning. The term afterburning is applicable when the treatment process is located downstream of a primary combustion process.

Incineration is often applied to effluent streams containing combustible gases in which the volume flow rate of gas is large and the concentration of the combustible contaminant is relatively low. Incineration can be used to control (1) malodorants such as mercaptans and H_2S, (2) organic aerosols and visible plumes such as those produced by coffee roaster and enamel bake ovens, (3) combustible gases produced by refineries, and (4) solvent vapors produced by a variety of industrial processes.

Three different types of systems are utilized for pollution control: direct flame, thermal, and catalytic incineration systems. The use of these systems requires a knowledge of the flammability range of the pollutant air/gas mixture. The concentration of combustible contaminants and O_2 relative to the flammability range may determine the type of incineration system used and requirements for supplemental fuel and/or air. Four different types of effluents are amenable to treatment by incineration (Table 10.4).

Type I and II effluents can be treated in afterburners. They both require the use of supplemental fuel. Type II requires supplemental air. Type II effluents are

Table 10.4   Incinerable Effluents

Type I	25% LEL, >15% O_2
Type II	25% LEL, <15% O_2
Type III	Between LEL and UEL
Type IV	>UEL

usually treated in direct flame or flare incinerators. Type IV effluents are above the upper flammability limit and therefore cannot be combusted without significant dilution.

Direct Flame Incinerators. Direct flame systems are employed when effluent gases are between the LEL and UEL. The flare incinerator is a direct flame type system used in petrochemical plants and refineries. It is usually used for combustible waste gases that cannot be conveniently treated any other way. The flare unit is an open-ended combustion system pointed upward (Figure 10.10).

Thermal Incineration. Thermal incineration is normally utilized for the treatment of effluent gases which have a combustible concentration below the LEL. Thermal incineration requires that effluent gases be heated to high temperatures (750 to 800°C). Efficient combustion can be achieved by providing the necessary reaction time, optimal retention or dwell time, and turbulent mixing of reaction gases. A thermal incinerator is illustrated in Figure 10.11.

Efficient thermal incineration requires the provision of supplemental fuel to bring the reaction chamber temperature to 750 to 800°C. This supplemental fuel

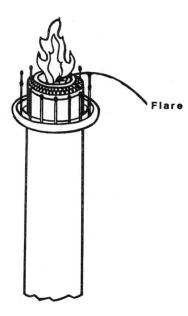

Figure 10.10 Flare incinerator.

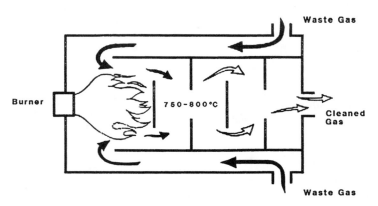

Figure 10.11 Schematic diagram of a thermal incinerator.

requirement can be reduced considerably by passing hot exhaust gas through a heat exchanger which extracts heat energy from exhaust gases and transfers it to the influent gases entering the combustion chamber. The heat exchanger may provide up to 75% of the energy requirements necessary for the incineration of waste gases.

Thermal incineration is used to treat effluent streams from a number of industrial processes including varnish cooking, paint bake ovens, meat smokehouses, fat rendering plants, paint application, and resin manufacturing. The major advantage of thermal incineration in these applications is its dependable performance. However, because of fuel requirements, operating costs are relatively high.

Catalytic Incineration. Effluent gases with combustible concentrations below the LEL can also be treated by incineration that employs catalytic substances. A catalytic incineration system is illustrated in Figure 10.12. The effluent gas stream is initially heated to temperatures of 300 to 475°C in a gas-fired preheat chamber. The preheated gases are discharged to a catalytic bed, where combustion occurs. The most effective and commonly used catalysts are platinum and palladium, although other metals or metal salts may be used. As effluent gases enter the catalytic bed, they are adsorbed on the surface of the catalyst, increasing the concentration of reacting substances. Combustion efficiencies of 95 to 98% can reasonably be expected in industrial applications. The heat released may increase the temperature of treated gases by 100 to 300°C. As was the case for thermal incineration, these hot exhaust gases can be passed through a heat exchanger to provide some of the preheat energy necessary to promote efficient combustion.

Catalytic incineration has several advantages compared to thermal incineration. First, retention times for effluent gases in catalytic units are significantly less. The retention requirement for thermal incineration is 20 to 50 times greater. As a result, catalytic incinerators are much smaller than thermal units. The smaller size is reflected in lower capital equipment costs. The most attractive advantage, however, is the lower supplemental fuel requirement and lower operating costs. Despite these advantages, catalytic incineration does not have the wide acceptance

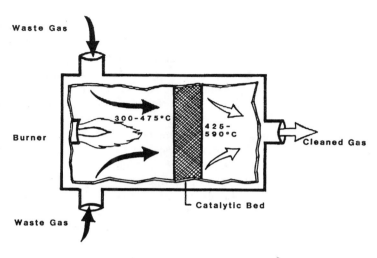

Figure 10.12 Schematic diagram of a catalytic incinerator system.

and use of thermal incinerators. This is primarily because catalytic systems may be "poisoned" by a variety of substances, including lead, phosphorus, zinc, and iron oxide. This poisoning or fouling of catalytic surfaces decreases performance and increases catalyst maintenance.

Catalytic incineration is used to treat effluent gases from a variety of industrial processes, including emissions from varnish cooking, paint and enamel bake ovens, asphalt oxidation, phthalic anhydride manufacturing, printing presses, coke ovens, and formaldehyde manufacture.

Adsorption

A variety of contaminant gases can be removed from waste gas streams by being passed through beds or media containing solid sorbents. Gases coming into contact with solid sorbents may bond to sorbent surfaces by forces similar to van der Waals forces. This phenomenon is described as physical adsorption. If a chemical reaction takes place between the sorbent surface and contaminant gas (adsorbate), the process is called chemical adsorption or chemisorption. Both processes are exothermic as adsorbed chemicals lose their molecular motion and release heat. In physical adsorption, heat release is close to the magnitude of condensation. The process can be reversed by adding an equivalent amount of heat.

Sorbents are available for different commercial and industrial applications. They include activated carbons, simple and complex oxides, and impregnants. Activated carbons such as charcoal are very porous materials with high surface areas available for sorption. Because they contain neutral atoms, there are no electrical gradients. As a result, polar compounds are not attracted in preference to nonpolar ones. This makes them effective in adsorbing organic compounds in humid air streams.

Oxide adsorbents include silicaceous materials, such as silica gel, Fuller's and diatomaceous earth, molecular sieve (aluminum silicates), synthetic zeolites, and activated alumina. They have an inhomogeneous molecular-scale charge distribution, and as a result have a preference for polar compounds. They are useful when the sorption of specific chemical species is desired. Because of their high affinity for H_2O vapor, moisture must be removed first.

Sorbents can be prepared for special uses by impregnating them with reactant substances or catalysts. Impregnants such as bromine are used to collect ethylene on charcoal (by conversion to 1,2-dibromoethane), iodine for mercury, lead acetate for hydrogen sulfide, and sodium silicate for hydrogen fluoride. Catalytic impregnants include metal salts containing chromium, copper, silver, palladium, or platinum. These metals catalyze oxidation reactions.

Sorbents are produced as granular materials of small size to provide limited resistance to gas flow. They may be used in disposable or rechargeable canisters, fixed regenerable beds, traveling beds, and fluid beds. Disposable and rechargeable canisters and regenerable beds have received the greatest industrial and commercial application.

Disposable and rechargeable canisters are commonly used for low-volume exhaust flows and air waste streams with low contaminant concentrations. These are in the form of carbon-impregnated fibrous media or permanent containers. Disposable and rechargeable canisters are commonly used for vapor recovery from storage tanks.

Fixed regenerable beds are typically used when the quantity of gas treated or adsorbate concentration is high. They are particularly attractive for solvent recovery when the cost of regeneration is less than the cost of the sorbent.

Both thin bed [several inches (centimeters) thick] and thick bed (1 to 3 ft; 0.3 to 1.0 m) regenerable systems are available. The latter because of their size and capacity are most appropriate for industrial applications. They are attractive when effluent concentrations of solvent vapors exceed 100 ppmv and when exhaust flows exceed 10,000 CFM (4.7 m³/sec).

A two-unit bed regenerable adsorption system of the type used in vapor recovery is shown in Figure 10.13. In a dual adsorber system, one unit can be used in an on-stream adsorbing mode, the other in regenerating. They can also be used simultaneously when vapor concentrations are relatively low and regeneration can be done at the end of the work shift. Parallel operation doubles the air handling capacity of the system. Thick bed systems can be oriented vertically or horizontally. In either case gas flow is generally downward.

In regenerative systems, desorption is achieved by heating, evacuation, stripping with an inert gas, displacing with a more adsorbable material, or a combination of these. Heating is the most common desorption method. The adsorbate can be subsequently collected and reused. Stripping can be used to concentrate contaminants for incineration. Steam displacement is also commonly used. As the adsorbate is displaced, water vapor is adsorbed. The desorbed vapors are condensed along with excess steam. The bed is then regenerated by removing water by passing hot air over it. Desorption and regeneration by steam can be seen in a dual adsorber system in Figure 10.13.

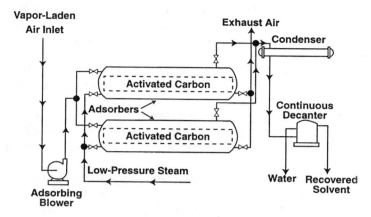

Figure 10.13 Dual adsorption system. (From Theodore, L. and A.J. Buonicore (Eds.). 1994. Springer-Verlag, New York. With permission.)

A variety of factors may affect the efficient operation of adsorption systems. These include adequate sorbent capacity, ample contact time, low resistance to air flow, and uniform airflow distribution. It may also be desirable to pretreat the waste gas stream to remove particles and competing vapors and to cool it. Ample contact time can be achieved by utilizing relatively low flow velocities and extended bed depths.

Though significant control results when adsorption is used for solvent recovery, pollution control has historically been only a secondary concern. Adsorption for solvent recovery is used for economic reasons. Its major application for pollution control occurs when contaminants must be reduced to low concentration ranges (ppbv to ppmv). This is particularly the case with malodors where concentrations in the ambient atmosphere of less than 100 ppbv may not be acceptable. Odor applications include food processing (rendering, fish processing, etc.), chemical and process manufacturing (glue, tanning, pulp, and paper), and other operations (painting and coating, foundries, and animal laboratories).

Adsorption is a useful control method where low emissions of very toxic substances and carcinogens are required. These include such substances as mercury in mercury cell chloralkali plants and radioactive gases in nuclear facilities. Its use is likely to increase as facilities begin to comply with new "toxics" control requirements under the 1990 CAA Amendments. Many of the substances controlled under these regulations are carcinogenic.

Absorption

Pollution control systems which utilize liquid or liquid/solid media to remove selected gases from a waste gas stream are called scrubbers. Gaseous pollutants are absorbed when they come into contact with a medium in which they are soluble or with which they can chemically react. Though water has been the most common scrubbing medium, it is almost always amended with substances such as lime, limestone, ethanolamine, etc., to increase its collection efficiency and absorptive capacity. Collection efficiency is also affected by the following factors: gas solubility, gas and liquid flow rates, contact time, mechanism of contact, and the type of collector.

The fixed-bed packed tower scrubbing system is predominantly used to control gaseous pollutants in industrial applications. It can achieve removal efficiencies of 90 to 95%.

Packed tower scrubbers can be operated as countercurrent, cocurrent, and cross-flow systems. Countercurrent systems are the most widely used. The scrubbing medium is typically introduced at the top and flows downward; waste gases, on the other hand, flow upward. Countercurrent systems are used for difficult to control gases because the packing height is not restricted, thus allowing for higher removal efficiencies. It also has the highest pressure drop and least capacity to handle solids. A schematic cross-section through a countercurrent fixed-bed packed tower can be seen in Figure 10.14.

Cocurrent systems have both the scrubbing media and waste gases flowing in the same direction. They have a finite limit on absorption efficiency but are useful when the scrubbing medium has a high solids content. They can operate at relatively high scrubber liquid flow rates without causing plugging.

Cross-flow designs are used for moderately or highly soluble gases or when chemical reactions are rapid. Waste gas flows horizontally while the scrubbing liquid flows downward. An attractive feature of such systems is that they have lower pressure drops.

As indicated previously, the absorption of a gas in a scrubbing medium depends on its solubility or chemical reactivity. Solubility depends on gas type, concentration (mole fraction or vapor pressure), system pressure, the scrubbing liquid, and temperature.

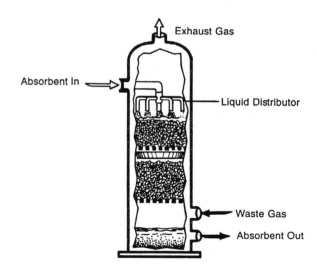

Figure 10.14 Schematic diagram of a fixed-bed packed scrubbing tower.

Figure 10.15 Scrubber packing materials. (From Wark, K. and C.F. Warner. 1981. Harper & Row Publishers, New York. With permission.)

As with adsorbents, a scrubbing medium is limited as to the amount of a gas it can absorb. On saturation it cannot absorb additional gases. In most packed tower scrubbers, the liquid is recirculated from a sump. Such recirculation has to be limited, or the medium will reach saturation. Fresh scrubbing liquid is added, and used liquid is withdrawn continuously to prevent saturation from occurring.

Collection efficiency in scrubbers is enhanced by using a packing medium. Packing media may include steel, ceramic, or thermoplastic materials. These materials come in a variety of shapes (Figure 10.15) which have a high surface area. By providing a larger surface over which the scrubbing liquid is spread, a greater contact area and collection efficiency can be achieved. Thermosetting plastics are the most widely used packing materials. They are less subject to corrosion and pressure drop than metal or ceramic types.

Scrubbers experience a number of operating and maintenance problems. These include corrosion of the scrubber shell and the plugging of packed beds. To reduce corrosion, internal scrubber surfaces need to be constructed with stainless steel, lined steel, exotic metals, polyvinyl chloride, or fiberglass-reinforced plastic. Because of cost, fiberglass or lined steel are commonly used. Plugging may occur as a result of the deposition of undissolved solids, deposition of insoluble dust, precipitation of dissolved salts, or the precipitation of salts produced on absorption. Scaling often occurs when calcium and magnesium react in alkaline scrubbing media with CO_2 to produce carbonates. Such scaling may require flushing with an acidic solution.

Scrubber Wastes. After the medium is removed from the scrubbing system, absorbed materials must be treated and disposed. For acidic gases waste scrubbing liquids may require neutralization. Solids removal may require the use of settling ponds. Solids generally must be landfilled.

CONTROL OF SULFUR OXIDES

As indicated in Chapters 4 to 6, sulfur oxides (SO_x) generated in the combustion of coal, high-sulfur fuel oils, and smelting of metallic ores have enormous impacts on air quality and the environment. As a result, considerable regulatory attention and allocation of economic resources has focused on the reduction of SO_x emissions to the atmosphere.

When emission reduction requirements were first promulgated by state and local agencies to achieve air-quality standards for SO_x, most sources chose to comply by using low-sulfur fuels. Because of the severe economic impact that fuel-use policies (particularly low-sulfur coal) had and would have on the high-sulfur midwestern and eastern coal fields of the U.S., Congress in the 1977 CAA Amendments mandated that USEPA promulgate new source performance standards (NSPS) for large coal-fired power plants that required the use of flue gas desulfurization (FGD) systems.

A variety of SO_x control technologies have been developed and are being used to reduce SO_x emissions to the atmosphere. Some technologies are widely used and others much less so. Some SO_x technologies such as coal beneficiation, coal gasification, and solvent refining, remove sulfur prior to combustion, others such as fluidized bed combustion (FBC) during the combustion process, and FGD systems after combustion has been completed and gases are to be emitted to the atmosphere.

Coal Beneficiation

Coal is a very heterogeneous mineral consisting of both combustible organic and noncombustible inorganic matter. Both fractions may contain considerable sulfur. The inorganic fraction which contains pyritic sulfur can be removed by coal washing. In coal washing, inorganic minerals are separated from the organic portion by suspending coal fragments in water where differences in specific gravity result in the sinking of heavy, inorganic mineral matter (which contains sulfur) and the floating of the lighter, organic coal. Coal cleaning can on average remove 50% of pyritic sulfur and 20 to 30% of total sulfur.

Coal washing is widely used by coal producers to reduce the ash content of coal to both enhance its combustibility and meet ash specifications of coal users. Sulfur reduction is a side benefit associated with the enhanced coal quality produced by washing. Sulfur reduction by coal beneficiation or cleaning is usually not adequate to meet emission reduction requirements. However, it may prove useful when combined with other control technologies such as FGD.

Solvent Refining

Solvent refining or chemical cleaning is a technology which has been undergoing development for several decades. It is based on the principle that sulfur and other impurities can be removed by solvent extraction. Such a technology has the potential for obviating the need for FGD systems. However, it is presently not economically viable.

Coal Gasification and Liquefaction

Coal can be precleaned on conversion to gases and liquids. In converting coal to synthetic fuels (synfuels), sulfur and ash can be removed. These synfuel processes are, unfortunately, significantly less energy efficient that direct coal combustion.

Considerable development efforts were focused on synfuels in the U.S. in the late 1970s and early 1980s when energy concerns were high. These concerns have eased considerably, and interest in synfuel development projects has waned. Coal gasification received major development attention in South Africa which was under international economic sanctions because of its former apartheid policies.

Fluidized Bed Combustion

In conventional combustion systems, fuel is burned in a fixed bed or by suspension firing. In FBC, air at high velocities is passed upward from a distribution grid (Figure 10.16). The solids containing fuel, inert granular bed materials, and dolomite/limestone (sulfur sorbents) are held in turbulent suspension by the upward flow of air. This turbulent suspension appears diffuse, behaving like a liquid in that bubbles of gas rise through the bed as if it were boiling.

Combustion within the bed is intense. Coal burns so rapidly that the bed at any given time is composed almost entirely of the inert materials and sorbents present before combustion was initiated. Because of the intense combustion, high

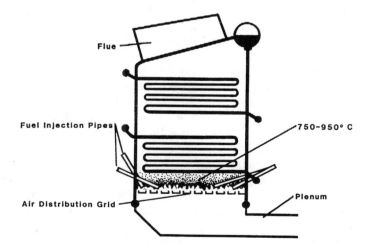

Figure 10.16 Fluidized bed combustion system. (From PAT Report. 1974. *Environ. Sci. Technol.* 8:978–980. With permission.)

volumetric heat release is achieved. High heat transfer rates are also obtained by the horizontal boiler tubes immersed in the fluidized bed. As a result, FBC can operate at lower temperatures. Optimum FBC temperatures range from 750 to 950°C, whereas conventional utility boilers operate at temperatures as high as 1400 to 1500°C. These lower operating temperatures result in lower NO_x emissions.

As coal burns in the FBC system, it releases SO_2 which reacts with limestone in the suspension. Collection efficiencies up to 90% or better can be achieved. New limestone is periodically added to maintain sorbent reactivity.

Fluidized bed combustion has become an established option for large industrial boilers at scales equivalent to 10 to 25 MW. More than 300 atmospheric fluidized bed systems are in use in Europe and the U.S. to supply heat to industrial processes, municipalities, etc. Fluidized bed technologies in the 75 to 350 MW scale necessary for the generation of electricity have been developed and are now in operation in several states in the U.S.

The atmospheric FBC system operates roughly the same as a conventional boiler in driving a steam turbine, except that it produces significantly lower emissions of SO_x and NO_x. Pressurized FBC systems operate at pressures 6 to 16 times higher than atmospheric pressure. Because of the increased energy in the high-pressure gases entering the boiler, such systems can drive both a gas turbine and a steam turbine — an arrangement known as a combined cycle. Power generation efficiencies well above 40% can be obtained (compared with 30 to 35% for conventional coal-fired technology).

Flue Gas Desulfurization

Flue gas desulfurization systems are widely used to remove SO_x from a variety of large and small sources including utility boilers, industrial boilers, and primary and secondary metals smelting. Multiple hundreds of FGD systems have been installed and operated in the U.S. in the past two decades. FGD systems are required for all new or modified large coal-fired utility boilers given permits to construct after June 1979 under USEPA-revised NSPS. These NSPS required 90% reduction of SO_x for utility boilers using coal with sulfur contents greater than 1.5% and 70% reduction for coals less than 1.5%.

Flue gas desulfurization systems can be described in terms of the nature of the sorbent used. In wet systems flue gases are scrubbed in a liquid and/or liquid/solid slurry. Solid sorbents are used in dry systems. Flue gas desulfurization systems can also be described in terms of the use/reuse of sorbents. In regenerable systems, generally expensive sorbents are recovered by stripping SO_x from the scrubbing medium. Useful products such as elemental sulfur, sulfuric acid, liquefied SO_x, and gypsum may be recovered depending on the FGD system used and regeneration technology employed. In nonregenerable (throwaway) systems, the scrubbing medium is relatively inexpensive and so recovery is not economically desirable. Throwaway systems do, however, have significant waste disposal concerns and costs.

Wet Systems. There are six major wet FGD systems used in the U.S. and other parts of the developed world. These include the Wellman-Lord system and lime, limestone, sodium carbonate, dual alkali, and magnesium oxide scrubbing systems. Lime and limestone scrubbing systems are widely used to meet SO_x reduction requirements in coal-fired utility boilers. Both can remove 90+% of SO_x in flue gases, and both are nonregenerable.

In lime scrubbing, the scrubbing medium consists of a slurry of 15 to 20% solids. It is produced by adding calcium oxide (CaO) to water which on reaction produces calcium hydroxide [$Ca(OH)_2$] (Equation 10.3). Reactions with SO_2 to produce calcium sulfite and calcium sulfate in the scrubbing liquid are summarized in Equations 10.4 to 10.6.

$$CaO + H_2O \rightarrow Ca(OH)_2 \tag{10.3}$$

$$SO_2 + H_2O \leftrightarrow H_2SO_3 \tag{10.4}$$

$$H_2SO_3 + Ca(OH)_2 \rightarrow CaSO_3 \cdot 2H_2O \tag{10.5}$$

$$CaSO_3 \cdot 2H_2O + 1/2O_2 \rightarrow CaSO_4 \cdot 2H_2O \tag{10.6}$$

Limestone scrubbing is more widely used in the coal-fired utility industry. The limestone scrubbing medium consists of a slurry of pulverized limestone. Removal of SO_2 from flue gases occurs as a result of the following reactions:

$$CaCO_{3(S)} + H_2O + 2SO_2 \rightarrow Ca^{+2} + 2HSO_3^- + CO_2 \tag{10.7}$$

$$CaCO_{3(S)} + 2HSO_3^- + Ca^{+2} \rightarrow 2CaSO_3 + CO_2 + H_2O \tag{10.8}$$

As in lime scrubbing, an excess of O_2 in flue gases will result in the conversion of some of the $CaSO_3$ to $CaSO_4$ (Equation 10.6).

In lime and limestone FGD systems, flue gases pass into and through the scrubbing unit as seen in Figure 10.17. Gas flow through the scrubbing tower is countercurrent to the flow of the scrubbing medium. After contact with flue gases, the scrubbing medium contains calcium sulfite, calcium sulfate, and unreacted sorbents. Because it has considerable sorbent ability remaining, it will be recirculated from an effluent holding tank where calcium sulfite and calcium sulfate are actively precipitated. The scrubber effluent not recycled is pumped to a thickener or concentrating tank from which it will be pumped to a holding pond to be stabilized and ultimately disposed. Though scrubbing liquids are reused, they must also be continually replenished with unexposed scrubbing media.

Both lime and limestone scrubbing are similar in process flow and equipment used. Lime is, however, much more reactive than limestone, with a better utilization of the scrubbing medium. However, the cost of lime is high compared to

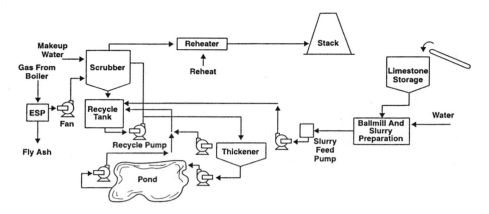

Figure 10.17 Flue gas desulfurization system. [From Theodore, L. and A.J. Buonicore (Eds.). 1994. Springer-Verlag, New York. With permission.]

that of limestone. Lime scrubbers can routinely achieve 95% removal of SO_2; limestone, 90%.

Though lime and limestone scrubbing systems can achieve high FGD efficiencies, their operation is technically demanding. Several factors can reduce their performance and reliability. One of these is scaling and plugging of scrubber equipment by insoluble compounds. Proper control of pH is required to reduce the occurrence of scaling problems.

High desulfurization efficiencies require high scrubbing medium-to-gas ratios. However, high ratios result in higher operating costs because of the pressure drop in the scrubber and higher pumping energy needs. Energy consumption by FGD systems is 3 to 6% of the power generated by the facility. In addition to these high operating costs, FGD systems used on coal-fired utility boilers also have high capital costs. Capital costs on an 1000 MW unit typically exceed $100 million (U.S.).

Dry Scrubbing Systems. Dry scrubbing is an FGD technology which involves the injection of alkaline sorbent materials such as lime, limestone, sodium carbonate, etc., directly into boilers or flue gases. Sorption occurs on the surface of solid particles which are captured on dust collectors. Dry systems are not regenerable.

Dry systems have several important advantages over wet systems, including (1) freedom from scaling and plugging, (2) no sludge handling requirements, (3) less corrosion and maintenance, and (4) lower energy and water requirements. The major disadvantage is that they cannot achieve the 90% efficiencies that have been required for high-sulfur coal; they can, however, achieve the 70% reductions required for low-sulfur coals under the revised 1979 NSPS. Therefore, they are particularly advantageous in the low-sulfur coal-rich western states, where the availability of water is limited.

CONTROL OF NITROGEN OXIDES

Because of concerns associated with the role of NO_x in photochemical oxidant formation, as well as acidic deposition, stationary sources, particularly large coal-fired power plants, are increasingly being required to reduce NO_x emissions.

A variety of control techniques for reducing emissions from utility and industrial boilers have been developed. One of the most widely used is the low NO_x burner, now a standard part of new utility boiler designs.

Low NO_x burners inhibit the formation of NO_x by controlling the mixing of air and fuel. In these burners which are located in pulverized coal boiler walls, the amount of excess air used to insure good combustion can be reduced. Reducing excess air from 20 to 14%, for example, has been shown to reduce NO_x production by 20%. Reducing excess air reduces the amount of O_2 available to combine with N_2. Additional reductions can also be achieved by staged combustion (fuel is combusted in a primary zone which is fuel rich, followed by secondary and following zones which are fuel lean). Combustion staging methods reduce NO_x formation by either reducing available O_2 or providing excess O_2 to cool the combustion process.

A variety of flue gas treatment techniques are available for NO_x emission control. One of these, selective catalytic reduction, is widely used in Japan. Catalysts such as mixtures of titanium and vanadium used in combination with reducing gases such as NH_3, H_2, CO, and H_2S can convert NO_x to N_2. The following reactions involving NH_3 are illustrative.

$$4NO + 4NH_3 + O_2 \rightarrow 4N_2 + 6H_2O \qquad (10.9)$$

$$2NO_2 + 4NH_3 + O_2 \rightarrow 3N_2 + 6H_2O \qquad (10.10)$$

READINGS

1. Boubel, R.W., D.L. Fox, D.B. Turner, and A.C. Stern. 1994. *Fundamentals of Air Pollution*. 3rd ed. Academic Press, New York.
2. Bubenick, D.V. 1984. "Control of Fugitive Emissions." 745–764. In: S. Calvert and H. Englund (Eds.) *Handbook of Air Pollution Technology*. Wiley, New York.
3. Buonicore, A.J. 1980. "Air Pollution Control." *Chem. Eng.* June 30:81–101.
4. Cooper, C.D. and F.C. Alley. 1986. *Air Pollution Control: A Design Approach*. PWS Engineering, Boston.
5. Croker, B.B. and K.B. Schnelle, Jr. 1984. "Control of Gases by Absorption, Adsorption and Condensation." 135–201. In: S. Calvert and H. Englund (Eds.). *Handbook of Air Pollution Technology*. Wiley, New York.
6. Danielson, A.F. 1973. *Air Pollution Engineering Manual*. 2nd ed. USEPA Publication No. AP-40.
7. First, M.W. 1977. "Control of Systems, Processes and Operations." 4–38. In: A.C. Stern (Ed.). *Air Pollution. Vol. IV. Engineering Control of Air Pollution*. 3rd ed. Academic Press, New York.

8. Freeman, H., T. Harten, J. Springer, P. Randall, M.A. Curran, and K. Stone. 1992. "Industrial Pollution Prevention: A Critical Review." *JAWMA* 42:618.

9. Gottschlich, C.F. 1977. "Combustion." 365–376. In: A.C. Stern (Ed.). *Air Pollution. Vol. IV. Engineering Control of Air Pollution.* 3rd ed. Academic Press, New York.

10. Henzel, D.S., B.A. Laseke, E.O. Smith, and D.O. Swenson. 1982. *Handbook for Flue Gas Desulfurization — Scrubbing with Limestone.* Noyes Data Corporation, Park Ridge, NJ.

11. Hesketh, H.E. 1991. *Air Pollution Control, Traditional and Hazardous Pollutants.* Technomic Publishing Co., Inc., Lancaster, PA.

12. Higgins, T.E. (Ed.). 1995. *Pollution Prevention Handbook.* CRC/Lewis Publishers, Boca Raton, FL.

13. Hudson, J.L. and G.T. Rochelle (Eds.). 1982. *Flue Gas Desulfurization.* ACS Symposium series No. 188. American Chemical Society, Washington, DC.

14. Kaplan, N. and M.A. Maxwell. 1977. "Removal of SO_2 from Industrial Waste Gases." *Chemical Engineering* (deskbook issue). Oct. 17:127–135.

15. Kilgroe, J.D. 1984. "Coal Cleaning." 419–435. In: S. Calvert and H. Englund (Eds.). *Handbook of Air Pollution Control Technology.* Wiley, New York.

16. PAT Report. 1974. "Fluidized Bed Steam Generators for Utilities." *Environ. Sci. Technol.* 8:978–980.

17. Special Report. 1975. "The Building of Tall Stacks and Not So Tall Stacks." *Environ. Sci. Technol.* 9:522–527.

18. Spencer, D.F., M.J. Gluckman, and S.B. Alpert. 1982. "Coal Gasification for Electric Power Generation." *Science* 215:1571–1576.

19. Theodore, L. and A.J. Buonicore (Eds.). 1994. *Air Pollution Control Equipment: Selection, Design, Operation, and Maintenance.* Springer-Verlag, New York.

20. Tomany, J.P. 1975. *Air Pollution: The Emissions, the Regulations and the Control.* American Elsevier Publishing Co., Inc., New York.

21. Turk, A. 1977. "Adsorption." 329–363. In: A.C. Stern (Ed.). *Air Pollution. Vol. IV. Engineering Control of Air Pollution.* Academic Press, New York.

22. Turner, L.H. and J.D. McKenna. 1984. "Control of Particles by Filters." 249–283. In: S. Calvert and H. Englund (Eds.). *Handbook of Air Pollution Technology.* Wiley, New York.

23. U.S. DOE. 1990. *Clean Coal Technology, The New Coal Era.* DOE/FE-0193P.

24. USEPA. 1979. *Sulfur Emissions: Control Technology and Waste Management.* EPA/600/9-79-019.

25. USEPA. 1992. *Opportunities for Pollution Prevention Research to Support the 33/50 Program.* EPA/600/R-92/175.

26. USEPA. 1993. *Alternative Control Techniques Document — NO_x Emissions from Process Heaters (Revised).* EPA/453/R-93/034.

27. Wark, K. and C.F. Warner. 1981. *Air Pollution: Its Origin and Control.* 2nd ed. Harper & Row Publishers, New York.

28. Weinstein, N.J. 1984. "Unconventional Fuels." 435–484. In: S. Calvert and H. Englund (Eds.). *Handbook of Air Pollution Control Technology.* Wiley, New York.

29. White, H.J. 1984. "Control of Particulates by Electrostatic Precipitation." 283–317. In: S. Calvert and H. Englund (Eds.). *Handbook of Air Pollution Control Technology.* Wiley, New York.

QUESTIONS

1. Discuss the application of pollution prevention principles to air pollution control.

2. In fabric filtration, particle collection is less efficient after the filter material has been cleaned. Why?

3. What are the advantages and disadvantages of using tall stacks to "control" air pollution?

4. When would it be most desirable to use baghouses as opposed to electrostatic precipitators to collect flyash from coal-fired utility boiler flue gas?

5. Describe in detail how an electrostatic precipitator (EP) collects particles entrained in a waste gas stream. What variables affect the performance of the EP?

6. Describe approaches used to control fugitive emissions.

7. Describe operational and maintenance problems associated with scrubbing systems.

8. Describe the relationship between collection efficiency and particle size.

9. What factors determine collection efficiency in an activated carbon adsorption system?

10. Under what circumstances would flare incineration be employed? Thermal incineration?

11. A source exceeds opacity standards because of the large number of small particles (less than 1 μm) which comprise its plume. To reduce emissions sufficiently to meet the standard, what dust collection system or systems would be the most appropriate? Why?

12. What is coal beneficiation? What are its limitations relative to meeting SO_2 emission standards for coal-fired utility boilers?

13. What collection/control principles serve as the basis for the following gas-cleaning equipment: (1) scrubber, (2) baghouse, (3) cyclone, (4) deep-bed activated carbon column, and (5) thermal incinerator?

14. Why are packing materials used in flue gas desulfurization systems?

15. What are the relative advantages and disadvantages of using thermal incineration or adsorption for the control of HC vapors in a given control application?

16. Describe how fluidized bed combustion works and how SO_x and NO_x emissions are reduced compared to conventional coal-fired boiler units.

11

Over the past three decades, the U.S. has invested hundreds of billions of dollars to enhance and protect air quality in our cities, metropolitan areas, and country-sides. Most pollution control efforts have focused on protecting public health. These efforts to protect human health from ambient pollutant exposures have for the most part failed to consider other airborne exposures to toxic pollutants: most notable are exposures occurring indoors, in our homes, and in our nonindustrial work environments.

Numerous field investigations and scientific studies have shown that air inside public access buildings (office, commercial, and institutional) and residences is contaminated by a variety of gas- and particulate-phase contaminants which may be present in sufficient concentrations to adversely affect the health of those exposed. In addition, indoor exposures may vary from 8 hours/day in office, commercial, and institutional buildings to upwards of 16 to 24 hours/day in residences. In contrast, the average time that an individual spends outdoors is approximately 2 hours/day. Thus longer exposure periods are likely to occur in building environments.

There have been numerous reported cases of illness complaints in offices and other public access buildings, as well as homeowner complaints of building-related illness. Major long-term health concerns have also been expressed for cancer risks associated with exposures to common indoor contaminants such as asbestos, radon, and environmental tobacco smoke.

INDOOR/OUTDOOR RELATIONSHIPS

Historically, we have thought that the indoor environment protected us from the polluted air of the ambient environment. Indeed, in episode control plans, environmental protection authorities advise citizens to remain indoors during "warning" and "emergency" stages. This is particularly true for individuals who

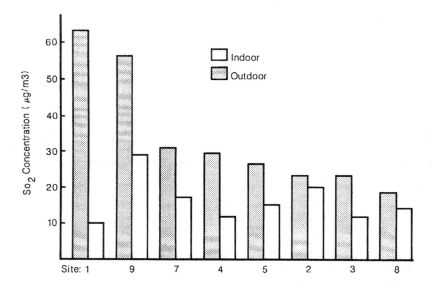

Figure 11.1 Concurrent measurements of indoor and outdoor concentrations of SO_2 at eight sites. (From Ferris, B.G., Jr., et al. 1978. *Proceedings First International Indoor Climate Symposium.* Danish Building Research Institute, Copenhagen.)

may be at special risk, for example, those who have existing respiratory or cardiovascular disease.

To a limited extent, the indoor environment does protect us from polluted ambient air. For reactive gases such as sulfur dioxide (SO_2) and ozone (O_3), indoor concentrations may only be a fraction of those outdoors. A relationship between indoor and outdoor levels of SO_2 is illustrated in Figure 11.1. In general, where ambient levels of SO_2 are moderate to high, indoor/outdoor (I/O) ratios range from 0.3 to 0.5. However, where ambient SO_2 levels are low (I/O ratios of 0.7 to 0.9), the depletion of SO_2 by the building envelope and indoor environment is relatively small. Indoor/outdoor ratios for O_3 appear to range from 0.1 to 0.3, but may be as high as 0.7. Although indoor levels of O_3 are significantly less than those reported for the ambient environment, the latter is the primary source of O_3 indoors. Indoor sources may include electronic dust cleaners, ion generators and, in office environments, photocopy machines.

Nitrogen dioxide (NO_2) is also a relatively reactive gas, and in the absence of indoor sources, I/O ratios are usually less than 1.0. For example, in residences with electric stoves, an I/O ratio of 0.38 has been reported. However, residences with gas cooking appliances or unvented gas or kerosene heaters commonly have NO_2 levels that exceed those outdoors. With indoor sources NO_2 exposures may exceed ambient exposures by a factor of two or more.

Carbon monoxide (CO) is a relatively unreactive gas, so that in the absence of indoor sources the ratio of indoor/outdoor CO concentrations typically approaches unity. Consequently, during an air pollution episode, the indoor envi-

Table 11.1 Indoor/Outdoor Ratios of
Respirable Particles Related to
Tobacco Smoking

Study 1	I/O ratio
Residences with smokers	4.4
Residences without smokers	1.4
Office buildings	1.1
Study 2	**I/O ratio**
Residences with 1 smoker	1.7
Residences with 2+ smokers	3.3
Residences without smokers	1.2

From Yocum, J.E. 1982. *JAPCA* 32:500–520.
With permission.

ronment will not be associated with reduced CO exposures. There are exceptions; for example, the use of gas cooking appliances or unvented gas or kerosene space heaters can produce indoor CO levels significantly above those in the ambient environment; malfunctioning gas furnaces may result in life-threatening household CO contamination.

A variety of nonmethane HCs (NMHCs) are known to contaminate both ambient and indoor air. Because of the presence of a variety of solvent-containing materials, indoor NMHC levels are significantly higher (I/O ratios 1.5 to 1.9) than ambient levels.

A broad range of I/O ratios (e.g., 0.3 to 3.5) have been reported for particulate matter (PM). Particle levels indoors are highly dependent on indoor activities, such as tobacco smoking and cooking. They are also affected by the presence or absence of gas cooking appliances and unvented space heaters, air filtration systems, and ventilation air. The single most important contributor to elevated respirable particles is tobacco smoking. The effects of tobacco smoking on I/O ratios for respirable particles are summarized for two studies in Table 11.1. Note that respirable particle concentrations are often higher indoors even in the absence of tobacco smoking.

Although radon is released into the ambient environment from soil and rock sources, outdoor concentrations are very low. Because of certain physical phenomena, radon concentrations in indoor environments are often several orders of magnitude higher than outdoors.

Formaldehyde (HCHO) is a ubiquitous air contaminant found both in ambient and indoor environments. Elevated ambient concentrations are due to direct emissions from motor vehicle exhaust and to secondary chemical reactions in the atmosphere. During smog episodes, peak HCHO levels may reach 0.1 ppmv. However, urban levels rarely exceed 0.05 ppmv, with concentrations in the range of less than 0.01 ppmv more common. Indoor HCHO levels in residences, depending on sources present and environmental conditions at the time of testing, may range from 0.02 to 0.40 ppmv or more. As a result, the most significant population exposure to HCHO occurs indoors.

Indoor and outdoor environments differ significantly in both types and levels of contaminants common to each. Contaminants with sources predominantly outdoors include SO_2, O_3, pollens, and a variety of organic chemicals. Contaminants generated primarily from indoor sources include HCHO, radon, asbestos, ammonia (NH_3), acrolein, a variety of organic chemicals, and viable organisms or matter of biological origin. Contaminants common to both indoor and outdoor environments include CO, NO_x, CO_2, PM, organic chemicals, and mold spores.

Differences in indoor/outdoor levels of pollutants result from the chemical reactivity of some ambient pollutants and their ability or inability to pass through the building envelope, the presence of strong sources indoors, and building construction and maintenance practices that reduce air exchange between indoor and outdoor environments.

PERSONAL AIR POLLUTION EXPOSURE

The primary air quality standards for CO, NO_x, SO_2, O_3, PM, and lead reflect national goals for permissible ambient air pollution exposure levels that would in theory safeguard public health. With the exception of short averaging times associated with 1-hour standards for CO and O_3, long exposure durations are implicit in most primary standards. However, average population exposure is relatively short, on the order of 2 hour/day (Table 11.2). Significantly longer exposures occur indoors, particularly in our homes. Significant exposure can also occur in transit to and from one's employment or other destination (Figure 11.2).

Personal exposure to criteria pollutants regulated by ambient air-quality standards may be considerably different from the continuous exposures implicit in such standards. Total outdoor, indoor, and personal exposures to respirable particles are summarized in Figure 11.3 for 46 subjects studied in Topeka, KS. It is evident that personal respirable particle exposures were significantly greater than either indoor or outdoor exposures.

The results summarized in Figure 11.3 suggest that reliance on ambient air-quality standards to protect public health from pollution exposures may be unrealistic. They also bring into question whether it is necessary to continue to require additional expensive controls on sources of ambient air pollutants when the incremental health protection afforded by such controls may be small.

Table 11.2   Potential Exposure to Air Pollutants Associated with Different Environments

	Employed men	Homemakers	Overall population
Indoors	21.7	22.8	20.8
Transit	0.7	0.4	1.4
Outdoors	1.6	0.8	1.8

From Moschandreas, D.J. 1981. *Bull. N.Y. Acad. Med.* 57:845–860. With permission.

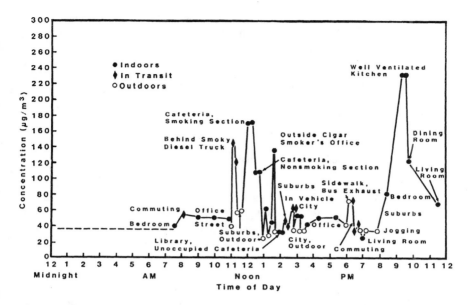

Figure 11.2 Personal exposure to respirable particles in different environments. (From Budiansky, S. 1980. *Environ. Sci. Technol.* 14:1023–1027. With permission.)

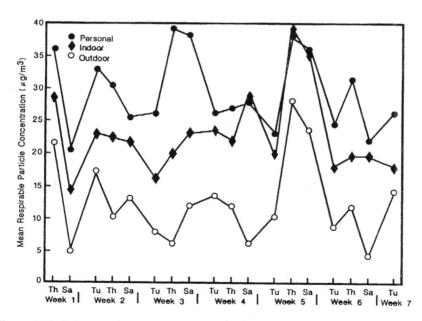

Figure 11.3 Indoor, outdoor, and personal exposure of Topeka, KS residents to respirable particles. (From Budiansky, S. 1980. *Environ. Sci. Technol.* 14:1023–1027. With permission.)

It is more logical to assess health risk from air pollutants on a basis of total air pollution exposure to specific pollutants and to estimate exposures on a population basis. Such assessment dramatically indicates how important the indoor environment is as a source of pollutant exposure. In many instances, the health significance of indoor exposures may exceed that of outdoor exposures. For pollutants primarily generated indoors, the health risk is almost entirely due to breathing contaminated indoor air.

INDOOR AIR-QUALITY PROBLEMS

Though many pollutants are common to most nonindustrial building environments, indoor air-quality problems differ somewhat in large nonresidential office, commercial, and institutional public access buildings as compared to the environments of our homes. As a result, air-quality problems are described below within the context of these buildings types.

Public Access Buildings

Office, commercial, and institutional buildings open to the public are described as public access buildings. They vary in size, the nature of their construction, how they are heated, cooled, and ventilated, and the nature of activities which occur within them. In many cases they are relatively large buildings serving hundreds to thousands of occupants. In most cases they are mechanically ventilated. Because such buildings often have a large number of occupants, a diversity of activities which occur within them, and mechanically controlled ventilation systems, they experience unique indoor air-quality (IAQ) problems and concerns. These are often expressed in the context of health and/or comfort complaints.

Building-Related Illness. When causal factors for health complaints have been identified, the air-quality problem is described as building-related illness (BRI) or specific building-related illness (SBRI). Building-related illness is characterized by what are often a unique set of symptoms accompanied by clinical signs, laboratory test results, and specific pollutants. Included in BRI are nosocomial infections, the hypersensitivity diseases (e.g., common allergy, asthma, hypersensitivity pneumonitis, humidifier fever), fiberglass dermatitis, and toxic effects associated with contaminants such as carbon monoxide, ammonia, and formaldehyde.

Sick Building Syndrome. "Sick building syndrome" (SBS) is used to describe a spectrum of nonspecific symptoms which on investigation appear to have no identifiable cause. These commonly include mucous membrane symptoms (irritation of the eyes, nose, throat, sinuses), general symptoms (headache, fatigue, lassitude), skin irritation (dryness, rashes), and to a lesser extent respiratory symptoms (cough, shortness of breath).

Sick Buildings/Problem Buildings. Buildings subject to health, comfort, and/or odor complaints can be best described as being problem buildings. They comprise a building population in which some form of dissatisfaction with air quality has been expressed by occupants sufficient to require an IAQ investigation. In many cases the cause of complaints can be readily identified and remediated. In many other cases, there appears to be no single causal factor, and mitigation efforts prove to be ineffective. When they include a relatively largely percentage of occupants with complaints, they are often described as "sick buildings."

Field Investigations and Systematic Building Studies. An apparent relationship between illness complaints and building environments has been reported in a large number of problem building investigations conducted by a variety of governmental, occupational, and public health agencies, and private consultants responding to occupant/building management requests. In the U.S., hundreds of such investigations have been conducted by health hazard evaluation teams of the National Institute of Occupational Safety and Health (NIOSH). Symptoms reported have included eye, nose, throat and skin irritation; headache and fatigue; respiratory symptoms such as sinus congestion, sneeze, cough, and shortness of breath; and less frequently, nausea and dizziness. Reported potential causal/contributory factors to complaints for 529 investigations conducted between 1971 and 1988 are summarized in Table 11.3.

Field investigations have an important role in initially identifying and defining problem building phenomena. The inherent bias involved in conducting studies of buildings subject to occupant health and comfort complaints and the relatively unsystematic approaches used in many investigations makes them of limited scientific usefulness. As a result, investigators in several countries have attempted to conduct systematic studies of a variety of buildings (mostly office-type). These have included a study of 47 office buildings in the U.K., 14 town halls in Denmark, 60 office buildings in the Netherlands, the large office building

Table 11.3   Problem Types Identified in NIOSH Building Investigations

Problem type	Buildings investigated	%
Contamination from indoor sources	80	15
Contamination from outdoor sources	53	10
Building fabric as contaminant source	21	4
Microbial contamination	27	5
Inadequate ventilation	280	53
Unknown	68	13
Total	529	100

From Seitz, T.A. 1989. *Proceedings Indoor Air Quality International Symposium: The Practitioner's Approach to Indoor Air Quality Investigations.* American Industrial Hygiene Association, Akron, OH. With permission.

Table 11.4 Symptom Prevalence Among Occupants of 14 Danish Town Hall Buildings

	Prevalence rates (%)	
	Males (n = 1093–1115)	Females (n = 2280–2345)
Symptoms		
Eye irritation	8.0	15.1
Nasal irritation	12.0	20.0
Blocked, runny nose	4.7	8.3
Throat irritation	10.9	17.9
Sore throat	1.9	2.5
Dry skin	3.6	7.5
Rash	1.2	1.6
Headache	13.0	22.9
Fatigue	20.9	30.8
Malaise	4.9	9.2
Irritability	5.4	6.3
Lack of concentration	3.7	4.7
Symptom groups (males and females)		
Mucous membrane irritation	20.3	
Skin reactions	4.2	
General symptoms	26.1	
Irritability	7.9	

From Skov, P. et al. 1987. "The Sick Building in the Office Environment: The Danish Town Hall Study." *Environ. Int.* 13:349–399. With permission.

study in northern Sweden, and studies of 12 apparently healthy buildings in California.

Systematic building studies have served to characterize symptom prevalence rates in mostly nonproblem buildings and in some cases sick buildings. Symptom prevalence rates for both males and females in 14 noncomplaint Danish town halls and 11 Swedish sick buildings are summarized in Tables 11.4 and 11.5.

Systematic research studies have shown significant prevalence rates of mucous membrane, general (headache, fatigue, lassitude), and skin symptoms in complaint and noncomplaint office buildings. They have shown that building/work-related health complaints occur in all buildings surveyed with complaint rates varying from building to building. As indicated in Tables 11.4 and 11.5, complaint rates are generally higher in sick/problem buildings compared to those in which problems have not been previously reported.

Risk Factors. Investigators conducting systematic building studies have attempted to identify potential risk/contributory factors to building/work-related health complaints. These have included people-related, work/building environment, and contaminant-related concerns.

A variety of people-related risk factors have been shown to be related to symptom prevalence. These include gender, allergy, a variety of psychosocial factors, and tobacco smoking. All studies have shown that symptom-reporting rates are several times higher among females compared to males. Most studies report that a predisposition to allergy or atopy is a significant risk factor for SBS-

Table 11.5 Symptom Prevalence Among Occupants
of 11 "Sick" Swedish Office Buildings

Symptom	Total mean prevalence (%)	Range
Eye irritation	36	13–67
Swollen eyelids	13	0–32
Nasal catarrh	21	7–46
Nasal congestion	33	12–54
Throat dryness	38	13–64
Sore throat	18	8–36
Irritative cough	15	6–27
Headache	36	19–60
Abnormal tiredness	49	19–92
Sensation of getting a cold	42	23–77
Nausea	8	0–23
Facial itch	12	0–31
Facial rash	14	0–38
Itching on hands	12	5–31
Rashes on hands	8	0–23
Eczema	15	5–26

From Norback, D. et al. 1990. *Scand. J. Work Environ. Health* 16:121–128. With permission.

type symptoms. Symptom prevalence is also significantly associated with psychosocial variables such as work stress and job satisfaction. Symptom prevalence increases with increased workplace stress and decreases with increasing job satisfaction. Nonsmoker exposure to tobacco smoke has reportedly been observed to increase symptom prevalence rates.

A variety of environmental factors (temperature, humidity, air movement, etc.) have been investigated as potential risk factors for SBS symptoms. Significant relationships have been reported for elevated temperature (21 to 25°C) and low relative humidity (<30% RH) and increased SBS prevalence rates. The adequacy of ventilation or the relationship between outdoor ventilation rates and SBS symptoms and outdoor ventilation rates and occupant dissatisfaction with air quality has been studied by a variety of investigators. Most studies have failed to show a significant relationship between SBS symptom prevalence rates and outdoor ventilation. Several studies have indicated that low ventilation rates (<10 L/sec · person) are risk factors for SBS symptoms.

Various office materials and equipment have been implicated as potential contributory factors to work/building-related health complaints. These include handling no-carbon-required copy paper and other papers and working with office copy machines such as wet process photocopiers, electrostatic photocopiers, laser printers, diazo-copiers, microfilm copiers, spirit duplicators, as well as video display terminals.

A variety of gas- and particulate-phase substances have been evaluated as potential risk factors for occupant health complaints. These have included bioeffluents (such as CO_2), HCHO, volatile organic compounds (VOCs), and dust. At this time there is little evidence to implicate bioeffluents as contributing to health complaints. Formaldehyde levels in public access buildings are relatively low, and most studies do not indicate that such exposures are sufficient to cause symptoms which are commonly reported. Human exposure studies and some epidemiological studies suggest a potential causal role of total volatile organic compounds (TVOCs) in SBS symptoms. A variety of epidemiological studies have implicated either airborne or settled dust as a risk factor for SBS symptoms. Most notable has been the association between the macromolecular or organic fraction of floor dust.

Residences

A variety of IAQ problems have been reported for residential environments. These typically differ from those experienced in public access buildings.

Residential environments in the U.S. are characterized by large population size (tens of millions), ownership status, differences in construction characteristics, differences in age and condition, limited complexity of building systems (e.g., no mechanical ventilation), and relatively unique exposure concerns. They may be single or multifamily, one to two stories or multistory, site-built, or manufactured.

Exposure concerns in residences are unique in that exposures to contaminants may for some individuals vary from approximately 12 to 24 hours/day. Exposed populations include infants and children, healthy adults, and those with existing illness. Under closure conditions which occur in the heating and cooling seasons, contaminant dilution is limited to air exchange which is associated with infiltration.

Though exposures to major identified indoor air contaminants may occur in all building environments, such exposures are often much greater in residences. Because of higher exposure concentrations, longer exposure durations, and the nature of exposed populations described above, a number of contaminants in residential environments are of significant public health concern. These include radon, HCHO, biological contaminants such as dust mite and cockroach antigens, mold and animal danders, environmental tobacco smoke (ETS), unvented combustion appliances, and pesticides. More limited concerns are exposures associated with craft activities and business activities conducted in the home.

Known health risks associated with exposures to contaminants in residential environments include acute mucous membrane and central nervous system symptoms associated with exposure to irritant vapors such as HCHO and possibly TVOCs; upper respiratory symptoms characteristic of chronic allergic rhinitis associated with exposures to dust mite, cockroach, cat, and dog antigens, mold spores and particles; pulmonary symptoms which may be associated with exposures to ETS or combustion byproducts such as NO_2 from unvented space heaters; asthma associated with exposures to a variety of allergenic substances that cause common allergic rhinitis, as well as nonspecific irritants that may induce asthmatic attacks; flue gas spillage that may result in high CO exposures which cause central nervous system (CNS) symptoms as well as hundreds of deaths annually; acute CNS and mucous membrane symptoms associated with the misapplication of pesticides; mucous membrane and central nervous system symptoms associated with solvent exposures from crafts and hobbies; and cancer that may be associated with exposures to radon, ETS, chlorinated termiticidal pesticides, and a variety of VOCs.

POLLUTANTS/POLLUTION PROBLEMS OF CONCERN

Combustion Byproducts

Combustion byproducts are released into indoor air from a variety of sources. These include open cooking fires fueled by wood, charcoal, coal, dung, etc. in third-world countries; unvented gas cooking appliances; gas and kerosene space heaters; unvented gas fireplaces; and to a lesser extent, vented combustion systems such as fireplaces, wood-burning appliances, gas and oil furnaces, and hot water heaters. They also include tobacco smoking, the most ubiquitous source of combustion byproducts in indoor environments.

Gas Cooking Appliances. Gas cooking stoves and ovens can be significant sources of CO, CO_2, NO, NO_2, aldehydes, respirable particles, and a variety of VOCs. Levels of CO in kitchens with operating gas stoves have been reported to range from 10 to 40 ppmv. Gas stoves and ovens appear to be the major source of NO_2 and human exposures in residential environments. Nitrogen dioxide levels in gas cooking homes have been reported to range from 18 to 35 ppbv, compared to 5 to 10 ppbv in nongas cooking residences.

Exposure to elevated levels of NO_2 associated with gas cooking has been proposed as a potential causal factor for respiratory symptoms and pulmonary function changes in school-age children living in gas cooking homes. Although several epidemiological studies have shown significant relationships between gas cooking and respiratory symptoms, others have not. As a result, the relationship among gas cooking stoves, elevated NO_2 levels, and respiratory symptoms/pulmonary function changes has not been definitively demonstrated.

Unvented Space Heaters. Unvented gas and kerosene heaters are widely used to heat residential spaces. In the U.S., gas space heaters are commonly used in southern states where winter temperatures are relatively moderate. This is also true for Australia. Kerosene heaters are widely used in Japan and were once popular (late 1970s to early 1980s) in the U.S. as a space heating alternative. Though unvented space heaters emit CO, such emissions are low and do not pose a CO poisoning concern.

Although the asphyxiation hazard has been eliminated from unvented kerosene and gas space heaters, they nevertheless have the potential for significantly contaminating indoor spaces with NO, NO_2, CO, CO_2, SO_2, aldehydes, VOCs, and respirable particles. Emissions depend on the type of heater or burner design, fuel used, and heater maintenance. For example, radiant kerosene heaters burn at lower temperatures and emit higher concentrations of CO, VOCs, and respirable particles than convective heaters. Because convective heaters burn hotter, emissions of NO_x may be 5 to 10 times higher than radiant heaters. Emissions of NO_x from radiant heaters are almost entirely NO_2, whereas convective heaters produce approximately equal quantities of NO and NO_2.

Kerosene heaters can be a significant source of SO_2 emissions. Kerosene fuels classified as Grade 1-K (S content of 0.04% or less) have low SO_2 emissions; those classified as Grade 2-K (S content up to 0.3%) may have SO_2 emissions 5 to 10 times greater.

Steady-state contaminant concentrations associated with kerosene heater operation in an experimental chamber (equivalent to the volume of a typical bedroom) under a range of air exchange conditions are illustrated in Figure 11.4. Under these worst-case conditions, kerosene heater operation can result in exposures to SO_2 and NO_2 that exceed air-quality standards. For purposes of reference, air exchange rates in residences typically range from 0.25 to 1.0 ACH. Concentrations of SO_2 in these chamber studies are in the range of those shown to cause asthmatic attacks in sensitive individuals (see Chapter 5).

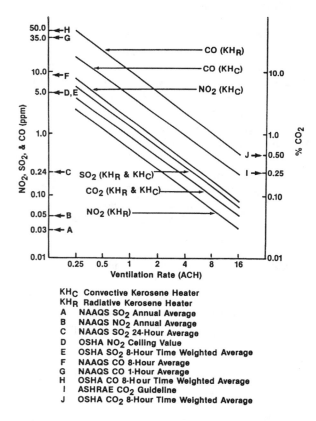

KH$_C$ Convective Kerosene Heater
KH$_R$ Radiative Kerosene Heater
A   NAAQS SO_2 Annual Average
B   NAAQS NO_2 Annual Average
C   NAAQS SO_2 24-Hour Average
D   OSHA NO_2 Ceiling Value
E   OSHA SO_2 8-Hour Time Weighted Average
F   NAAQS CO 8-Hour Average
G   NAAQS CO 1-Hour Average
H   OSHA CO 8-Hour Time Weighted Average
I   ASHRAE CO_2 Guideline
J   OSHA CO_2 8-Hour Time Weighted Average

Figure 11.4 Emissions and steady-state concentrations of NO_2, SO_2, CO, and CO_2 associated with radiant (R) and convective (C) space heaters. (From Leaderer, B.P. 1982. *Science* 218:1113–1115. With permission.)

Wood-Burning Appliances. During the late 1970s to the mid-1980s, wood-burning appliances such as stoves and furnaces became popular as residential space heating alternatives. An estimated 8 to 10 million units were used in the U.S. Though no recent-use surveys have been conducted, it appears that with fuel price stabilization, wood-heating appliance use has dramatically decreased.

Combustion byproducts may be emitted from wood-burning appliances because of improper installation, negative air pressure and downdrafts, leaks in appliance parts and flue pipes, and starting, reloading, stoking, and ash-removing operations. Wood-burning appliances have been reported to emit CO, NO_x, SO_2, aldehydes, fine particles, and VOCs. Some of the 200+ VOCs identified in wood smoke are known carcinogens, such as polycyclic aromatic hydrocarbons (PAHs).

Significant contamination of indoor spaces by wood heat emissions has been reported for conventional units. Airtight appliances, on the other hand, emit only trace quantities of combustion gases and particles into indoor spaces. The effect of heater type on PAH concentrations both indoors and outdoors is illustrated in Figure 11.5.

Contamination of indoor air may also occur in the operation of wood-burning fireplaces. Because they are open to room air, fireplaces have the potential for being a significant source of indoor air contamination. However, they

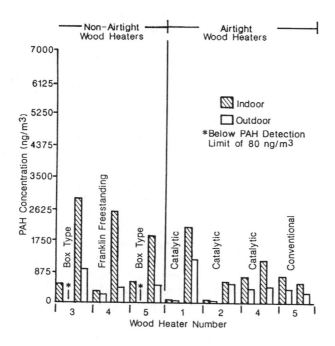

Figure 11.5 PAH concentrations associated with wood heater types. (From Knight, C.V. 1986. *Proceedings IAQ '86*. ASHRAE, Atlanta, GA. With permission.)

are primarily employed for their aesthetic appeal, and the frequency of use is often quite limited.

In parts of the U.S. where wood supplies are abundant and firewood inexpensive, the use of wood-burning appliances may be so extensive that their emissions may result in ambient concentrations which exceed air-quality standards (e.g., PM_{10}). This has been the case in mountain valley communities such as Missoula, MT, Corvallis, OR, Aspen, CO, and Danbury, VT. As a result, the U.S. Environmental Protection Agency (USEPA) has promulgated performance new source standards for new wood-burning appliances. These performance standards result in decreased emissions to both the ambient and indoor environments.

Environmental Tobacco Smoke. The use of tobacco products (and presumably other weeds) by approximately 40 million smokers in the U.S. results in significant indoor contamination from combustion byproducts that pose significant exposures to millions of others who do not smoke but who must breathe contaminated indoor air. Several thousand gas- and particulate-phase compounds have been identified in tobacco smoke. Some of the more significant include respirable particles, nicotine, nitrosamines, PAHs, CO, CO_2, NO_x, acrolein, HCHO, and hydrogen cyanide.

Environmental tobacco smoke (ETS) is a combination of sidestream (SS) smoke (released between puffs) and exhaled mainstream (MS) smoke. Environmental tobacco smoke exposures depend on the number and type of cigarettes consumed in a space, room volume, ventilation rate, and proximity to the source. Concentrations of major tobacco-related contaminants measured in a variety of building environments are summarized in Table 11.6. From these data as well as other studies, tobacco smoke (where smoking is occurring) appears to be the major source of respirable particles (RSP) in many indoor environments.

The health hazards of MS to smokers are well known. It is the major cause of lung cancer (and other cancers as well) and respiratory diseases such as chronic bronchitis and pulmonary emphysema. It is also a major risk factor for cardiovascular disease.

A number of gas-phase (and presumably particulate) constituents of MS and ETS are potent mucous membrane irritants, causing irritation of the eyes and upper respiratory system in both smokers and nonsmokers. Most notable of these irritants are the aldehydes acrolein and HCHO. Exposure to acrolein in a smoke-filled room is believed to be the major cause of eye irritation experienced. In addition to irritant effects, ETS exposures are reported to cause an increase in urinary nicotine and cotinine levels and pulmonary resistance in nonsmokers. Studies of nonsmoking office workers exposed to ETS have observed reductions in small airway function comparable to those in smokers consuming 1 to 10 cigarettes per day. Other workplace studies indicate that nonsmokers exposed to ETS experience more cough, phlegm production, eye irritation, chest colds, and days lost from work than those who were not exposed to ETS.

Respiratory symptoms in infant children have been associated with parental smoking. These effects are particularly significant when exposure occurs *in utero.*

Table 11.6   Tobacco-Related Contaminant Levels in Buildings

Contaminant	Type of environment	Levels	Nonsmoking controls
		ppmv	
CO	15 restaurants	4	2.5
	Arena (11,806 people)	9	3.0
	44 work rooms	2.8	2.0
		$\mu g/m^3$	
RSP/TSP	Bar and grill[a]	589	
	Bingo hall[a]	1140	
	Fast food restaurant[a]	109	
	44 work rooms	117	
	Restaurant	200	
	Cocktail lounge	400	
	Restaurant	240	
	Restaurant	144	
		ppbv	
NO_2	Restaurant	63	50
NO	Bar	21	48
	44 work rooms	84	62
		$\mu g/m^3$	
Nicotine	Restaurant	5.2	
	Cocktail lounge	10.3	
	44 work rooms	1.1	
		ng/m^3	
Benzo(α)pyrene	Arena	9.9	0.69

Note: RSP/TSP, respirable particles/total suspended particulates.
[a] Short-term sampling (<0.5 hour).

Exposure to ETS from parental smoking is a major risk factor for asthma in children. The incidence of asthmatic attacks in children has been reported to decrease dramatically when parents quit smoking or quit smoking indoors.

Involuntary exposure to ETS appears to be a major risk factor for lung cancer. In a review of a number of epidemiological studies, USEPA has concluded that ETS poses a significant lung cancer risk, with an estimated 3000 deaths per year associated with such exposure. The Occupational Safety and Health Administration (OSHA) has identified ETS as a presumed human carcinogen and has initiated rulemaking to regulate workplace exposures.

Environmental tobacco smoke may also increase the lung cancer risk associated with exposures to radon. Tobacco smoke particles serve as foci for radon progeny deposition, suspension in air, and transport and deposition in lung tissue.

Recent epidemiological studies have indicated that the risk of cardiovascular disease mortality associated with ETS exposure constitutes a more serious public health problem than ETS-associated lung cancer. An estimated 35 to 40,000 deaths annually in the U.S. have been attributed to ischemic heart disease associated with ETS.

Asbestos

Asbestos became a major indoor air-quality concern in the U.S. in the late 1970s when public health and environmental protection authorities concluded that friable (crushable under hand pressure) asbestos-containing building materials (ACBM) in school buildings had the potential for posing a cancer risk to children and other building occupants. Prior to 1978, ACBM was widely used in schools and other large buildings.

Asbestos is a collective term for a number of mineral silicates which are comprised of heat- and fire-resistant fibers of high tensile strength. The most widely used asbestos mineral in the U.S. was chrysotile or white asbestos. It accounted for more than 90% of the fibrous mass used in the 3000 or so different asbestos product applications. Less commonly used was amosite or brown asbestos. Typical asbestos applications included sprayed-on fireproofing, acoustical plaster, thermal system insulation, friction products, fibrocement board and pipes, roofing products, spackling compounds, mastics, and vinyl asbestos floor tile.

Asbestos has been regulated as a hazardous air pollutant by USEPA since 1973. Occupational asbestos exposures have been causally associated with asbestosis (a fibrogenic, debilitating lung disease), lung cancer, mesothelioma (a cancer of the lining of the chest and abdominal cavities), and scarring of the pleural (chest) cavity called pleural plaques. Asbestosis is caused by heavy, long-term exposures to asbestos fibers and is unlikely to be a health risk to building occupants. Because there is apparently no threshold or safe level of exposure, most public health concern has focused on risks of developing lung cancer or mesothelioma. Because children would be potentially exposed earlier in life, providing for a longer latency period, their risk of developing mesothelioma sometime in their life would be increased. Though relatively rare (an estimated 1200 to 1500 cases per year in the U.S.), mesothelioma is not curable and as a consequence has a mortality rate of 100%.

The presence of asbestos fibers in ACBM does not of itself pose a risk of human exposure. ACBM must in some way be disturbed before fiber release and exposure can occur. In many products asbestos fibers are held in some type of matrix. Fiber release occurs when this matrix is broken by hand pressure (hand friable) or mechanical force (mechanically friable). Because hand friable ACBM poses the greatest risk of fiber release and building air contamination, USEPA in 1973 began to regulate building demolition and renovation and banned sprayed-on fireproofing materials (Figure 11.6). In 1978 USEPA banned sprayed-on and troweled-on acoustical plaster and premolded asbestos products used in thermal system insulation (Figure 11.7). These materials continue to remain in place in hundreds of thousands of buildings in the U.S.

Over time, friable ACBM may become damaged and/or lose its adhesion to the substrate to which it was applied. Thermal system insulation applied to steam and hot water lines, boilers, etc. may become friable when damaged by maintenance or service activities. Damage to ACBM increases the likelihood that fibers will become airborne and cause human exposure. Exposure may be increased by

Figure 11.6 Sprayed-on friable asbestos fireproofing.

custodial activities such as dusting/cleaning and activities of occupants which may cause resuspension of settled fibers.

In response to asbestos health concerns in schools, USEPA initiated a guidance and technical assistance program in 1979. This was followed by legislation in 1983 which required schools to be inspected and tested for asbestos. Because of inadequacies in the Asbestos-In-Schools Program, Congress in 1986 passed

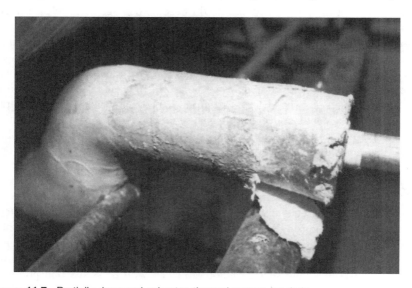

Figure 11.7 Partially damaged asbestos thermal system insulation.

the Asbestos Hazard Emergency Response Act (AHERA) which required that all schools be inspected by accredited inspectors and that a management plan be prepared by each school corporation for each building under its jurisdiction. Requirements for inspection and management plans are limited to school buildings. Under OSHA rules designed to protect custodial and service workers, materials that presumably contain asbestos are to be identified and workers notified and trained as to the potential health risk associated with its disturbance.

In the late 1980s, much public health attention was given to potential asbestos fiber exposures and health risks to the general building population such as school children, faculty, and administrative staff. There is increasing consensus in the scientific and regulatory communities that exposure risks to the general building population are very small. Increasing scientific evidence indicates that building custodial and service staff may be at significant risk of developing asbestos-related disease. Studies of custodial workers, sheet-metal workers, and carpenters have all shown significantly elevated prevalence rates of asbestos-related pleural plaques. Such scarring of chest cavity tissue results in impaired pulmonary function.

The presence of friable or damaged-bound ACBM does not itself constitute an exposure hazard to building occupants or health hazard to those who are at greatest risk (custodial staff and service workers). Factors which may increase exposure and potential health risks include (1) location and accessibility, (2) occupant activities, (3) ACBM age and condition, and (4) water damage. The accessibility of ACBM (typically acoustical plaster) in school corridors and gymnasia increases the likelihood of student-induced damage. Vibration associated with athletic events may contribute to releasing fiber from acoustical materials. As sprayed- or troweled-on ACBM ages, it loses its cohesiveness and adhesiveness, increasing the probability that it will become damaged and release fibers. Delamination may occur particularly if the material has been damaged by water associated with leaking roofs or some other similar occurrence. Service worker exposure typically occurs when ACBM is damaged by work-related activities.

The presence or absence of asbestos fibers and the mineral type present can be determined from bulk samples using polarized light microscopy. Abatement worker exposures can be determined from fiber counts using phase-contrast microscopy (PCM) of collected personal samples. Achievement of clearance guidelines after an asbestos abatement is determined from aggressive sampling with fiber concentrations determined by PCM or transmission light microscopy.

Formaldehyde

The widespread contamination of residences and, to a much lesser extent, public-access buildings by HCHO came to the attention of public health authorities in the late 1970s and early 1980s as a result of health and odor complaints associated with urea-formaldehyde foam insulation (UFFI), mobile homes, and conventional homes constructed and/or furnished with pressed wood products. Like asbestos, HCHO was a widely used industrial and commercial chemical. As a result, it was and continues to be a ubiquitous contaminant of indoor air.

Formaldehyde is a potent mucous membrane irritant causing, on acute exposure, eye and upper respiratory system symptoms. It is also a potent dermal irritant. On relatively chronic exposure, it appears to cause neurological symptoms such as headache, fatigue, disturbed sleep, and depression. It is a sensitizing substance causing allergic reactions of the skin, whole-body reactions from renal dialysis exposures, and asthma. There are suggestions that it may cause allergy-type symptoms from sensitization responses as well as its irritant properties. Animal studies, as well as epidemiological studies of exposed workers and residents of mobile homes, indicate that it can cause cancer in humans.

A variety of field investigations, as well as systematic epidemiological studies, have demonstrated strong associations between HCHO exposures and health effects commonly found in residences during the early 1980s. Statistically significant relationships have been reported for residential HCHO exposures and eye irritation, runny nose, sinus congestion, phlegm production, cough, shortness of breath, chest pain, headache, fatigue, unusual thirst, sleeping difficulty, dizziness, diarrhea, rashes, and menstrual irregularities.

Sources and Levels. Although HCHO has been and continues to be used in a variety of products, only a relatively few emit free HCHO at levels that can cause significant indoor air contamination. Problem products have included pressed wood materials such as particleboard (underlayment, paneling, cabinetry, furniture), medium-density fiberboard (furniture, cabinetry), hardwood plywood (paneling, cabinetry), UFFI, and acid-catalyzed finishes (applied to furniture and cabinetry). In each case the product used as an adhesive, foamed material, or resin finish contained urea-formaldehyde copolymer.

Urea-formaldehyde resins, though they make excellent adhesives and other products, are nevertheless chemically unstable. They release HCHO from the volatilizable, unreacted HCHO in the resin and from the hydrolytic decomposition of the resin copolymer itself. It is the release of the former which is responsible for the initially high HCHO levels that were associated with new mobile homes, conventional homes with particleboard underlayment, and homes that had been recently insulated with urea-formaldehyde foam. Emissions from pressed wood products now commercially available are less than 10% of those used in the late 1970s and early 1980s. Relatively low-emission pressed wood products have low emissions of unreacted HCHO that plagued early products, building environments, and, of course, occupants.

Formaldehyde-emitting products differ in their emission potential. As a result, indoor HCHO levels reflect the nature and potency of sources present. Indoor levels are also affected by the quantity of HCHO-emitting material present. Buildings with high load factors (surface area of HCHO-emitting products to air volume) such as mobile homes have experienced the highest reported HCHO levels. Formaldehyde sources also interact. Depending on individual circumstances, interactions may result in the suppression of emissions from one source by another, complete additivity, or where sources of equal potency are present, a less than additive augmentation of HCHO levels.

Most research on HCHO levels in residences was conducted in mobile homes, UFFI houses, and conventional houses with a variety of HCHO sources in the late 1970s and early to mid-1980s. As a result, data on HCHO levels reflect that period which can be described as historically high for indoor HCHO.

Highest levels of HCHO were reported for mobile or manufactured homes. Concentrations varied as a function of mobile home age, environmental factors at test, and geographical area. Concentrations varied from 0.05 to 1 ppmv or greater, with average concentrations in the range of 0.09 to 0.18 ppmv. In general, concentrations of HCHO in mobile homes manufactured today are much lower. Some new mobile homes may have initial HCHO levels as high as 0.40 ppmv, but most are less than 0.20 ppmv. Residences with particleboard underlayment had HCHO concentrations which at their extreme exceeded 0.50 ppmv but, depending on age and environmental conditions, were in the range of 0.05 to 0.20 ppmv. The use of particleboard underlayment in conventional stick-built houses today is very uncommon. It has been replaced by oriented-strandboard, a material that has very low HCHO emissions. Residences with UFFI initially experienced HCHO levels of 0.4 to 0.5 ppmv or greater but decreased rapidly to a range of 0.03 to 0.12 ppmv with an average of 0.06 ppmv a year or two after application. Though still widely used in Great Britain, very few houses in the U.S. are insulated with UFFI today. Residences with one or more miscellaneous sources (e.g., pressed wood furniture, hardwood plywood, pressed wood cabinets) had HCHO concentrations which ranged from 0.03 to 0.15 ppmv with average concentrations of 0.03 to 0.05 ppmv.

Formaldehyde concentrations decrease significantly with time. As a result, highest concentrations and exposures will occur in building environments where HCHO-emitting products are new. As previously indicated, pressed wood products emit significantly less HCHO than they once did. Though they may still cause excessive exposure levels and health problems, the combination of lower emissions and time-dependent decreases associated with older, once potent products have greatly diminished the public health significance of HCHO contamination in building environments.

Environmental Factors. The level of HCHO that occurs in a building environment at a given time depends on a variety of environmental factors, including indoor temperature and relative humidity, indoor/outdoor temperature differences, and wind speed. Formaldehyde levels increase in response to increases in either temperature or relative humidity and, of course, both. An increase in indoor temperatures of 5 to 6°C (8 to 10°F) will result in a doubling (Figure 11.8) of HCHO levels, and an increase of relative humidity of 40% will result in an approximate increase of 40% in HCHO levels. Increased differences in indoor/outdoor temperatures and increased wind speed result in higher building exchange rates associated with infiltration. As a result of this fact and lower relative humidities, low HCHO levels occur on cold winter days.

Safe Levels of Exposure. What level of HCHO exposure is safe? As with most chemical exposures, this question cannot be easily answered. OSHA, which

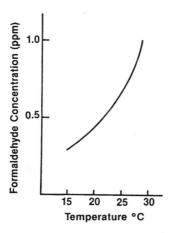
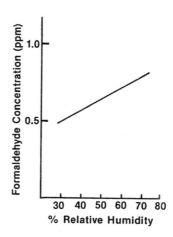

Figure 11.8 Effect of temperature and relative humidity on indoor formaldehyde concentrations.

regulates worker exposure, has a permissible exposure level (PEL) of 0.75 ppmv time-weighted average (TWA) 8-hour exposure with an action level requiring medical surveillance and training of 0.50 ppmv TWA. Taking cancer risks into account, NIOSH has a recommended exposure limit of 0.016 ppmv.

In response to building-related problems associated with residential HCHO exposures, a number of countries have adopted air-quality guidelines for HCHO exposures in residences. Countries with guideline values of 0.10 to 0.12 ppmv include Australia, Germany, Denmark, the Netherlands, and Italy. Canada and California have adopted guideline values of 0.10 ppmv as an action level (need of remediation), with a target level of 0.05 ppmv. The World Health Organization (WHO) recommends a guideline value of 0.082 ppmv.

Compliance with these guideline values may not be sufficient to protect the health of sensitive populations such as asthmatic children. Studies conducted in Arizona have shown significant pulmonary function deficits in asthmatic children with average household exposure levels of 0.06 ppmv. Other studies in Canada have observed dose–response relationships between HCHO levels and eight symptoms at average exposure concentrations of 0.045 ppmv.

Organic Compounds

A large variety of organic compounds (in addition to HCHO) have been reported to be present in public access buildings and residential environments. Organic compounds found in indoor air include: very volatile organic compounds (VVOCs) with boiling points that range from less than 0 to 50–100°C, volatile organic compounds (VOCs) with boiling points in the range of 50–100 to 240 – 260°C, semivolatile organic compounds (SVOCs) with boiling points which range from 240 – 260 to 380 – 400°C, and organic compounds which are solids and have boiling points in excess of 400°C. Because of their great variety, relative abundance in indoor air (2 to 100 times higher indoors), potential to cause both sensory irritation and central nervous system symptoms, and gas-phase presence, VOCs have received major research attention as potential causal factors of building-related symptoms, particularly in office environments.

Volatile organic compounds include aliphatic hydrocarbons (HCs) which may be straight, branched-chained, or cyclic; aromatic HCs; halogenated HCs (primarily chlorine); and oxygenated HCs such as aldehydes, alcohols, ketones, esters, ethers, and acids. They are emitted by a variety of sources including building materials and furnishings, consumer products, building maintenance materials, humans, office equipment, and tobacco smoking.

Studies in Danish residences and office buildings have reported the presence of 40 different VOCs consisting mainly of C_8–C_{13} alkanes, C_6–C_{10} alkylbenzenes, and terpenes. In a study of 42 commonly used building materials, Danish investigators identified emissions of an average 22 different compounds associated with each material. The ten compounds with the highest steady-state concentrations were toluene, n-xylene, terpene, n-butylacetate, n-butanol, n-hexane, o-xylene, ethoxyethylacetate, n-heptane, and p-xylene. Organic compounds identified in 40 east Tennessee homes included toluene, ethyl benzene, n-xylene, p-xylene, nonane, cumene, benzaldehyde, mesitylene, decane, limonene, undecane, naphthalene, dodecane, tridecane, tetradecane, pentadecane, hexadecane, and 2,2-methylnaphthalene.

Concentrations of individual VOCs measured in indoor air are typically several orders of magnitude lower than occupational threshold limit values (TLVs) or permissible exposure limits (PELs). As a result, it is not likely that individual VOCs (other than HCHO) are responsible for building-related health complaints.

A potential causal relationship between SBS-type symptoms and exposures to a combination of VOCs at relatively low individual concentrations has been proposed by Danish investigators and supported by human exposure studies. It is described as the total volatile organic compound (TVOC) theory. In the TVOC theory, sensory irritation and possibly neurological (headache and fatigue) symptoms in problem buildings may be due to the combined effect of the many VOCs found in indoor air.

Effects have been reported at exposure concentrations as low as 5 mg/m^3 (1 ppmv) toluene equivalent TVOC to a mixture of 22 different compounds as well as 3 submixtures. A proposed dose–response relationship between TVOC exposures and discomfort/health effects is indicated in Table 11.7.

The relationship between TVOC concentrations and building-related health complaints has been evaluated in a number of systematic building studies. Several major studies have failed to show a dose–response relationship between area TVOC levels and retrospectively reported symptoms. In one study, however, significant associations were observed between VOC levels determined at each building occupant's workstation and simultaneously reported symptom intensity. In another study of 11 buildings, significant positive correlations were observed between symptom prevalence rates and log TVOC concentrations. In a study of problem schools, symptom prevalence was not only associated with log TVOC but also to concentration of terpenes, n-alkanes (C_8–C_{11}), and butanols. In the 11-

Table 11.7 **Proposed Dose–Response Relationship Between Exposures to TVOC Mixtures and Health/Comfort Effects**

Total volatile organic compound concentration (mg/m³)	Response	Exposure range
<0.20	No effects	Comfort range
0.20–3.0	Irritation/discomfort possible	Multifactorial exposure range
3.0–25.0	Irritation and discomfort; headache possible	Discomfort range
>25.0	Neurotoxic effects in addition to headaches	Toxic range

From Molhave, L. 1990. *Proceedings Fifth International Conference on Indoor Air Quality and Climate.* Vol. 5. Toronto.

building study, TVOC concentrations varied from 0.05 to 1.38 mg/m³; in the problem school study, 0.07 to 0.18 mg/m³. In the former case, TVOC levels were in the multifactorial range in Table 11.7; in the latter case, the no-effects range.

Concentrations of individual VOCs measured in indoor air are typically in the low $\mu g/m^3$ (up to 30 $\mu g/m^3$ or so) or ppbv (up to 25 ppbv) range. Concentrations of TVOCs are typically in the low mg/m³ range (typically <5 mg/m³) or ppmv (typically <1 ppmv) range.

Pesticides

Pesticides represent a special case of chemical contamination of buildings. These are substances with well-known toxicities that are deliberately introduced (and in many cases inadvertently) into building environments for controlling pests such as moths, fleas, cockroaches, flies, ants, etc.; rodents such as rats and mice; mold; and bacteria.

Insecticides and antimicrobials are the most commonly used pesticides indoors. Antimicrobial substances commonly used include hypochlorites, ethanol, isopropanol, pine oil, glycolic acid, 2-phenylphenol, and 2-benzyl-4-chlorophenol.

A variety of insecticides are used in buildings or have been used to treat building materials. These include diazinon, chlordane, heptachlor, lindane, malathion, chlorpyrifos, propoxur, aldrin, dieldrin, folpit, bendiocarb, methoxychlor, ronnel, dichlorvos (DDVP), pentachlorophenol, paradichlorobenzene, naphthalene, methyl demeton, and boric acid. Insecticides may be applied in a variety of ways: as an emulsion spray, by fogging devices (bug bombs), impregnated in pest strips or flea and tick collars, and as bait poison. They may be released directly into the air (fly spray, bug bombs), sprayed on surfaces (cockroach, flea, and ant control), applied to the substructure and/or surrounding ground surfaces (termites), or placed in solid form around materials to be protected (moth balls/flakes).

Most insecticides are SVOCs with vapor pressures in the range of 10^{-2} to 10^{-4}. As a result, indoor air concentrations, with the exception of misapplications, are usually quite low with mean concentrations less than 5 $\mu g/m^3$. Highest concentrations are reported for home foggers that use DDVP (vapor pressure of 1.2 × 10^{-2} mm Hg). Within the first hour, DDVP concentrations may be in the range of 4 to 6 mg/m^3 and then rapidly dissipate.

Major indoor pesticide contamination and exposure concerns have been expressed relative to the use of chlordane/heptachlor formulations for termiticidal application and the use of such substances as pentachlorophenol and lindane as wood preservatives. Chlordane as a termiticidal application is no longer permitted in the U.S. However, it was widely used to treat ground surfaces around substructure perimeters in new houses and injected under existing slab houses. Slab pressure injection under military houses resulted in widespread occupant health complaints. Significant air contamination occurred in many houses as the applied pesticide and the organic solvent vehicle penetrated heating ducts in slabs.

Chlorpyrifos, more commonly known by its trade name Dursban, is now the most widely used termiticidal pesticide in the U.S. It is also applied to the ground around building substructures or within slabs. Indoor contamination from chlorpyrifos occurs in the same way that chlordane did; volatilized pesticide becomes entrained in soil gas which is drawn into the building by infiltration.

Pentachlorophenol (PCP) has been used to treat timbers for log houses and foundation lumber in the U.S. Because PCP is a chlorinated HC and accumulates in fat tissue, PCP blood serum levels have been reported to be seven times higher in occupants of PCP timber-treated log homes compared to individuals living in nonlog homes.

Health concerns associated with pesticide use indoors, treated building materials, and termiticidal applications have focused on acute effects reporting by building occupants immediately after pesticidal applications and the longer-term carcinogenic potential of pesticides such as chlordane, heptachlor, and PCP.

Acute effects have most notably been reported with the use of home fogging devices. Principal complaints have included nausea, dizziness, headaches, and skin irritation. It has been estimated that from 1966 to 1977, occupants of 2.5 million households experienced acute pesticide-related symptoms. Acute effects have also been reported for the misapplication of termiticides and the use of wood preservatives. In some cases it is likely that acute symptoms may be due to exposures to high VOC solvent vehicle concentrations rather than the active pesticide itself.

Chlordane, heptachlor, lindane, dieldrin, and PCP are chlorinated HCs which accumulate in fat tissue and are suspected human carcinogens. Because of their potential human carcinogenicity, building-related exposures may pose a long-term health risk.

Radon

Radon is a noble, nontoxic gas produced in the decay of radium-226. Radium is found in high concentrations in uranium and phosphate ores and mill tailings and to a lesser extent in common minerals such as granite, schist, limestone, etc. As a result, radon is ubiquitously present in the soil and air near to the surface

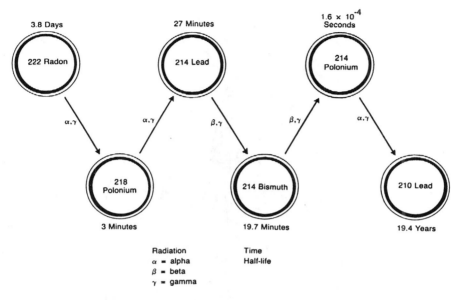

Figure 11.9 Radioactive decay of radon and its progeny.

of the earth. As radon undergoes radioactive decay, it releases an alpha particle, gamma ray, and progeny that quickly decay (Figure 11.9) to release alpha and beta particles and gamma rays. Because radon progeny are electrically charged, they readily attach to particles, producing a radioactive aerosol. These aerosol particles may be inhaled and deposited in the bifurcations of respiratory airways (Figure 11.10). Irradiation of tissue at these sites poses a significant risk (based on exposure dose) of lung cancer tumor development.

Radon Concentrations and Sources. Radon concentrations measured in buildings in the U.S. are calculated and reported as picocuries per liter (pCi/L). A picocurie is equal to a trillionth of a Curie or 2.2 radioactive disintegrations per minute. A working level (WL) is a quantity of radon which will produce 1.3×10^5 million electron volts (MeV) of potential alpha particle energy per liter of air. One WL is equal to the concentration of radon progeny in equilibrium with 100 pCi/L radon.

Because of the relatively common occurrence of radium-226 in the earth's crust, radon is present in ambient air with average concentrations ranging from 0.20 to 0.25 pCi/L. Radon levels in buildings may in many cases be several times or orders of magnitude higher than ambient levels. This is particularly the case in single family dwellings. In one of the worst cases ever reported, radon concentrations as high as 2600 pCi/L (13 WL) were measured in a southeastern Pennsylvania home. Unusually high residential radon levels have been associated with the Reading Prong, a small geological formation that extends through southeastern Pennsylvania through New Jersey and into Connecticut.

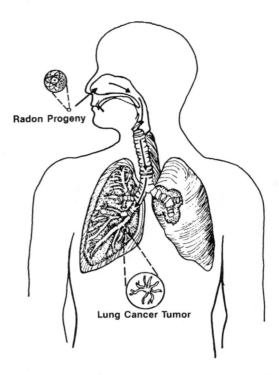

Radon Progeny

Lung Cancer Tumor

Figure 11.10   Inhalation and deposition of radon progeny in humans and lung cancer tumor formation.

Significant geographical differences in radon levels in the U.S. have been reported. Geometric mean values based on 3-month averages for 30,000 nonrandomly tested houses were reported as 3.43 pCi/L in the Northeast; 2.36, Midwest; 0.64, Northwest, 2.40, mountain states; and 1.43, Southeast. The geometric mean for the entire sample population was 1.74 pCi/L, and 0.5% exceeded 100 pCi/L. Other studies have reported a lower geometric mean value of approximately 1.0 pCi/L.

The major source of indoor radon is soil gas which is transported into buildings by pressure-induced convective flows. Other sources include well water and masonry materials. Buildings constructed on soils that have high radon release and transport potential typically have the highest radon levels. Radon levels in a house vary in response to temperature-dependent and wind-dependent pressure differentials and to changes in barometric pressure. Radon transport is enhanced when the base of a building is under significant negative pressure.

Health Effects/Risks. The major health risk which has been definitively linked to radon exposures is lung cancer. This association has been well documented for uranium miners. Extrapolations from miner radon exposure/health data indicate that humans exposed to the lower concentrations characteristic of residences and other buildings may also be at risk. Risk assessments conducted

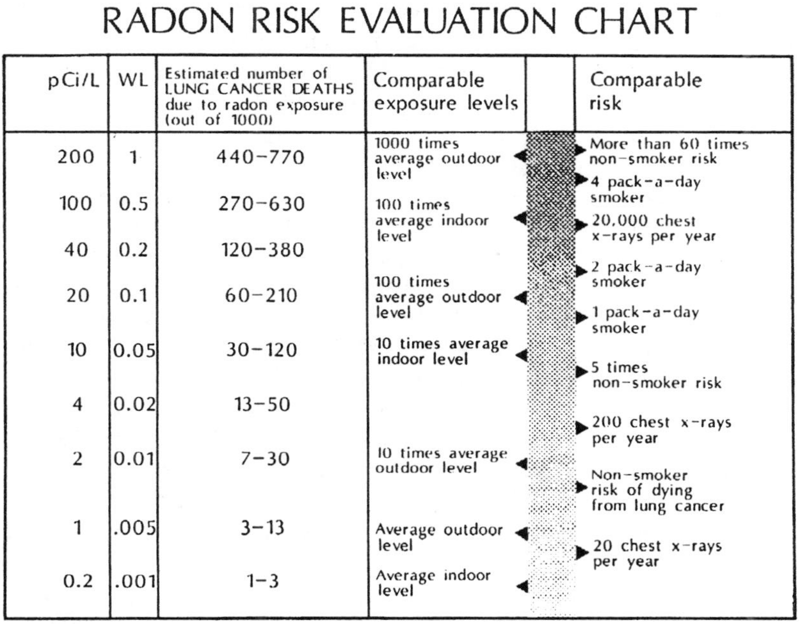

RADON RISK EVALUATION CHART

pCi/L	WL	Estimated number of LUNG CANCER DEATHS due to radon exposure (out of 1000)	Comparable exposure levels		Comparable risk
200	1	440–770	1000 times average outdoor level		More than 60 times non–smoker risk
					4 pack–a–day smoker
100	0.5	270–630	100 times average indoor level		20,000 chest x–rays per year
40	0.2	120–380			2 pack–a–day smoker
20	0.1	60–210	100 times average outdoor level		1 pack–a–day smoker
10	0.05	30–120	10 times average indoor level		5 times non–smoker risk
4	0.02	13–50			200 chest x–rays per year
2	0.01	7–30	10 times average outdoor level		Non–smoker risk of dying from lung cancer
1	.005	3–13	Average outdoor level		20 chest x–rays per year
0.2	.001	1–3	Average indoor level		

Figure 11.11 USEPA radon risk estimates. (From USEPA. 1986. EPA-86-004.)

by the National Council on Radiation Protection (NCRP) predict 9000 lung cancer deaths in the U.S. per year from indoor radon exposures; those conducted by the National Academy of Sciences' Committee on the Biological Effects of Radiation (BEIR IV) predict 13,500 radon-related lung cancer deaths per year. USEPA estimates approximately 15,000 to 16,000 lung cancer deaths annually from indoor radon exposures. USEPA lifetime risk assessments associated with different average radon exposures are shown in Figure 11.11. Equivalent risks are also shown for tobacco smoking and exposure to chest X-rays. Note the relatively large lifetime lung cancer risk of 1 to 3 per 1000 population for outdoor levels of 0.2 pCi/L (0.001 WL) and 13 to 50 per 1000 for an average lifetime exposure to 4 pCi/L (0.02 WL) — the USEPA action level for remediation. In the latter case, an annual average lifetime exposure to 4 pCi/L would be equivalent to the lung cancer risk associated with smoking five cigarettes per day. Using these risk estimates or the more conservative NCRP one, radon appears to be one of the most hazardous substances humans are exposed to. Because of these lung cancer risks, USEPA and the U.S. Surgeon General in 1988 issued a public health advisory recommending that all homes be tested and that remediation be undertaken when long-term (3 mo to 1 year) test results are excessive, that is, above the action guideline of 4 pCi/L.

There is also a limited amount of epidemiological evidence to suggest that radon exposures may be a significant risk factor for acute myeloid leukemia, cancer of the kidney, melanoma, and certain childhood cancers. In Great Britain, approximately 6 to 12% of acute myeloid leukemia cases are estimated to be associated with radon exposures; for regions where radon levels are higher, the predicted range is 23 to 43%. For the world average radon exposure of 1.3 pCi/L, approximately 13 to 25% of acute myeloid leukemias are predicted to be associated with radon.

Biological Contaminants

As described in previous pages, exposures to toxic chemicals such as combustion byproducts, HCHO, VOCs, and pesticides, hazardous particles such as asbestos, and radioactive radon gas have the potential for significant adverse health effects. A variety of biological contaminants can also cause significant illness and health risks. These include infections from airborne exposures to viruses that cause colds and influenza and bacteria that cause Legionnaires' disease and tuberculosis (TB). They also include respiratory ailments such as hypersensitivity pneumonitis, humidifier fever, asthma, chronic allergic rhinitis, and SBS-type symptoms. Such ailments may be caused by exposures to mold, fungal glucans, mycotoxins, thermophilic actinomycetes, bacterial endotoxins, antigens (produced by mold, dust mites, cockroaches, and mammalian pets), microbial volatiles, or organic dust.

Legionnaires' Disease. The outbreak of Legionnaires' disease among some of those that attended the Pennsylvania American Legion convention in Philadelphia in 1976 was a dramatic example of an infectious disease outbreak associated with a building. After an intensive investigation, Centers for Disease Control staff isolated the causal organism, the bacterium *Legionella pneumophila*; and identified the most probable mode of transmission, contaminated air entrained in one of the air handling systems that served the hotel lobby. Legionnaire's disease causes pneumonia-like symptoms. Though it has a low attack rate (5%), mortality among those affected is high (15 to 20%).

Legionella pneumophila is widely present in the environment, being commonly isolated from surface waters and soil. It is relatively resistant to chlorine and passes through most water treatment systems. Its optimum growth temperature is at or above 33°C (90°F). As a result, significant growth can occur in cooling tower waters, evaporative condensers, institutional and domestic hot water heaters, hot tubs, and spas. Drift from cooling towers/evaporative condensers can become entrained in building air supply. Disease outbreaks occur when a susceptible population is exposed to high bacteria levels. Risk factors identified for Legionnaires' disease include middle age, smoking, alcohol consumption, treatment with immunosuppressive drugs, and travel.

Legionella pneumophila has also been identified as the cause of a nonpneumonic disease known as Pontiac fever. It is an ailment characterized by fever, malaise, myalgia, and headaches. It has a short incubation period (2 days), a high attack rate (95%), and is not fatal.

Legionnaire's disease and Pontiac fever appear to be caused by exposures to *L. pneumophila* which has become aerosolized. Outbreaks have been reported throughout the U.S. and many other parts of the developed world in hotels, office buildings, student unions, country clubs, cruise ships, hospitals, industrial facilities, etc.

Tuberculosis. Despite a many decades-long sense that TB is no longer a problem, it has become an increasing public health concern in the U.S. The incidence of TB has been increasing in the U.S., in part because of immigration and also because the bacterium that causes the disease has become increasingly drug resistant.

The disease is spread when the bacterium becomes airborne. Its spread is a major health concern in hospitals and places where people congregate such as homeless shelters and prisons.

Hypersensitivity Pneumonitis and Humidifier Fever. Outbreaks of illness in large buildings and residences occur as a result of air contamination by a variety of microorganisms and microbial products. Most notable are illness syndromes hypersensitivity pneumonitis and humidifier fever. Both are characterized by symptoms of fever, chills, myalgia, malaise, shortness of breath, and cough. Onset usually occurs 4 to 6 hours after exposure with recovery within 18 hours. Symptoms may diminish in severity on continued exposure with recurrence after a period of nonexposure (e.g., over weekends). In hypersensitivity pneumonitis, long-term exposures to causative agents causes irreversible lung damage. Chronic lung damage does not, however, occur in cases of humidifier fever reported in Europe.

Outbreaks of hypersensitivity pneumonitis have been reportedly associated with airborne exposures to thermophilic actinomycetes, high mold levels, nonpathogenic amoeba and flavobacterium species and their endotoxins. In many outbreaks of hypersensitivity pneumonitis, the sources of exposures to causal agents included components of heating, ventilating, and air conditioning systems such as duct work, humidifiers, air washers, and fan coil units. In other cases mold or bacterially infested building materials (such as ceiling tile) or furnishings (carpeting) were identified as the source. The major cause of humidifier fever appears to be the contamination of cool-mist humidifiers (commonly used in Great Britain) by bacterial endotoxins.

A common potential source of air contamination that may result in outbreaks of hypersensitivity pneumonitis is microbial slime that grows on organic dust entrained in cooling coil condensate waters that are not properly drained. As this microbial material dries, it tends to aerosolize and contaminate building supply air.

Asthma. Asthma is a serious disease of the pulmonary or lower respiratory system, characterized by episodal constriction of the respiratory airways resulting in shortness of breath, wheeze, cough, and chest tightness. Symptoms vary in severity. Onset may occur within an hour of exposure or be delayed 4 to 12 hours. The prevalence of asthma in the U.S. is estimated to be 10 to 15 million cases affecting 4 to 5% of the population. Most notable is the approximate 60% rise in asthmatic cases since 1982 and a doubling of asthma-related mortality (now approximately 5000 deaths per year). Much of the observed increase in asthma has occurred in children.

Asthma has been shown to be strongly associated with airborne allergens such as house dust (contaminated with dust mite fecal waste), mold spores, animal danders (particularly cat and dog), pollen, and cockroach excretions and body parts. Asthma is initiated in sensitive individuals by exposures to substances which induce an immunological sensitization. Subsequent asthmatic attacks can occur on reexposure to the asthmagenic substance or a variety of nonspecific irritants.

Most asthma cases occur as a result of exposures to airborne allergens. In most cases such exposures occur in residential environments. Because of the seriousness of the disease and the very large population of affected individuals, factors that cause asthma and initiate asthmatic attacks represent a very significant IAQ problem.

Chronic Allergic Rhinitis. Chronic allergic rhinitis or common allergy is a relatively nonserious ailment that may affect 20% of the U.S. population. It affects the upper respiratory system with characteristic symptoms of sinus inflammation, runny nose, congestion, sneezing, and eye irritation. It is caused by immunological sensitization to antigens found in house dust, mold spores and hyphal fragments, animal danders, saliva, urine, insect excreta and body parts, pollen, and allergens found in food. As with asthma the primary cause of allergy symptoms is exposure to inhaled allergens. Because of the nature of the exposure, common allergy is a major IAQ problem.

DIAGNOSING INDOOR AIR-QUALITY PROBLEMS

As indicated in previous sections, there are a number of significant IAQ problems in buildings. Except in severe cases of acute illness, most of these problems (or potential problems such as radon) go unrecognized. In the case of problem buildings, the need to conduct a building investigation to diagnose the nature of a suspected problem only occurs when occupant complaints are sufficiently intense to convince building management that air-quality problems need investigation.

Illness symptoms associated with HCHO, combustion byproducts such as CO, and exposures to indoor inhalant allergens and elevated radon levels are major IAQ problems associated with residential environments. As a result, it is incumbent that homeowners recognize that they may have such problems and seek professional assistance.

The task of investigating building-related health and sometimes odor complaints may fall on a variety of local, state, and federal public or occupational health agencies, or private consultants. A variety of protocols are used to conduct such investigations in problem public access buildings. Protocols have not been developed to conduct residential investigations.

USEPA, in cooperation with NIOSH, has developed a model protocol for conducting health and comfort complaints in public access buildings. Though designed for in-house personnel, it can be readily adapted for use by public agencies and private consultants.

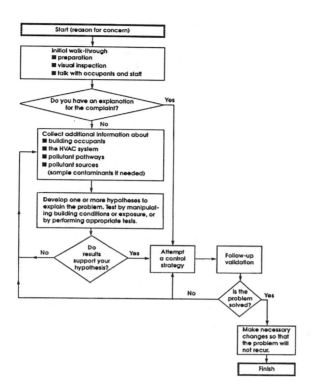

Figure 11.12 USEPA/NIOSH building investigation protocol. (From USEPA/NIOSH. 1991. EPA/400/1-91/003. DDHS Publication No. 91-114.)

The USEPA/NIOSH investigative protocol is characterized by systematic information gathering and hypothesis testing (Figure 11.12). It includes an initial walk-through of problem areas to gather information on building occupants, ventilation systems, contaminant sources, and pollutant pathways. The acquisition of additional information is recommended if the walkthrough does not identify the cause or causes of reported problems. Information gathering is intended to provide the basis for developing a hypothesis as to what may be the cause or causes. Because many complaints occur in poorly ventilated buildings and spaces, it is desirable to inspect and evaluate components of ventilation systems that may serve problem areas and other parts of the building.

The USEPA/NIOSH investigative protocol gives limited attention to air sampling, suggesting that such sampling may not be required to solve most problems and may even be misleading. Air sampling is only recommended after all other investigative activities have been exhausted. Exceptions include CO_2 to determine ventilation adequacy and temperature, humidity, and air movement for thermal comfort.

Relatively speaking, the diagnosis of an IAQ problem in residences is a relatively simple undertaking. Residences are small, contain few individuals, and have relatively simple mechanical systems. Additionally, relatively few contaminants are responsible for most residential health and odor complaints, and causal connections are more easily made. Commonly encountered IAQ problems in residences include exposure to allergens such as mold, dust mite excreta, cockroach excreta and body parts, and animal danders which can cause allergy and asthma, HCHO that causes irritant-type symptoms, CO that causes CNS symptoms, and a variety of other problems including sewer gas odors, misapplied pesticides, etc. Diagnosis involves information gathering on occupant symptoms (type and onset patterns) and assessment of the building environment for possible sources. It may also include air testing or collection of surface samples to confirm a hypothesis of what the potential causal factor or factors may be.

Air Testing

Air testing is widely used in air-quality investigations in buildings as screening measurements for contaminants such as radon. A variety of air-quality test procedures have been developed for common indoor contaminants such as HCHO, CO, CO_2, biological contaminants, and radon.

Air testing requires knowledge of test procedures, their limitations, factors that affect test results, and how to interpret them.

PREVENTION AND CONTROL MEASURES

A variety of approaches are available to both prevent IAQ problems before they occur and to control and mitigate them once they are manifest. These can be described as source management, exposure control, and contaminant control. Source management includes source exclusion, removal, or treatment. Exposure control focuses on scheduling activities to reduce occupant exposures. Contaminant control focuses on reducing contaminant levels once they become airborne. Contaminant control is usually achieved by the use of ventilation and/or air cleaning.

Source Management

Exclusion. In new building construction and remodeling, IAQ problems can be reduced or minimized by the use of low emission products that are increasingly becoming available. Carpeting and mastic products are now being produced and sold which have relatively low emissions of VOCs and the odor-producing substance called 4-PC. Formaldehyde emissions from pressed wood products are much lower than they once were as lower emission products have been developed. Low levels of HCHO can be attained/maintained by avoiding the use of HCHO-emitting products such as particleboard, hardwood plywood paneling, medium-density fiberboard, and acid-cured finishes. Alternative products include softwood plywood, oriented-strand board, decorative gypsum board and hardboard panels, and in the case of finishes, a variety of varnishes and lacquers.

Contamination of indoor air by combustion byproducts can be avoided by using electric rather than gas appliances for cooking, using only electric or vented gas and oil space heating appliances, and restricting smoking. The use of pesticides indoors can also be largely avoided by implementing integrated pest management practices. Integrated pest management can also be used to minimize cockroach infestations.

High radon levels can be avoided by selecting homesites which have a high clay content. However, such soils tend to have moisture problems which increase health risks associated with mold and dust mites.

Source Removal. The single most effective measure to reduce contaminant levels is to identify and remove the source. It is an effective control measure for HCHO. Its effectiveness depends on the correct identification of the most potent source or sources in a multiple source environment. Because of interaction effects, HCHO levels are determined by the most potent source/sources present. Removal of minor sources would not reduce HCHO levels.

In buildings with mold infestation, the mitigation of hypersensitivity reactions such as hypersensitivity pneumonitis, asthma, and allergy will in many cases require the removal and replacement of mold-infested construction materials and furnishings.

Source Treatment. Sources may be treated or modified to reduce contaminant emissions. For example, one can use encapsulants to prevent the release of asbestos fibers from acoustical plaster. Formaldehyde-emitting wood products can be treated with scavenging coatings or encapsulated with vinyl materials. Radon treatment measures include sealing cracks in masonry walls and slabs and gaps around concrete floor-based plumbing.

CONTAMINANT CONTROL

Ventilation

Contaminant concentrations in building spaces can be reduced by diluting indoor air with less contaminated outdoor air. Such dilution may occur in part or in whole by infiltration/exfiltration, natural ventilation, and forced or mechanical ventilation. Ventilation air exchange in residential buildings occurs by natural ventilation, that is, by opening windows and/or doors during mild to warm seasonal conditions, and by infiltration/exfiltration under cold or hot seasonal closure and building climate control conditions. Infiltration/exfiltration also plays a role in building air exchange in mechanically ventilated office, commercial, and institutional buildings.

Infiltration/Exfiltration. All building structures "leak"; that is, the building envelope contains many avenues whereby air enters and leaves. These include cracks or gaps around windows, doors, electrical outlets, ceiling light fixtures, and the building base. It also includes exhaust outlets. The infiltration of outside air and the exfiltration of inside air is affected by the tightness of the building envelope and by environmental factors such as indoor and outdoor temperature differences and wind speed. The effects of these environmental factors on building

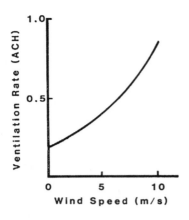 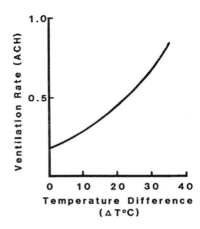

Figure 11.13 Effects of wind speed and indoor/outdoor temperature differences (ΔT) on building air exchange.

air exchange can be seen in Figure 11.13. Note that high air exchange occurs on cold, windy days and that low air exchange occurs on calm, moderate days when temperature differences are small. During the heating season, indoor and outdoor temperature differences result in pressure differences that draw air into the base area of the building (infiltration) and force air out the top (exfiltration). This is the so-called stack effect. It occurs in all buildings (including single story) but is most pronounced in tall buildings.

Natural Ventilation. A building is "naturally" ventilated when occupants open windows and, in some cases, doors. Air exchange can be considerable, as can contaminant dilution. The degree of air exchange, or ventilation depends on the magnitude of window and door opening, the position of open windows and doors relative to each other, duration, and the same environmental factors that affect infiltration/exfiltration. In the absence of year-round climate control, natural ventilation is the primary means by which residential and some nonresidential building spaces are made more comfortable during warm weather conditions.

Mechanical Ventilation. Mechanical ventilation is the most widely used contaminant control measure in large office, commercial, and institutional buildings. Mechanical systems are used to provide general dilution ventilation and local exhaust ventilation.

General dilution ventilation is used in most large buildings to dilute and remove human bioeffluents (contaminants emitted by humans) which cause odor and comfort complaints. It is also applied as a generic measure to reduce overall building contaminant levels and in attempts to mitigate SBS-type symptoms and IAQ complaints.

The use of mechanical ventilation for contaminant control is based on general dilution theory wherein contaminant levels are expected to decrease by 50% each time air volume/exchange is doubled. It is likely to be relatively effective in contamination problems that are episodic such as tobacco smoking, emissions from office copiers, or the relatively constant emissions of human bioeffluents. It is less effective in reducing levels of contaminants (such as HCHO and VOCs) that are emitted by diffusion processes.

A variety of studies have attempted to evaluate the effectiveness of general dilution ventilation in reducing both contaminant levels and occupant health complaints. The results have been mixed. In general it appears that increasing ventilation rates to 20 cubic feet per minute (10 L/sec) per person results in decreased symptom prevalence with higher ventilation rates being relatively ineffective.

A consensus has developed over the years that adequate ventilation is essential for the maintenance of comfortable and low human odor building environments. As a result, guidelines have been developed to help building designers and operators provide adequate ventilation rates to meet the needs of building occupants. In North America, ventilation guidelines are developed and recommended by the American Society of Heating, Refrigerating and Air-Conditioning Engineers (ASHRAE). In Europe and East Asia, such guidelines have been developed by governmental agencies, regional collaboration, and the WHO.

Ventilation guidelines specify a rate of outdoor air ventilation that can be reasonably expected to provide acceptable air quality related to human odor and comfort. Ventilation guidelines specify quantities of outdoor air that will maintain building steady-state CO_2 levels (typically 800 or 1000 ppmv) at maximum design capacity. These specifications are in cubic feet per minute per person or liters per second per person. Under the design occupancy, the provision of the specified minimum ventilation rate would maintain CO_2 levels below guideline values. ASHRAE guideline ventilation values for different building spaces or conditions of use are summarized in Table 11.8.

Local exhaust ventilation is used in special circumstances to minimize levels of certain contaminants in buildings. It is particularly well suited to situations where sources are known, emissions levels are high, and localized and general dilution ventilation would not be acceptable. Its major application is to control odors from lavatories. It is also used to control combustion byproducts and odors from cafeteria kitchens, vapors, and gases from institutional laboratories, and ammonia from blueprint machines.

Air Cleaning

As with the case of ventilation, air cleaning systems are designed to reduce contaminant levels after the contaminant is airborne. Air cleaning for the control of building contaminants has been less widely used than ventilation. Air cleaning systems can be designed to control PM (dust), gas-phase contaminants, or both.

Particle or dust cleaning systems have a long history of use. Particle control has been widely applied. A minimum level of air cleaning (dust-stop filters) is used in most building HVAC systems to protect mechanical equipment from soiling. Less often, filtration systems are employed to maintain clean building spaces. A variety of filters and filtration systems are available for dust control in buildings. These vary from low-efficiency dust-stop filters, to medium-efficiency

Table 11.8 American Society of Heating, Refrigerating and Air-Conditioning
Engineers (ASHRAE) Ventilation Guidelines

Facility type	Estimated occupancy (persons/1000 ft²)	Outdoor air requirements	
		CFM/person	L/sec/person
Food and beverage			
Dining rooms	70	20	10
Bars, cocktail lounges	100	30	15
Hotels, motels, resorts, etc.			
Lobbies	30	15	8
Assembly rooms	120	15	8
Dormitory sleeping areas	20	15	8
Office spaces	7	20	10
Conference rooms	50	20	10
Smoking lounges	70	60	30
Sports and amusements			
Spectator areas	150	15	8
Game rooms	70	25	13
Ballrooms/discos	100	25	13
Theaters/auditoriums	150	15	8
Education			
Classrooms	50	15	8
Libraries	20	15	8
Hospital			
Patient rooms	10	25	13

pleated-panel filters, to high-efficiency pleated panel filters and electronic air cleaners, to very high-efficiency HEPA filters which have the capability of removing 99.97% of particles at the hardest to control diameter of 0.3 µm. Particle filtration systems can be used relatively effectively in reducing indoor particle and mold spore levels.

Gas-phase contaminant control by air cleaning is more difficult, expensive, and usually less effective than particle control. Such air cleaning typically employs sorbents or specially impregnated sorbents in panel beds or in pleated filters. The most widely used sorbent is activated carbon (charcoal). It can be effective in removing VOCs which have a relatively high molecular weight (such as toluene, benzene, xylene, methyl chloroform, etc.). It is relatively ineffective in removing low molecular weight substances such as ethylene, HCHO, etc. Such substances may be removed from air by potassium permanganate on activated alumina and specially impregnated charcoal.

Many air cleaners are commercially available (particularly in the residential market) whose manufacturers claim a high degree of effectiveness in removing a broad range of indoor contaminants. Most claims have not been validated. Based on design features and operating parameters, it is unlikely that such claims are true.

INDOOR AIR QUALITY AND PUBLIC POLICY

In the U.S., the identification of a potential health-threatening environmental problem has historically been followed by the enactment of federal legislation which provides a regulatory framework for control and/or the promulgation of specific regulations by USEPA. In the case of IAQ, a variety of federal agencies have limited authority under existing legislation. These include the Consumer Product Safety Commission (CPSC) which regulates dangerous and defective products, USEPA which has specific authority relative to radon and general authority to conduct IAQ research, the Department of Housing and Urban Development (HUD) which sets standards on the habitability of manufactured houses and federally subsidized housing, OSHA which sets and enforces workplace health standards, and the National Institute of Occupational Safety and Health (NIOSH) which has the authority to recommend workplace health standards, conduct research, and perform workplace health hazard evaluations (including IAQ).

Under its statutory authority, the CPSC attempted to ban the use of UFFI. This ban was subsequently overturned by a federal appeals court. The CPSC also evaluated emissions of HCHO from wood products and emissions from kerosene and gas space heaters for possible regulatory action. Under its statutory authority, HUD issued product standards for emissions of HCHO from particleboard and hardwood plywood used in manufactured housing. It also required manufacturers to prominently post warning signs on their products about the health dangers of HCHO.

USEPA has been the most active federal agency in actions involving IAQ. It has the authority to promulgate and enforce regulations under the Asbestos Hazard Emergency Response Act (AHERA) which requires schools to conduct asbestos inspections, prepare management plans, and use accredited personnel for those functions, as well as conducting abatement operations.

In 1988, USEPA, in conjunction with the U.S. Surgeon General, issued a public health advisory recommending that all homeowners test their homes for radon and to abate high radon levels. This advisory was subsequently extended to schools. USEPA has been authorized by Congress to promulgate rules for radon testing and abatement activities.

USEPA has maintained a leadership role in the broad area of IAQ in part because of their expertise related to air quality and the agency's research and public policy interest. Over the past decade, USEPA has increasingly emphasized voluntary efforts in controlling pollution. Because IAQ is not in many instances (houses) in the public domain, it naturally lends itself to voluntary efforts and the leadership experience of an agency such as USEPA.

OSHA has also attempted to deal with IAQ concerns associated with both industrial and nonindustrial workplaces. It has identified ETS as a human carcinogen. Such a finding is often a prelude to regulation of workplace exposures. OSHA has also issued proposed rules that would regulate smoking and ventilation in nonindustrial workplaces. These proposed rules are highly controversial, and their promulgation in final form is in doubt at the time of this writing.

There are inherent limitations on the ability of government to regulate IAQ problems. As a result, it is apparent that a significant public education program is needed to provide individuals with the appropriate information so they can make informed choices about the quality of air which they wish to breathe indoors.

READINGS

1. American Conference of Governmental Industrial Hygienists. 1984. "Evaluating Office Environmental Problems." *Annals ACGIH. Vol. 10.*

2. ASHRAE Standard 62-1989. *Ventilation for Acceptable Air Quality.* American Society of Heating, Refrigerating and Air-Conditioning Engineers, Atlanta, GA.

3. Budiansky, S. 1980. Indoor Air Pollution. *Environ. Sci. Technol.* 14:1023–1027.

4. Burge, H.A. (Ed.). 1995. *Bioaerosols.* CRC/Lewis Publishers, Boca Raton, FL.

5. Cone, J.E. and M.Q. Hodgson (Eds.). 1989. "Problem Buildings: Building-Associated Illness and the Sick Building Syndrome." *State of the Art Reviews — Occupational Medicine.* Hanley and Belfus, Inc., Philadelphia, PA.

6. Ferris, B.G., Jr., F.E. Speizer, J.D. Spengler, D. Dockery, Y.M.M. Bishop, M. Wolfson, and S. Colome. 1978. "Relationships Between Outdoor and Indoor Air Pollution and Implications on Health." 25–37. In: P.O. Fanger and O. Valbjorn (Eds.). *Proceedings on First International Indoor Climate Symposium.* Danish Building Research Institute, Copenhagen.

7. Godish, T. 1989. *Indoor Air Pollution Control.* Lewis Publishers, Inc., Chelsea, MI.

8. Godish, T. 1995. *Sick Buildings: Definition, Diagnosis and Mitigation.* CRC/Lewis Publishers, Boca Raton, FL.

9. Knight, C.V. 1981. "Indoor Air Quality Related to Wood Heaters." In: *Proceedings of IAQ '86: Managing Indoor Air for Health and Energy Conservation.* American Society of Heating, Refrigerating and Air-Conditioning Engineers, Atlanta, GA.

10. Leaderer, B.P. 1982. "Air Pollutant Emissions from Kerosene Space Heaters." *Science* 218:1113–1115.

11. Molhave, L. 1990. "Volatile Organic Compounds, Indoor Air Quality and Health." 15–34. In: *Proceedings Fifth International Conference on Indoor Air Quality and Climate. Vol. V.,* Toronto.

12. Morey, P.R., J.C. Feeley, and Q.A. Othen (Eds.). 1990. *Biological Contaminants in Indoor Environments.* ASTM STP 1071. American Society of Testing Materials, Philadelphia, PA.

13. Moschandreas, D.J. 1981. "Exposure to Pollutants and Daily Time Budgets of People." In: Symposium on Health Aspects of Indoor Air Pollution. *Bull. N.Y. Acad. Med.* 57:845–860.

14. National Research Council. Committee on Indoor Pollutants. 1981. *Indoor Air Pollutants.* National Academy Press, Washington, DC.

15. Norback, D. and C. Edling. 1990. "Indoor Air Quality and Personal Factors Related to the Sick Building Syndrome." *Scand. J. Work Environ. Health* 16:121–128.

16. Samet, J.M. and J.D. Spengler (Eds.). 1991. *Indoor Air Pollution — A Health Perspective.* John Hopkins University Press, Baltimore, MD.

17. Seitz, T.A. 1989. "NIOSH Indoor Air Quality Investigations, 1971-1988." pp. 163–171. In: D.M. Weekes and R.B. Gammage (Eds.). *Proceedings of the International Indoor Air Quality Symposium: The Practitioner's Approach to Indoor Air Quality Investigations.* American Industrial Hygiene Association, Akron, OH.

18. Skov, P. and O. Valbjorn. 1987. "The Sick Building in the Office Environment: The Danish Town Hall Study." *Environ. Int.* 13:339–349.

19. Spengler, J.D. and K. Sexton. 1983. "Indoor Air Pollution: A Public Health Perspective." *Science* 221:9–17.

20. USEPA. 1986. *A Citizen's Guide to Radon.* EPA-86-004.

21. USEPA/NIOSH. 1991. *Building Air Quality: A Guide for Building Owners and Facility Managers.* EPA/400/1-91/1003. DDHS (NIOSH) Publication Number 91-114.
22. U.S. Surgeon General. 1986. *The Health Consequences of Involuntary Smoking.* DDHS (CDC) 87-8398.
23. Walsh, P.J., C.S. Dudney, and E.D. Copenhaver (Eds.). 1984. *Indoor Air Quality.* CRC Publishers, Boca Raton, FL.
24. Weekes, D.M., and R.R. Gammage (Eds.). 1989. *Proceedings of the Indoor Air Quality International Symposium: The Practitioner's Approach to Indoor Air Quality Investigations.* American Industrial Hygiene Association, Fairfax, VA.
25. Yocum, J.E. 1982. "Indoor-Outdoor Relationships. A Critical Review." *JAPCA* 32:500–520.

QUESTIONS

1. Compare the health risks associated with indoor air pollution to those from ambient pollution. How comparable are they?

2. What are the health risks associated with radon exposures in buildings? What factors contribute to these risks?

3. Indoor/outdoor ratios of various contaminants vary considerably. Discuss indoor/outdoor ratios for CO, O_3, SO_2, and RSP and factors which affect them.

4. Describe advantages and disadvantages of using general dilution ventilation to control contaminants in an office building.

5. Formaldehyde is a major indoor contaminant. Describe its health effects and factors which affect indoor levels.

6. Describe three major people-related risk factors for SBS-type symptoms.

7. Describe differences between Legionnaires' disease, hypersensitivity pneumonitis, and asthma and factors which cause or contribute to these ailments.

8. What is the TVOC theory? What is its relationship to SBS-type symptoms?

9. Describe three major ways that occupants of residential environments are exposed to pesticides.

10. Describe control measures that could be effectively used to control (1) human bioeffluents, (2) CO, (3) HCHO, (4) lavatory odors, and (5) respirable particles.

11. Describe steps you would employ in conducting an investigation in an office or school building.

12. From a public policy perspective, how are ambient and indoor air pollution problems different?

13. In controlling contaminant levels in building spaces, ventilation is more commonly used than air cleaning. Why?

14. What are the major health risks associated with environmental tobacco smoke?

15. There are a variety of indoor air-quality problems. Based on health statistics and the nature of the illness, which is the most important? Make the case for your answer.

16. Asbestos-containing products occur in many buildings. What are the health risks to general building occupants? How do these compare to service personnel?

17. Though health complaints may be associated with a building or work environment, symptoms may not be due to exposure to airborne contaminants. How may this be so?

18. How do indoor pollution problems differ in residences as compared to public access buildings?

19. What are differences (if any) between problem buildings and sick buildings?

20. What are the advantages of source control over contaminant control?

12   NOISE POLLUTION

Noise can be characterized as unwanted sound. It is unwanted because it can cause hearing losses, interfere with speech communication, disturb sleep, and interfere with the performance of complex tasks. It may also be unwanted because it is annoying, disturbing moods, relaxation, and privacy. Noise will be discussed here in the context of air pollution, as air is the principal medium through which it is transmitted and received by the human ear.

SOUND

Sound is a form of energy produced by a vibrating object or an aerodynamic disturbance. The former includes our vocal chords, operating motors, whistles; the latter, thunder, sonic booms, and ventilating fans.

Sound energy is propagated in the form of waves which represent a compression and decompression of molecules of air, water, or solids. The disturbance of air molecules by sound energy produces variations in normal atmospheric pressure. As these pressure variations reach our eardrums, they cause them to vibrate. The transmission of these vibrations to the inner ear and their interpretation by the brain result in our sensory perception of sound.

Sound can only be transmitted through a medium that contains molecules — it cannot move through a vacuum. The speed of sound depends on the density of the transmission medium, with increased speed associated with increased density. The speed of sound in air, water, and steel are 1125, 4860, and 16,500 ft/sec (343, 1482, and 5030 m/sec), respectively.

Sound Energy

When an object vibrates, it radiates acoustical energy. The rate at which a source produces and radiates acoustical or sound energy is described as sound power, expressed in energy units per unit time. In conventional use, sound power values are expressed in dimensionless units called decibels (dB), calculated from the following equation:

$$S_P = 10 \log_{10} \frac{W_e}{W_o} dB \qquad (12.1)$$

where W_e = measured sound energy, watts (w) and W_o = reference power, 10^{-12}w.

The sound power is the logarithm of the ratio of the measured power and a reference power (10^{-12} w) multiplied by 10.

When sound energy passes through a unit area, its flow is described as sound intensity. It is typically expressed as watts per square meter or square centimeter. sound intensity values are expressed in decibels, and their calculation is similar to sound power.

$$S_1 = 10 \log_{10} \frac{1_e}{I_o} dB \qquad (12.2)$$

where I_e = measured sound intensity, w/m^2 and I_o = reference intensity, 10^{-12}w/m^2.

When sound energy radiating from a source strikes a surface, it produces sound pressure. This pressure can be measured by instruments known as sound pressure level meters. Sound pressure values are also expressed in decibels. Calculations of sound pressure levels differ slightly from those for sound power and sound intensity. Since pressure in sound waves varies sinusoidally, the root mean square values (square root) of the pressure changes are used to calculate sound pressure levels.

$$Spl = 10 \log_{10} \frac{P_e^2}{P_r^2} dB \qquad (12.3)$$

$$Spl = 20 \log_{10} \frac{P_e}{P_r} dB \qquad (12.4)$$

In calculations of sound pressure, the reference pressure (P_r) is the threshold of human hearing, 2×10^{-4} μbars or 2×10^{-5}N/m^2. P_e is the sound pressure being measured. For purposes of illustration, let us calculate the sound pressure level when P_e = 2.0 μbar.

$$Spl = 20 \log_{10} \frac{2.0}{0.0002} dB \qquad (12.5)$$

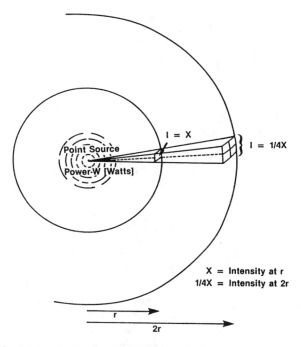

Figure 12.1 Sound propagation from a vibrating source.

$$\text{Spl} = 20 \log_{10} 10,000 \text{ dB}$$

$$\text{Spl} = 20\,(4)\,\text{dB}$$

$$\text{Spl} = 80 \text{ dB}$$

As an object vibrates, sound waves radiate outward in an ever-expanding sphere. At increasing distances, the sound power remains the same but sound intensity declines as the area through which sound energy passes increases. In a free field (space where sound is not reflected) sound intensity and sound pressure are reduced by 6 dB (factor of four) each time the distance from the source is doubled. This relationship is illustrated in Figure 12.1. At a distance of two radii (2r), the area of the sound sphere is four times as large as the area at a distance of one radius (r). The change in sound intensity and pressure follows the inverse square law (Equation 12.6).

$$S_1 = \frac{w}{4\pi r^2} \qquad (12.6)$$

where: w = watts and $4\pi r^2$ = area of the sound sphere.

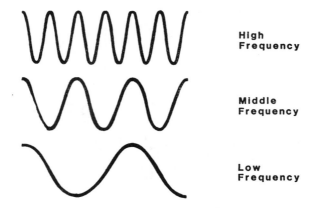

High
Frequency

Middle
Frequency

Low
Frequency

Figure 12.2 Sound waves of varying frequencies.

Each time the radius doubles, the area increases by $4\pi r^2$ of its previous value and sound intensity decreases inversely with this increase in area.

As sound is propagated by a source, it is radiated in the form of waves. These waves are formed by alternating compressions (wave peaks) and rarefactions (wave troughs) and vary in their length, frequency, and amplitude. The shorter the wavelength, the more frequent the waves are per unit time. The higher the wave, the more sound energy it has. Sound waves of varying length and frequency are illustrated in Figure 12.2. Sound waves of varying amplitude are illustrated in Figure 12.3.

Frequency is a major characteristic of sound and its discrimination by the human auditory organ constitutes the act of hearing. Frequency is expressed in cycles per second or hertz (Hz). If 1000 complete sound waves pass a point per second, it has a frequency of 1000 Hz

Sound generated by a source is not generally produced as discrete frequencies. Most sounds have a range of dominant frequencies by which they are characterized. For example, male human speech is characterized by dominant frequencies of less than 2000 Hz. For the speech of females, the range of

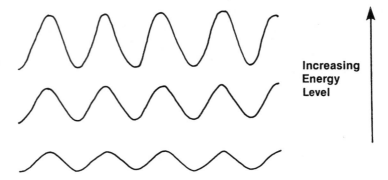

Increasing
Energy
Level

Figure 12.3 Sound waves showing increasing amplitude or intensity.

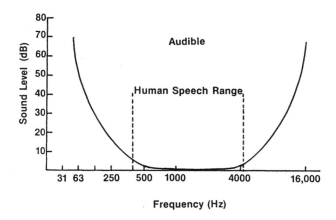

Figure 12.4 Frequency range of human hearing.

dominant frequencies is somewhat higher. Humans can hear sounds in the frequency range of 50–20,000 Hz. Frequencies in this range are not heard equally well. As can be seen in Figure 12.4, humans hear best those frequencies that correspond to human speech. Human hearing in the very low and high frequencies is relatively limited.

Whether a human can hear a sound or not depends on its spectrum of frequencies and its intensity. Decibel readings for a variety of sound sources experienced in the home, work, and ambient environments are summarized in Table 12.1. Note that 0 dB corresponds to the threshold of hearing. Sounds below 0 dB occur but cannot be heard. Sound pressure levels above 70 dB become increasingly objectionable as they interfere with speech communication. Brief exposures to sound levels above 140 dB may be painful, and long-term exposures to sound levels above 90 dB may result in hearing loss.

Sound Measurement

The instrument used to measure sound in decibels is described as a sound pressure level meter (Figure 12.5). It consists of a microphone, attenuator, amplifiers and an indicating meter. The sound pressure level meter measures sound pressure, converting it into electrical energy that may be displayed on the indicating meter by deflection of a needle or as a digital display. An integral part of sound measuring instruments are weighting networks or scales that emphasize or deemphasize sound in selected frequency ranges.

Weighting Networks

Weighting scales reflect our interest in noise effects on humans. Loudness, for example, not only varies with the sound pressure, but also with frequency, and the way it varies with frequency depends on sound pressure. It is therefore

Table 12.1 Sound Pressure Levels, Sources, and Human Responses

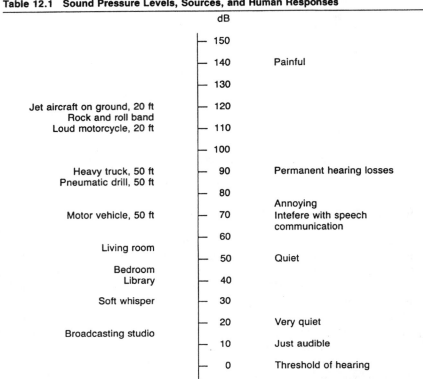

	dB	
	150	
	140	Painful
	130	
Jet aircraft on ground, 20 ft	120	
Rock and roll band		
Loud motorcycle, 20 ft	110	
	100	
Heavy truck, 50 ft	90	Permanent hearing losses
Pneumatic drill, 50 ft		
	80	
		Annoying
Motor vehicle, 50 ft	70	Intefere with speech communication
	60	
Living room		
	50	Quiet
Bedroom		
Library	40	
Soft whisper	30	
	20	Very quiet
Broadcasting studio		
	10	Just audible
	0	Threshold of hearing

useful to know something about the frequencies we are measuring, and weighting networks allow us, to some extent, to take into account human responses to sound.

Three weighting networks have been established by the American National Standards Institute (ANSI) for general purpose use. These include the A, B, and C scales. Figure 12.6 show how these three scales differ in their ability to measure sound levels across a frequency range. The A scale discriminates against frequencies less than 600 Hz. Low-frequency discrimination is more moderate on the B scale. The C scale corresponds more closely to the actual sound energy produced; its response is relatively uniform across the frequency spectrum. Readings on all three networks provide some indication of the frequency distribution of a recorded sound. If the level is greater on the C scale, this indicates that some of the sound is in frequencies less that 600 Hz. If no differences are observed among the three scales, this indicates that most of the measured sound is in frequencies greater than 600 Hz.

The sound reading should be expressed as a function of the networks used, for example, 20 dBA, 20 dBB, 20 dBC. At one time, it was customary to select the weighting networks according to level. For levels below 55 dB, the A scale

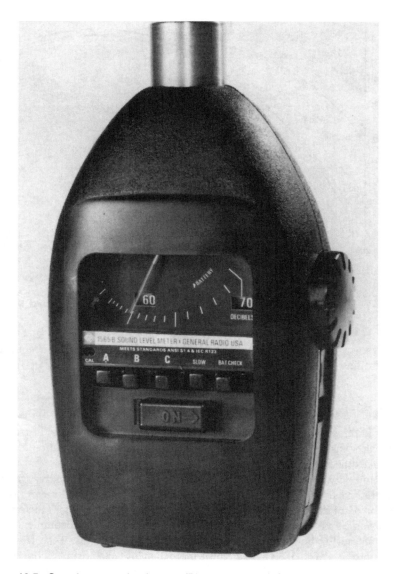

Figure 12.5 Sound pressure level meter. (Photo courtesy of GenRad, Inc., Concord, MA.)

was recommended, from 55–85 dB, the B scale, and above 85 dB, the C scale. It is now common practice to employ the A scale for most sound measurements. The dominance of the A scale is associated with its acceptance for measuring sound levels in occupational environments where compensable hearing impairments are limited to the frequency range of 500 to 2000 Hz. It comes closest to approximating hearing damage risk. Since humans show little response to fre-

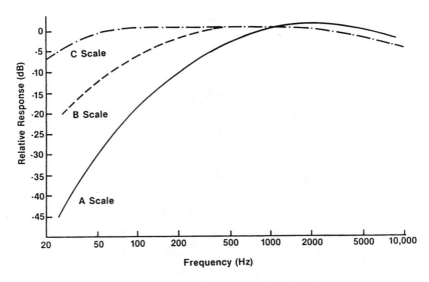

Figure 12.6 Frequency discrimination by A, B, and C weighting networks on a sound pressure level meter. (From Peterson, A.P. 1978. GenRad, Concord, MA. With permission.)

quencies less than 500 Hz, there is limited value in measuring such frequencies in controlling worker noise exposure.

Meter Response and Instrument Accuracy

In addition to the weighting networks, most sound pressure level instruments have a slow/fast option. The slow setting averages the rapid fluctuation of sound and makes the meter easier to read.

Commonly available sound pressure level meters are rated as type I, II, or III instruments. These designations reflect their accuracy. Type I meters have a precision of ±1 dB. These are primarily research instruments. Type II meters have a precision of ±2 dB. These are general purpose instruments recommended for use in measuring both industrial and community sound levels. Type III meters have a precision of ±3dB. These are little more than toys and are not recommended for any regulatory or technical use.

Measurement of Impulse Sound

General purpose sound pressure level meters cannot be used to measure impulse noises, i.e., those produced by a variety of industrial processes such as punch presses and pile drivers. These sounds consist of rapidly rising and falling sound pressure. Instruments must be specially designed to measure both peak and average impulse sound levels.

Table 12.2 Center Frequencies and Frequency Ranges of Octave Bands

Center frequency, Hz	Frequency range, Hz
31.5	18–45
63	45–90
125	90–180
250	180–355
500	355–710
1000	710–1400
2000	1400–2800
4000	2800–5600
8000	5600–11,200

Spectrum Analysis

Sound pressure levels can be determined for given frequency ranges. Instruments with these capabilities are called spectrum analyzers. The most common of these are octave band analyzers, which divide the frequency range into octaves. Octaves (Table 12.2) are characterized by a range of frequencies in which the upper limit is twice the lower limit. The center frequency of each octave is the geometric mean of these two limits.

Averaging Sound Measurements

It is often desirable to summarize multiple sound readings. For worker exposure, average levels are necessary to determine dose or exposure. When two or more sources are producing sound at the same time, we may wish to know their individual contributions as well as the total sound level. Decibels, because they are geometric values, cannot be added. Average values must be either determined from tables or calculated from an appropriate equation (Equation 12.7).

To calculate average exposures where multiple measurements are made, we must first convert decibel readings to arithmetic values, which are in turn averaged and converted back to a decibel value. This can be accomplished by calculating the Leq, or equivalent sound level, from Equation 12.7.

$$\text{Leq} = 10 \log \frac{\sum_{i=1}^{n} 10^{\frac{L_i}{10}}}{N} \qquad (12.7)$$

where L_i corresponds to each measure value in dB.

LOUDNESS

The human perception of increasing sound pressure describes the concept of loudness. As previously stated, loudness and sound pressure are related, but not equivalent. This is due in part to the fact that the human ear does not respond equally well across the sound frequency spectrum (Figure 12.3). More importantly, physiological processes in the human auditory system protect it from excessive sound stimulation. As a result, the relationship between sound level and auditory sensation is not a linear one across the audible range of frequencies. The relationship is close to being logarithmic rather than linear.

Loudness Ratings

Loudness, unlike sound pressure, cannot be measured directly. It can, however, be inferred from listener panel ratings of relative loudness. Figure 12.7 summarizes equal loudness contours for a range of frequencies and sound pressure levels obtained from panel studies. These equal loudness contours were obtained by asking listeners to rate the loudness of sound of various frequencies and sound pressure levels relative to a pure tone of 1000 Hz at a given sound pressure level. The loudness level is defined in units called phons, which at 1000 Hz are directly related to sound pressure level in decibels. This relationship, however, changes with frequency. Although the 40 phon contour line corresponds to 40 dB at 1000 Hz, 50 dB of sound pressure are required to produce a sound equally loud at 100 Hz.

Loudness can also be rated on the basis of judgments of listeners when one sound is one half, same, twice, four times, etc., as loud compared to another. Units of loudness developed from such ratings are called sones. The reference

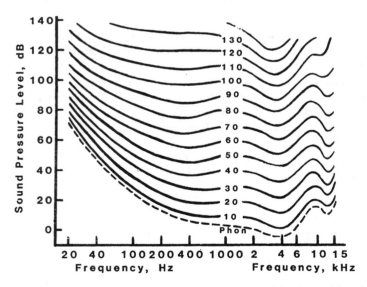

Figure 12.7 Relationship between sound pressure level and loudness (phons). (From Peterson, A.P. 1978. GenRad, Concord, MA. With permission.)

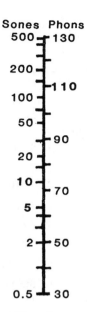

Figure 12.8 Phon/sone conversion nomogram.

sound for these ratings is a 1000-Hz pure tone with a sound pressure level of 40 dB. At 40 dB and 1000 Hz, the sound is characterized as having one sone of loudness and 40 phons. A sound twice this loud would be 2 sones; one four times as loud 4 sones.

Phons and sones at 1000 Hz can be interconverted by means of the nomograph in Figure 12.8. At 1000 Hz a 20-dB increase in sound pressure from 40 to 60 dB would result in a 20-phon increase in loudness and would be perceived as being four times as loud. The pressure/loudness ratio here would be 10:4, as a tenfold increase in sound pressure would be perceived as a fourfold increase in loudness. This can be roughly generalized across the frequency range of 500 to 2000 Hz.

NOISE EFFECTS

Annoyance

A major objectionable characteristic of sound or noise is the psychological annoyance it may create. The degree of annoyance may depend on the activity engaged in, previous conditioning, and the nature of the sound. Annoyance may often be related to loudness, as a loud noise is more annoying than one less so. A variety of noise annoyance criteria have been developed to judge sounds for their annoyance, noisiness, unacceptability, etc. From these criteria, it is evident that higher-frequency sounds (~5000 Hz) are more annoying than lower frequencies at the same sound level. Annoyance/acceptability levels may be determined from loudness calculations for broadband sound, the perceived noise level of Kryter, and noise criterion curves of Beranek.

In Kryter's noise rating scheme, observers compare noise based on their noisiness or acceptability. From these results, an effective perceived noise level (EPNdB) can be calculated. Units of noisiness are expressed as "noys." EPNdB ratings are widely used for rating aircraft noise.

The noise criteria curve rating system developed by Beranek was designed to achieve satisfactory background noise levels inside office buildings. It is useful in determining where and what noise reduction is desirable.

Interference with Speech Communication

In addition to a generalized factor of annoyance, sounds can interfere with human voice communication. Unwanted sounds can interfere with (or mask) our ability to hear human speech. This masking may make speech inaudible, change its quality, and reduce its clarity.

As background sound increases, we may not hear speech intelligibly. Consonants, although they contain most of the information in a word, are masked more than vowels, which are more intensely produced. Depending on the speech level, the source and relative frequency spectrum of speech and noise, certain noises may mask some speech sounds and not others.

Noise can be rated relative to its ability to interfere with speech communication. The preferred octave speech-interference level (PSIL) is based on the arithmetic average of sound pressure levels centered on octave bands 500, 1000, and 2000 Hz, which cover the human speech spectrum. One can use this rating system (Figure 12.9) to determine when speech communication and telephone use are easy, difficult, or impossible. One can also determine what changes in PSIL are necessary to decrease/increase communication difficulty. As evident in

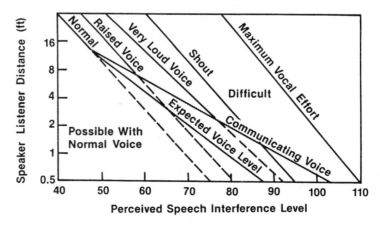

Figure 12.9 Relationship between listener distance and speech interference. (From National Bureau of Standards. 1971. NTID Publication No. 300.15.)

Figure 12.9, the distance of effective communication sharply decreases with increased PSIL levels.

HEALTH EFFECTS

Noise exposures have been alleged to be responsible for a variety of physical as well as psychological health effects. There has, however, been little good scientific evidence to support this thesis. Noise exposures can, however, adversely affect human hearing, that is, induce hearing loss. Noise-induced hearing losses are primarily associated with industrial noise exposures.

Mechanism of Hearing

To better understand how noise can cause hearing losses, let us examine how we hear. The organ of hearing of course is the ear. It consists of three regions: the outer, middle, and inner ear (Figure 12.10). The outer ear consists of the cartilaginous auricle and a short, funnel-shaped canal called the external auditory meatus. The outer ear canal is subtended by a thin membrane called the tympanic membrane or eardrum. The outer ear serves to funnel incoming sound waves to the eardrum, which vibrates at the same frequencies. The middle ear is an air-filled cavity that contains three bones in a series — the malleus (hammer), incus (anvil), and stapes (stirrup). The malleus is connected to the eardrum and the stapes to the oval window, the opening to the inner ear. These tiny middle ear bones transmit sound vibrations from the eardrum to the sound-sensitive portion of the inner ear. The change in the surface area from the eardrum (64 mm^2) to the oval window (3.2 mm^2) serves to amplify sound vibrations.

The inner ear consists of a series of passages known as the bony labyrinth. The bony labyrinth consists of the cochlea, vestibule, and semicircular canals. The cochlea (a snail-shaped structure) is the principal organ of the inner ear

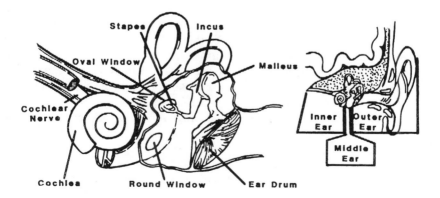

Figure 12.10 Anatomy of the ear.

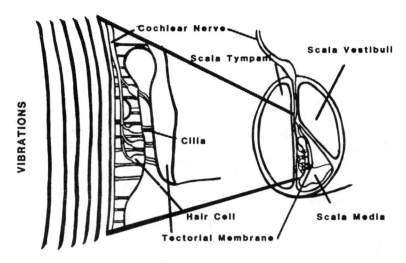

Figure 12.11   Cochlea and organ of Corti.

concerned with sound reception. The semicircular canals constitute the organ of equilibrium and body orientation.

The stapes of the inner ear transmits sound vibrations through the oval window to the fluid (perilymph) inside the cochlea. These continue to travel through the perilymph to the organ of Corti, which contains hair cells (Figure 12.11) that rest on frequency-sensitive fibers. Vibration of these fibers causes the hair cells to push/pull against the tectorial membrane, producing impulses that travel to the brain.

Noise-Induced Hearing Losses

Noise can result in three distinct forms of hearing loss. These, in increasing degree of severity, are: temporary threshold shift, permanent threshold shift, and acoustic trauma. In the first two, a change in hearing level relative to a previously measured level is indicated. A temporary threshold shift (TTS) is a reversible hearing loss which is of short duration, usually returning within a few seconds, minutes, or hours after exposure. In extreme cases, it may persist for weeks. Permanent threshold shift (PTS) is nonreversible; that is, it has no possibility for future recovery. Acoustic trauma is often caused by a short intense exposure of the auditory system to an extreme sound pressure rise associated with major explosions, fireworks, or gunfire. Acoustic trauma may result in ruptured ear drums and middle ear bones and/or damage to the organ of Corti. In less severe injuries associated with PTS caused by prolonged exposure to high-intensity noise, hair cells and associated structures of the organ of Corti may be damaged to varying degrees. Once the hair cells are destroyed, they do not regenerate; their function is lost forever.

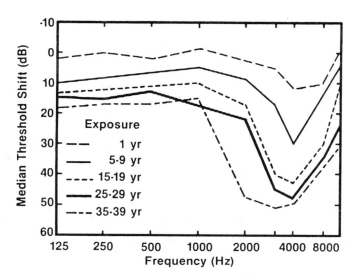

Figure 12.12 Noise-induced hearing changes in textile workers. (Reprinted from Taylor, W. et al. 1965. "Study of Noise and Hearing in Jute Weavers." *J. Acoust. Soc. Am.* 38:113–116. With permission.)

Occupational Exposure

Noise-induced hearing losses are most commonly associated with a variety of processes where prolonged exposure to high-intensity (greater than 90 dBA) noise levels are experienced. Changes in hearing acuity among jute weavers are depicted in Figure 12.12. Note the change in threshold level with increasing exposure duration. Also note that the earliest and most significant hearing losses occur at higher frequencies (approximately 4000 Hz), at the upper end of the speech range. With prolonged exposure, these initial changes in threshold broaden over a larger frequency range.

Under worker's compensation laws in the U.S., individuals with an industrially related hearing disability can obtain compensation for their hearing loss if it can be determined that their loss is compensable. This requires the establishment by audiometric evaluation of a 25-dB threshold shift in the frequency range of 500 to 2000 Hz. The large threshold shift at 4000 Hz observed for textile workers in Figure 12.12, although significant in terms of their ability to hear certain sounds, particularly consonants, would not be compensable.

The relationship between auditory damage/hearing losses and noise exposure is influenced by a variety of factors. These include sound pressure, exposure duration, sound frequency characteristics, individual susceptibility, susceptibility to specific frequencies, and characteristics of the time-dependent development in PTS. The sound pressure is important, as it can be related to increased hearing damage. With respect to frequency, noise in the band 300–600 Hz would affect

Table 12.3   U.S. Department of Labor Occupational
Noise Exposure Standards

Exposure duration (hour/day)	Permissible sound level (dBA)
8	90
6	92
4	95
3	97
2	100
1.5	102
1	105
0.5	110
≤0.25	115

hearing mainly at 1000 Hz; 600–1200 Hz at about 1500 Hz; 1200–2400, at about 3000 Hz; and 2400–4800, at 4000 and 6000 Hz.

Occupational Standards

Permissible noise exposure standards have been developed and promulgated by the Department of Labor to protect workers from overexposure to noise levels which have the potential for causing hearing losses. These standards are summarized in Table 12.3. In theory, a worker can be exposed to an average exposure of 90 dB, 8 hr/day, 5 days/week throughout a work career without risk of suffering a significant hearing loss. With increased sound level, the permissible exposure period per day decreases according to a 5-dB exchange ratio; that is, for every 5-dB increase in exposure level, the permissible exposure duration is halved. The USEPA, in contrast, recommends the 3-dB exchange ratio commonly used in Europe. It also has recommended a maximum average exposure of 85 dBA for 8 hr/day, 5-day work week, which it believes would more adequately protect workers. The difference between the OSHA/Department of Labor standards and those recommended by USEPA represents economic concerns that such standards would require significant economic burdens on industry to achieve compliance; it represents a compromise between worker protection from noise exposure and economic interests of industry.

COMMUNITY NOISE

Unlike occupational exposure, community noise does not apparently result in any damage to human hearing. It can, however, affect the quality of our lives by interfering with speech communication, disturbing sleep and relaxation, and intruding on our privacy. Because of the annoyance of a variety of community noise sources, pollution control agencies receive a significant number of noise-related complaints, more than all other environmental problems combined.

Major contributors to community noise levels include surface and air transportation, construction sites, industrial sources, and a variety of human activities.

The magnitude of these problems depends on the intensity and occurrence of the unwanted sounds and the proximity of the receiver to the sound source. A very noisy source in an industrial area may be less significant than one of less sound power in or near a residential neighborhood. Noise produced during the day may not be as annoying as at night, when it may interfere with sleep. Community noise problems usually result when an area or neighborhood has a mix of uses.

In residential neighborhoods, noise problems are often associated with domestic sources or activity. These may include air conditioners, lawn mowers, blaring radios or TVs, barking dogs, children playing, and loud parties. While not particularly intense, such domestic noise may for a variety of reasons (primarily social) be particularly annoying.

Construction sites and industrial plants can also be particularly noisy. These, however, tend to be less of a problem, as they are often isolated in time and space. Heavy construction activity generally occurs in commercial and industrial areas, and thus often does not significantly intrude on residential neighborhoods. Additionally, such activity is usually limited to the daytime hours.

Transportation

Dominant ambient noise sources are motor vehicle traffic, light-duty vehicles and trucks, heavy-duty trucks and buses, and motorcycles. Noise levels increase with increased vehicle numbers and speed. They also increase with vehicle size and weight, with large trucks responsible for the highest noise emissions. Highway noise results from the production of sound by the vehicle directly and by the interaction of vehicle tires and roadways. At speeds above 35 mph, tire noise dominates the noise emission picture of motor vehicles.

Aircraft noise produced by takeoff and landing operations at metropolitan airports is the second major source of community noise and the major source of citizen complaints. Complaints often result when aircraft fly over residential neighborhoods, with these flyovers being particularly disturbing at night. Aircraft noise is uniquely annoying because of the high-pitched whine of jet engines.

Assessment of Community Noise Exposures

Population exposures to community noise are determined from Ldn, a long-term annoyance quantity determined from a day-night equivalent sound level

Table 12.4 Urban Noise Exposures

Outdoor Ldn	Urban traffic	Aircraft	Construction sites	Freeway
		Number of people affected x 10⁶		
≥70 dB	4–12	4–7	1–3	1–4
≥65 dB	15–33	8–15	3–6	2–5
≥60 dB	40–70	16–32	7–15	3–6

From USEPA. 1974. Federal Register 39(121):22297–22299.

during a 24-hr period with a 10-dBA penalty added for nighttime occurrences to account for increased annoyance. Population exposures (U.S.) to noise from urban traffic, aircraft, construction sites, and freeways are summarized in Table 12.4. As a reference, USEPA has identified an outdoor Ldn level of 55 dBA as the level that will not result in adverse health or welfare effects.

Control of Community Noise

The abatement of community noise requires the identification of noise as a problem of sufficient magnitude to establish a regulatory program for its control. Regulatory programs for noise control exist at local, state, and federal levels. These programs vary considerably in form and scope. Many local noise ordinances are qualitative, prohibiting excessive noise or noise that results in a public nuisance. Because of the subjective nature of such ordinances, they are difficult to enforce.

Standards

The regulation of community noise requires a quantitative approach; that is, noise regulations should be based on well-defined standards, violations of which can be measured. Standards should be based on scientific criteria as to acceptable noise levels. Noise standards in Minnesota are summarized in Table 12.6. They are based on sound pressure and duration and are defined for five zones, including single- and multiple-family residential, commercial, manufacturing, and industrial areas. They are specified for daytime (7 a.m. to 10 p.m.) and nighttime (10 p.m. to 7 a.m.).

The L_{10} value indicates that 10% of the time this standard can be exceeded; the L_1 value 1% of the time. The L_{max} is the maximum acceptable instantaneous noise applied to sudden blaring noises, e.g., a fast motor vehicle. The EPNdB is used for aircraft noise. It is a single-event standard that heavily weighs the presence of very annoying discrete frequencies associated with the whine of jet engines. The impulse standard is designed for noise of very short duration.

Note that in Table 12.5 the most stringent noise standards are applied to residential neighborhoods for the nighttime hours. These standards assume that night levels are tolerable in commercial, manufacturing, and industrial zones. They also assume that sounds of short duration are the most easily controlled.

FEDERAL NOISE PROGRAMS

A variety of federal agencies have authority to regulate and control noise. These include the Federal Aviation Agency (FAA), Federal Highway Administration (FHA), Occupational Safety and Health Administration (OSHA), Housing and Urban Development (HUD), and USEPA. The FAA has authority to impose and enforce noise emission standards for commercial aircraft operation. The FHA is responsible for issuing standards for new and improved federal highway con-

Table 12.5 Minnesota Noise Standards

Zone	Daytime (0700–2200 hour)					Nighttime (2200–0700 hour)				
	L_{10}[a] (dBA)	L_1 (dBA)	L_{max} (dBA)	EPNdB (dBA)	Impulse (pst)[b]	L_{10} (dBA)	L_1 (dBA)	L_{max} (dBA)	EPNdB (dBA)	Impulse (psf)
Residential (single-family)	65	70	75	88	1.25	55	60	65	787	0.40
Residential (multiple-family)	70	75	80	93	2.22	60	65	70	83	0.70
Commercial	75	80	85	98	3.95	65	70	75	88	1.25
Industrial	85	90	95	105	12.50					

[a] Based on an hourly average
[b] Pounds per square foot

From Tyler, J.M. et al. 1974. *JAPCA* 24:130–135. With permission.

struction. OSHA has significant noise control responsibilities that are designed to protect workers from noise-induced hearing losses. HUD's authority involves the specification of noise control standards and techniques for HUD-assisted new housing construction.

The principal responsibility for ambient or community noise is vested in the Environmental Protection Agency. The USEPA role in noise abatement was authorized by Congress in its Federal Noise Control Act of 1972 and the Quiet Communities Act of 1978.

In the Noise Control Act of 1972, USEPA was required to develop and publish noise criteria and scientific information on noise effects, and to identify levels of environmental noise, the attainment and maintenance of which are necessary for the protection of public health and welfare. USEPA was also required to identify major noise sources which would need regulation. Under this authority, USEPA promulgated noise emission standards for interstate motor carriers and buses, medium- and heavy-duty trucks, and railroad carriers. Other noise standards were applied to construction equipment as well. The Noise Control Act of 1972 also authorized regulation of manufacturers of a variety of equipment and appliances requiring labeling as to sound levels produced.

The Quiet Communities Act of 1978 was designed to provide financial assistance to state and local governments and regional planning agencies to investigate noise problems, plan and develop a noise control capability, and purchase equipment. It also provided for assistance to facilitate community development and enforcement of community noise control measures, as well as to develop abatement plans for areas around major transportation facilities, including airports, highways, and railyards.

CONTROL MEASURES

The accomplishment of noise abatement objectives requires that control requirements be imposed on noise producers. Noise control is best achieved at the source. For new products, noise emissions standards can be specified and designed into products such as aircraft engines, motor vehicle tires, industrial machinery, and home appliances. The control of aerodynamically produced noise may require changing air flow to reduce turbulence, or placing sound-absorbent materials along its path. The control of noise from vibrating surfaces may require mounts to limit vibration, the use of flexible materials in joints, or materials of sufficient mass that they are not easily subject to vibration.

For existing sources, noise reduction may require retrofitting by a variety of engineering modifications, including use of sound-damping materials. The economics of retrofitting are such that it is far less expensive to initially purchase less noisy equipment than to pay for its control after installation and use. Some existing noise sources can also be isolated, and sound-absorbing enclosures placed around them.

An alternative to noise control at the source is the modification of the noise path, and/or its reception. For example, operational changes in commercial airline and flight patterns during takeoff and landing can significantly reduce noise exposures. Such operational requirements are already in effect at the direction of the FAA.

In the planning process for major highways and airports, the location and routing of these can minimize noise exposures in residential neighborhoods. Where major highways are in close proximity to residential neighborhoods, sound-absorbing walls, earthen mounds, and vegetation may be utilized to attenuate sound. Inside buildings, sound-absorbing furnishings (carpets, drapery) or construction materials (acoustical tile) can significantly reduce sound. In occupational environments, worker exposure can be reduced by using sound-attenuating ear plugs or earmuffs.

READINGS

1. Beranek, L.L. (Ed.). 1971. *Noise and Vibration Control*. McGraw-Hill Book Company, New York.
2. Bragdon, C.R. 1980. *Municipal Noise Legislation*. Fairmont Press, Atlanta, GA.
3. Burns, W. 1973. *Noise and Man*. 2nd ed. J.P. Lippincott Company, Philadelphia.
4. Chanlett, E.T. 1973. *Environmental Protection*. McGraw-Hill Book Company, New York.
5. Cunniff, P.F. 1977. *Environmental Noise Pollution*. John Wiley & Sons, New York.
6. Ellerbusch, F. (Ed.). 1990. *Guide for Industrial Noise Control*. Scitech Publishers, Matawan, NJ.
7. Hammond, R.E. 1981. "Human Hearing." *Carolina Tips* 44(3):13–15.
8. May, D.N. (Ed.). 1978. *Handbook of Noise Assessment*. Van Nostrand Reinhold Environmental Engineering Series, New York.
9. National Bureau of Standards. 1971. *Fundamentals of Noise: Measurement, Rating Schemes and Standards*. NTID Publication No. 300.15.
10. Pelton, H.K. 1993. *Noise Control Management*. Van Nostrand Reinhold, New York.
11. Peterson, A.P. 1978. *Handbook of Noise Measurement*. 8th ed. GenRad, Concord, MA.
12. Ropes, J.M. and D.L. Williamson. 1979. "EPA's Implementation of the Quiet Communities Act of 1978." *Sound & Vibration*, December: 10–12.
13. Taylor, W. et al. 1965. "Study of Noise and Hearing in Jute Weavers." *J. Acoust. Soc. Am.* 38:113–116.
14. Tyler, J.M., L.V. Hinton, and J.G. Olin. 1974. "State Standards, Regulations and Responsibilities in Noise Pollution Control." *JAPCA* 24:130–135.
15. USEPA. 1974. "Identification of Products as Major Sources of Noise." Fed. Regist. 39(121):22297–22299.

QUESTIONS

1. How are sound pressure and loudness related?

2. How do the concepts of sound and noise differ?

3. Given a measured sound pressure of 0.02 µbar, what is the sound pressure level in decibels?

4. Given a sound pressure level of 80 dB, what is the measured sound pressure (P_e) of µbars?

5. Calculate the average decibel value or Leq for the following sound measurements: 65 dB, 70 dB, 60 dB, 55 dB, 75 dB.

6. How does distance affect the pressure of sound measured at a receptor?

7. Why would permissible nighttime noise standards be more restrictive than those for daylight hours in residential neighborhoods?

8. What is the significance of the 5 dB rule?

9. What advantages are there to using the A scale for measuring human sound exposure? What disadvantages (if any) are there to using the A scale?

10. What is the most important noise problem in the ambient environment? Why?

11. How do exposures to excessively intense sounds cause hearing loss?

12. Why can one hear two sounds equally well even though they differ in sound pressure level by 30 decibels?

13. A receptor is exposed to two different sound sources at the same time. Exposure to source A is equal to 85 dB, to source B 80 dB. What is the average sound exposure of the receptor?

14. Why can't humans talk (with understanding) with whales and dolphins?

13   QUANTITATIVE ASPECTS

As seen in previous chapters, the science of air quality is multidisciplinary, involving physics, chemistry, meteorology, climatology, biology, and engineering. The practice of air-quality protection/enhancement or air pollution control includes aspects of public policy development and implementation, air monitoring, emissions assessment, data summarization, modeling, and design and installation of control equipment. A variety of quantitative principles and applications are used in day-to-day pollution control practice. They are the subject of this chapter. Special attention is given to the gas laws and their application to air-quality and emissions sampling/monitoring, emission assessment procedures, data summarization, air-quality modeling, and engineering controls.

GAS PROPERTIES AND LAWS

Depending on temperature and pressure, all matter can exist as a solid, liquid, or gas. Under the range of environmental conditions which exist in the ambient environment, the substances which comprise the atmosphere (see Chapter 1) are gas phase. Other substances which enter the atmosphere from either natural or anthropogenic sources may undergo phase changes. Volatile liquids may vaporize and condense; molten metals may vaporize and condense; small gas-phase molecules may react to produce higher molecular weight molecules which condense; heterogenous phase reactions may occur.

The measurement of solid and liquid phase substances is usually done gravimetrically. For gases, however, gravimetric determination is much more difficult because gases are so many times lighter than containers needed to hold them. As a result, gas-phase substances are measured in volume units. The volume occupied by a given quantity of a gas is sensitive to both temperature and pressure.

The relationship between gas volume and temperature, gas volume and pressure, and gas volume, temperature, and pressure can be explained by Charles' Law, Boyle's Law, and the Combined Gas Law, respectively. If the temperature of a given mass (m) of a gas is changed, then its volume (V) will change proportionately so that the ratio of V to the absolute temperature (T, 273+ °C) does not change if the pressure is held constant. In Charles' Law

$$\frac{V}{T} = k \qquad (13.1)$$

The constant (k) will be different for each pressure and quantity of gas. Once held constant, the ratio will not change.

Let us assume that a quantity of a gas such as sulfur dioxide (SO_2) at 20°C and atmospheric pressure of 760 mmHg occupies a volume of 1 L. If pressure is held constant, what will its volume be at 40°C?

Since the ratio of the volume to absolute temperature must remain the same after heating to 40°C:

$$\frac{V_2}{T_2} = \frac{V_1}{T_1} \qquad (13.2)$$

Solving for V_2 the final volume

$$V_2 = V_1 \frac{T_2}{T_1} \qquad (13.3)$$

$$V_2 = \frac{313°K}{293°K} \times 1\,L = 1.07\,L \qquad (13.4)$$

Let us assume that the same quantity of SO_2 occupies a volume of 1 L at a temperature of 20°C and a pressure of 760 mmHg. If we now change the pressure to 730 mmHg, what will the new volume be?

Using Boyle's Law:

$$P \times V = k \qquad (13.5)$$

As with Charles' Law, the constant (k) will be different for each temperature and gas quantity (m). It will remain the same as these are held constant.

If we change the pressure, then the volume will change:

$$P_2 V_2 = P_1 V_1 \qquad (13.6)$$

$$V_2 = \frac{P_1 V_1}{P_2} \qquad (13.7)$$

$$V_2 = \frac{760\,\text{mmHg}}{730\,\text{mmHg}} \times 1\,\text{L} = 1.04\,\text{L} \qquad (13.8)$$

If we combine Charles' and Boyle's Laws we have:

$$\frac{PV}{T} = k \qquad (13.9)$$

In one of the most common applications of the combined gas law in air-quality practice, we can "correct" for changes in gas volume that occur under ambient conditions of temperature and pressure from those of instrument calibration in the laboratory.

Let us assume that a sampling device has a flow rate of 1 L/min at a calibration temperature of 22°C and pressure of 730 mmHg. Sampling is conducted for 1 hour (60 min) at ambient conditions of 15°C and 760 mmHg. What is the sampling volume?

From the Combined Gas Law:

$$\frac{P_2 V_2}{T_2} = \frac{P_1 V_1}{T_1} \qquad (13.10)$$

$$V_2 = V_1 \frac{P_1}{P_2} \times \frac{T_2}{T_1} \qquad (13.11)$$

$$V_2 = 1\,\text{L}/\text{min} \cdot 60\,\text{min} \frac{730\,\text{mmHg}}{760\,\text{mmHg}} \times \frac{282°\text{K}}{295°\text{K}} \qquad (13.12)$$

$$V_2 = 55.1\,\text{L}$$

Let us now assume that we collect a sample of dry gas from a stack. The sample flows through a heated filter, cooled impingers, a pump, and a dry gas meter. The flow rate based on a calibration temperature of 22°C and 730 mmHg was 25 L/min. Sampling duration was 30 min. The stack gas temperature was 220°C, and the pressure was 790 mmHg. Determine the volume sampled.

From the Combined Gas Law:

$$V_2 = V_1 \frac{P_1}{P_2} \times \frac{T_2}{T_1} \qquad (13.13)$$

where $V_1 = $ flow rate $\times$ time.

$$V_2 = 25\,\text{L}/\min \times 30\,\min \frac{730\,\text{mmHg}}{790\,\text{mmHg}} \times \frac{473°\text{K}}{295°\text{K}} \qquad (13.14)$$

$$V_2 = 1110\,\text{L}$$

The total volume of gas sampled at stack conditions was 1110 L or 1.11 m³.

$$1110\,\text{L} \times \frac{1\,\text{m}^3}{1000\,\text{L}} = 1.10\,\text{m}^3$$

How do P, V, and T behave when the mass of the gas is not held constant? Experimentally, the ratio PV/T is found to be proportional to the mass (m) of the gas:

$$\frac{P \times V}{T \times m} = k \qquad (13.15)$$

This constant is fixed for an individual gas but differs for other gases. It is inversely proportional to the molecular weight. By including the molecular weight (M), we have what is described as the Ideal Gas Law:

$$\frac{P \times V \times M}{T \times m} = R \qquad (13.16)$$

Since mass (m) of a gas divided by its molecular weight equals the number of moles (1 mole = M in grams/liter) of the gas, m/M can be expressed as n, and the equation can be rewritten as

$$PV = nRT$$

$$\text{where } R = \frac{62.36\,\text{mmHg} - \text{L}}{\text{g mol} - °\text{K}} \qquad (13.17)$$

Let us assume that a gas with $m = 6$ g occupies a volume of 5 L at $T = 50°\text{C}$ and $P = 730$ mmHg. What is its molecular weight?

From the Ideal Gas Law

$$n = \frac{m}{M} \qquad (13.18)$$

therefore:

$$M = \frac{mRT}{PV} \tag{13.19}$$

$$M = \frac{6\,g \cdot 62.4 \cdot 323°K}{730\,mmHg \cdot 5\,L} \tag{13.20}$$

$$M = 33$$

The density of a gas equals the ratio of its mass divided by volume.

$$d = \frac{m}{V} \tag{13.21}$$

From the Ideal Gas Law:

$$PV = \frac{m}{M}(R \cdot T) \tag{13.22}$$

$$\frac{m}{V} = \frac{MP}{RT} = d \tag{13.23}$$

What is the density of SO_2 at a temperature of 760 mmHg and 25°C?

$$d = \frac{64 \cdot 760}{62.4 \cdot 298} = 2.6\,g/L \tag{13.24}$$

Assuming that air primarily consists of N_2, O_2, and argon in approximate proportions of 78%, 21%, and 1%, respectively, what is its average molecular weight?

$$\overline{MW} = 0.78\,(28) + 0.21\,(32) + 0.01\,(40) = 28.96 \tag{13.25}$$

Based on this average molecular weight, calculate the density of air at a temperature of 25°C and a pressure of 760 mmHg.

$$d = \frac{MP}{RT} \tag{13.26}$$

$$d = \frac{28.96 \cdot 760}{62.4 \cdot 298} \tag{13.27}$$

$$d = 1.184\,g/L$$

Gas concentrations can be expressed in a variety of units. As indicated in Chapter 2, it is common to express them as mixing ratios (ppmv, v/v) or micrograms per cubic meter ($\mu g/m^3$). These are related to each other through the Ideal Gas Law. Because volume will change with changes in temperature and pressure, it is desirable to use reference conditions; a reference temperature of 25°C (298°K) and pressure of 760 mmHg is commonly used. At these reference conditions, a gram molecular weight (mole) of a gas will occupy a volume of 24.45 L.

Air Sampling Applications

Concentrations of various gases as seen in Chapter 7 can be determined by real-time analyzers or by intermittent sampling procedures. For illustrative purposes let us sample a volume of air for SO_2 using a gas impinger and determine its concentration using a wet chemical procedure. Sampling conditions included 50 ml of sampling reagent, 0.2 L/min flow rate, and 24-hour sampling duration. On analysis with the pararosaniline method, the concentration of SO_2 in the sample was 0.8 $\mu g/ml$. From these data we can determine the concentration of SO_2 in air during the 24-hour sampling period.

The total mass (m) of SO_2 collected must first be determined. The concentration of SO_2 was 0.8 $\mu g/ml$ in a solution volume of 50 ml. Therefore $m = 0.8$ $\mu g/ml \times 50$ ml = 40 μg. The sampling rate was 0.2 L/min; sampling duration, 24 hours. Assuming ambient conditions of temperature and pressure, the total volume sampled was 0.2 L/min $\times$ 60 min/hour $\times$ 24 hours = 288 L.

Since ppmv = $\mu l/L$, we can convert the concentration of 40 $\mu g/288$ L accordingly:

$$\frac{40\,\mu g}{288\,L} \cdot \frac{\mu\,mole}{64\,\mu g} \cdot \frac{24.45\,\mu L}{\mu\,mole} = 0.053\,\mu l\,/\,L \tag{13.28}$$

$$= 0.053\,ppmv$$

Therefore, the 24-hour average SO_2 concentration was 0.053 ppmv or 53 ppbv. This compares to the 24-hour standard of 0.14 ppmv. Therefore, the concentration of a gas in air can be calculated from the following equation:

$$ppmv = \frac{\mu g \cdot 24.45}{M \cdot V} \tag{13.29}$$

where V equals the sample volume in liters standardized to 25°C and 760 mmHg $\left(\text{by } PV = nRT \text{ or } V_R = V \frac{P_1}{760} \cdot \frac{298}{T_1} \right)$. In the SO_2 sampling calculations above, volume was not standardized to reference temperature and pressure because of the uncertainty of these conditions over the 24-hour sampling period.

The ppmv concentration of a gas in air can be converted into its respective System International (SI) concentration by means of the following equation:

$$mg/m^3 = \frac{M}{24.45} ppmv \qquad (13.30)$$

If the ambient air quality standard for O_3 is 0.12 ppmv, what is the corresponding concentration in $\mu g/m^3$?

$$mg/m^3 = \frac{48}{24.45} \cdot 0.12$$

$$= 0.236\,mg/m^3 \qquad (13.31)$$

$$= 236\,\mu g/m^3$$

If the concentration of a gas in mg/m^3 is known, it can be converted into ppmv by using Equation 13.32.

$$ppmv = \frac{24.45}{M} mg/m^3 \qquad (13.32)$$

The 24-hour ambient air-quality standard for SO_2 is 365 $\mu g/m^3$; what is the equivalent standard in ppmv?

$$ppmv = \frac{24.45}{64} \cdot 0.365\,mg/m^3$$
$$\qquad (13.33)$$
$$= 0.14\,ppmv$$

Stack Sampling Applications

Stack sampling is conducted to determine mass emission rates of pollutants. The dry gas molecular weight, wet gas molecular weight, average absolute stack pressure, average stack gas velocity, and the stack gas flow rate must be determined to calculate mass emission rates. Each of these includes some aspect of gas laws.

The molecular weight of dry stack gases is determined from volumetric percentages of major stack gases obtained by using an Orsat analyzer. Let us assume that a stack sample had the following dry composition: 19% CO_2, 5% O_2, 0.1 % CO, and 75% N_2. The dry gas molecular weight (M_d) is determined by multiplying the volumetric fraction of each gas by its molecular weight.

$$
\begin{array}{lll}
CO_2 & 44\,(0.19) = 8.36 \\
O_2 & 32\,(0.05) = 1.60 \\
CO & 28\,(0.01) = 0.28 \\
N_2 & 28\,(0.75) = 21.00 \\
\hline
\end{array}
$$

therefore, $M_d = 31.24$

The next step in determining the stack gas flow rate is to measure the stack gas moisture content and determine the wet stack gas molecular weight (M_w). By measuring wet (T_{wb}) and dry bulb temperatures (T_{db}) in the stack and determining the specific volumes of the mixture and dry air from a high temperatures psychrometric chart, the percent moisture can be calculated from the following equation:

$$\% \text{ moisture} = 100 \ (V_m - V_d)/V_m \tag{13.34}$$

$$\text{where } T_{db} = 200°F \text{ and } T_{wb} = 120°F$$

Since in the U.S. such charts are in British units, let us use temperature °F and specific volume as ft^3/lb dry air. These can be converted to SI units by calculation or from conversion tables. From a high temperature psychrometric chart (Figure 13.1) the values of V_d and V_m would be 16.6 ft^3/lb (1.03 m^3/kg) and 18.3 ft^3/lb (1.14 m^3/kg), respectively. Therefore,

$$\% \text{moisture} = 100 \, (18.3 - 16.6) \, / \, 18.3$$
$$= 9.3\% \tag{13.35}$$

The wet gas molecular weight (M_w) can be calculated from the following equation:

$$M_w = 0.01 \ (M_d) \ (100 - \% \text{ moisture}) + 0.18 \ (\% \text{ moisture}) \tag{13.36}$$

therefore,

$$M_w = 0.01 \, (31.24)(90.7) + 0.18 \, (9.3)$$
$$M_w = 30.01 \tag{13.37}$$

The wet molecular weight is used to calculate the average stack gas velocity v_s using the equation

$$v_s = v_m \, K_p \ [(29.92)(28.95)/(P_s) \ (M_w)]^{1/2} \tag{13.38}$$

where v_m = uncorrected or measured velocity, actual feet per second (AFPS), K_p = type S pitot tube coefficient, P_s = average absolute stack gas pressure, inHg, 28.95 = molecular weight of air, and 29.92 inHg = pressure under reference conditions.

The average absolute stack gas pressure P_s can be calculated from the equation

$$P_s = P_b + (S.P.)/13.6 \tag{13.39}$$

where P_b = barometric pressure, inHg and $S.P.$ = static pressure, inH$_2$O.

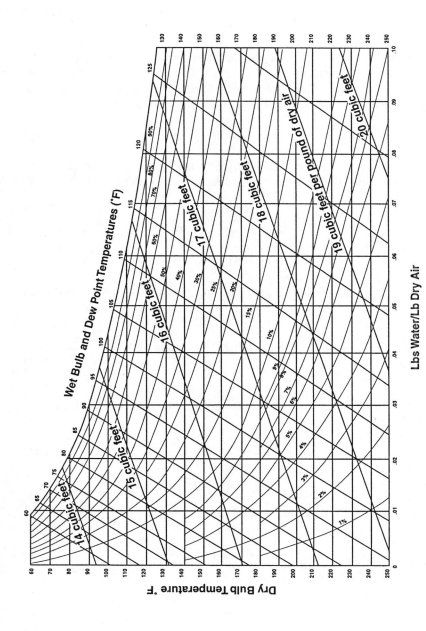

Figure 13.1 High temperature psychrometric chart.

If P_b = 29.0 inHg and $S.P.$ = 0.03 inH$_2$O then

$$P_s = 29.0 + (0.03)/13.6$$
$$P_s = 29.0 \, \text{inHg}$$

(13.40)

The average velocity (v_m) of stack gases measured across 20 traverse points in a 4 ft diameter stack is calculated from velocity pressure (ΔP, inH$_2$O) readings using a type S pitot tube. The uncorrected velocity (v_m) at each traverse point as well as the average can be calculated from the following equation:

$$v_m = 2.9 \, [\Delta P \, (T_{sdb})]^{1/2}$$

(13.41)

where T_{sdb} = °R (°F + 460) = absolute dry bulb temperature of stack gases.

Let us assume an average stack gas temperature of 650°R and ΔP of 0.30 inH$_2$O. Then

$$v_m = 2.9 \, [(0.30)(650)]^{1/2}$$
$$v_m = 40.49 \, \text{AFPS}$$

(13.42)

Now we can calculate the average stack gas velocity (v_s) where K_p = 0.855 (Figure 13.2).

$$v_s = 40.49 \, (0.855)[(29.92)(28.95)/(29.0)(30.01)]^{1/2}$$
$$v_s = 34.54 \, \text{AFPS}$$

(13.43)

The average actual stack gas flow rate in actual cubic feet/min (ACFM) can be calculated from:

$$Q = 60 \, (v_s) \, A$$

(13.44)

where Q = ACFM and A = cross sectional area of the stack.

Area (A) can be calculated from:

$$A = \pi D^2/4$$

(13.45)

where D = stack diameter, feet.

If the stack diameter is 4 ft, then

$$A = 3.14 \cdot (4)^2 4$$
$$A = 12.6 \, \text{ft}^2$$

(13.46)

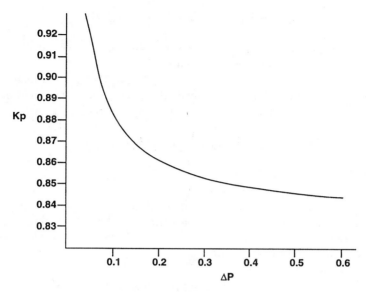

Figure 13.2 Calibration curve for type S pitot tube. (From Powals, R.J., et al. 1978. Technomic Publishing Co., Inc., Westport, CT. With permission.)

when $v_s = 34.54$ AFPS and $A = 12.6$ ft^2

$$Q = 60\,(34.54)(12.6)$$
$$Q = 26,110\,\text{ACFM} \tag{13.47}$$

In the above equation, v_s and A are multiplied by 60 to convert values from seconds to minutes.

The stack gas flow rate (Q) can now be expressed as standard cubic feet per minute (SCFM). Standard or reference conditions reflect regulatory requirements. In the U.S. they are 70°F (21°C) and 29.92 inHg (760 mmHg).

$$Q_s = (Q)(P_s)(T_s)(M_w)/(P_r)(T_r)(M_d) \tag{13.48}$$

From the above calculations,

$$Q_s = (26,110)(29.0)(530)(30.01)/(29.92)(660)(31.24)$$
$$Q_s = 19,522\,\text{SCFM} \tag{13.49}$$

Let us now collect a sample of dust from a stack using isokinetic sampling. The volume of air sampled determined from a dry gas meter was 120 ft^3 (3.4 m^3). We convert the meter volume (V_m) to standard conditions based on the following equation

$$V_s = V_m \, [(P_b - P_m - VP_c)/P_s](T_s \cdot T_m) \tag{13.50}$$

where P_m = meter pressure, inHg, VP_c = vapor pressure at dewpoint temperature, inHg, and T_m = absolute meter temperature, °R.

Measured data included V_m = 120 ft³, T_m = 540°R, P_m = 8.0 inHg, and T_c (absolute dewpoint temperature) = 495°R. Using a psychrometric chart or vapor pressure tables at T_c = 495°R, VP_c = 0.2111 inHg; therefore,

$$V_s = 115[(29.0 - 8.0 - 0.2111)/29.92](530/540)$$

$$V_s = 78.4 \, SCF \tag{13.51}$$

If water vapor is removed from sampled air, $VP_c = 0$.

In the sample of 78.4 SCF, a dust concentration (m) of 0.5670 g was measured during a sample collection period of 120 min. We can now calculate both the concentration and mass emission rate.

$$C = 0.5670 \, g \, / \, 78.4 \, SCF$$

$$C = 0.00721 \, g \, / \, SCF$$

$$\text{since } 1 \, SCF = 0.0283 \, m^3 \tag{13.52}$$

$$C = 0.254 \, g \, / \, m^3$$

$$C = 254 \, mg \, / \, m^3$$

The mass emission rate (ER) can be determined by multiplying the concentration of particles in the sample by the standardized mass flow rate.

$$ER = 0.0072 \, g \, / \, SCF \, (19,522 \, SCFM)$$

$$ER = 136.7 \, g \, / \, min \tag{13.53}$$

The emission rate can alternately be expressed as kilograms/hour by first dividing 139 g by 1000 and multiplying this value by 60

$$ER = (139/1000)(60) = 8.46 \, kg/hour \tag{13.54}$$

Because emission rates are often expressed as lb/hour, we can multiply the above by 2.2, the conversion factor from kg to lb.

$$ER = (8.46 \, kg/hour)(2.2 \, lb/kg) = 18.6 \, lb/hour \tag{13.55}$$

ACCURACY AND PRECISION

As indicated in Chapter 7, the quality and utility of ambient air-quality (as well as emissions) measurements are determined by the accuracy and precision of sampling methods used. Accuracy is the closeness of a measured value compared to the true or known value; precision the reproducibility of results.

All measurement techniques are subject to both determinate and indeterminate errors. In the former case, this includes errors associated with measuring devices, errors in the method of analysis, personal errors of those conducting measurements and a combination of these. Indeterminate errors result from extending a system of measurement to its maximum.

Sampling and analytical techniques for the same chemical species differ in their ability to accurately determine sample concentrations. It is imperative that the accuracy be known for the range of concentrations sampled. High accuracy (>90%) is desirable.

Accuracy can be described within the context of absolute or relative errors. The absolute error can be defined as

$$E = O - A \tag{13.56}$$

where O = observed value and A = actual value.

Let us assume that in making a series of measurements of a known concentration (1 ppmv) of SO_2 the average observed value is 0.925 ppmv; therefore,

$$E = 0.925 - 1.0$$
$$E = -0.075 \tag{13.57}$$

and the absolute error is -0.075 ppmv. Measured values less than the true value are expressed as negative error; those greater, as positive error.

We can calculate the relative error as a percentage using the following equation:

$$E_r = \frac{(E)(100)}{1\,\text{ppmv}} = -7.5\% \tag{13.58}$$

Accuracy is often reported as a percentage of the true or known value. Therefore the accuracy of the sampling method in this case was 92.5%. Accuracy can also be described by a range of error values such as $\pm10\%$.

The observed value used above was an average value determined from a series of 10 measurements which included 0.85, 1.0, 0.95, 0.96, 0.87, 0.90, 0.88, 0.92, 0.94, and 0.99 ppmv. Note that these varied around a mean of 0.925 ppmv. This variation reflects the precision of our measurements. Precision is determined by calculating the mean and standard deviation. In a normally distributed population of sample measurements, the standard deviation is a measure of the variability of the results around the sample mean. In a normal distribution (defined by a bell-shaped curve), 68.3% of sample data will vary around the mean within ± one standard deviation and 95% within ± two standard deviations. The standard deviation can be calculated from the following equation:

$$\sigma = \sum_{i=1}^{n} \frac{(x_i - \overline{x})^2}{N} \qquad (13.59)$$

where $\overline{x}$ = mean, x_i = individual value, and N = total number of measurements.

The standard deviation can easily be calculated on programmable scientific calculators or using statistical software packages.

From the above data, σ has been determined to be equal to 0.052. Therefore, the precision of our sampling results is ±0.052 ppmv.

Precision can be reported as a percentage by calculating the coefficient of variation (c.v.):

$$c.v. = \frac{\sigma}{\overline{x}}(100) \qquad (13.60)$$

$$= 5.62\%$$

Therefore, the precision of our measurements using one standard deviation as our reference was ±5.62%.

SUMMARIZING AIR POLLUTION DATA

Data collected in air-quality monitoring programs is usually unusable in the raw form that it is collected. It is therefore desirable to summarize it in ways that make it relatively easy to interpret and use for regulatory purposes.

Commonly used data summary techniques include graphical procedures and calculation of measures of central tendency (e.g., mean, median). For continuous data (which includes most air pollution data), commonly used graphical techniques include the histogram and cumulative frequency curve. These are illustrated for total suspended particulate (TSP) data (Table 13.1) in Figures 13.3 and 13.4. In the former case, frequency values were determined for 13 class intervals. An examination of the graph indicates that the data did not fit a bell-shaped curve and thus are not normally distributed.

In Figure 13.4, data have been plotted to illustrate a cumulative frequency curve. The cumulative frequency is determined by consecutively summing relative frequencies (absolute frequency/N) for each class interval until the cumulative frequency curve reaches a plateau of 1.0. The cumulative frequency curve can be used to determine the class interval below which 50%, 90%, etc., of the measured values lie by extending a straight line from 0.50 and 0.90 cumulative frequency until it touches the curve and then dropping a perpendicular to the class interval below. From Figure 13.4, 50% of measured values would be

Table 13.1 High Volume Sampling Data Collected at a Single Site

Sample #	TSP (μg/m³)	Sample #	TSP (μg/m³)	Sample #	TSP (μg/m³)
1	30.0	22	70.6	43	50.1
2	19.0	23	62.9	44	63.4
3	30.6	24	44.0	45	31.4
4	41.3	25	49.6	46	39.4
5	45.9	26	56.3	47	44.8
6	39.8	27	47.3	48	37.3
7	48.0	28	113.1	49	190.9
8	43.1	29	67.4	50	60.8
9	15.3	30	51.2	51	82.5
10	38.6	31	37.3	52	27.3
11	68.4	32	49.6	53	107.7
12	52.5	33	47.8	54	47.7
13	44.9	34	32.1	55	90.1
14	67.8	35	62.9	56	76.2
15	59.6	36	38.4	57	73.0
16	29.5	37	92.8	58	29.3
17	101.4	38	79.7	59	56.5
18	47.3	39	125.9	60	90.2
19	64.0	40	88.3	61	60.8
20	59.6	41	93.6	62	53.5
21	88.0	42	54.3		

Note: TSP, total suspended particulate.

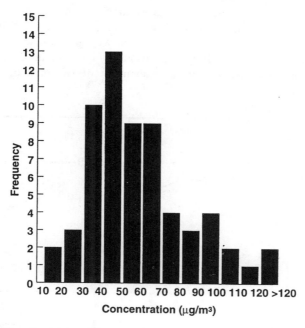

Figure 13.3 Histogram of TSP data.

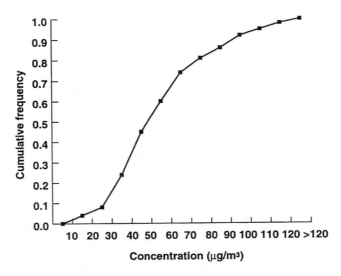

Figure 13.4 Cumulative frequency curve for TSP data.

expected to be ≤50 or >50 µg/m³; 90% of measured values would be expected to be ≤90, and 10%, >90 µg/m³.

Air pollution trends are often summarized as annual average concentrations over time (Figure 13.5). Box plots (Figure 13.6) may be used to indicate the distribution of data indicating percentiles.

Data are often summarized using the arithmetic mean if the data are normally distributed or the median or geometric mean if the data are not. When data are not normally distributed, the arithmetic mean will give a distorted picture of where most of the values lie. In the case of family income, the Bureau of Labor Statistics reports median values because about 5% of the population earns 40% of total annual income.

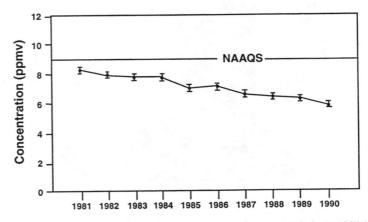

Figure 13.5 Trends in 8-hour average CO concentrations in the U.S. from 1981 to 1990. (From USEPA. 1991. EPA/600/8-90/045f.)

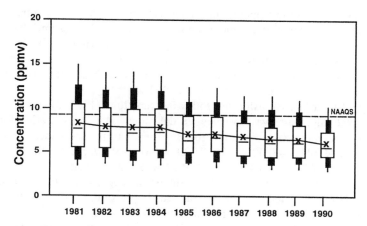

Figure 13.6 Box plot comparisons of trends in 8-hour average CO concentrations in the U.S. from 1981 to 1990. (From USEPA. 1991. EPA/600/8-90/045f.)

When data are not normally distributed as with TSP data, the median must be determined or geometric mean calculated. The median of a population of sequenced values (from smallest to largest) is the middle number when N is an odd number and the average of the two middle values when N equals an even number. From the data in Table 13.1, the median is 48.8 μg/m³. Summing all values and dividing by N, the arithmetic mean on the other hand is 59.9 μg/m³. As can be seen, the arithmetic mean is approximately 10 μg/m³ higher than the median.

Though the median is a good measure of central tendency (when data are not normally distributed), it cannot be used in statistical analyses. In such cases, it is desirable to calculate the geometric mean from the following equation

$$GM = antilog \frac{\log x_1 + \log x_2 + \log x_3 \cdots + \log x_n}{N} \qquad (13.61)$$

The log values can be obtained using a portable calculator. After summation and dividing by N, the antilog or arithmetic form must be determined. To obtain the antilog on your calculator, use the following steps: enter 10; press Y^x; enter average log value; press =. Using the procedure on the values in Table 13.1, the geometric mean is found to equal 50.8 μg/m³.

Because particulate matter (PM) concentrations determined from high volume sampler data were not normally distributed, the annual TSP air-quality standard was based on the geometric mean. PM_{10} data, however, are normally distributed; therefore, the arithmetic mean is calculated.

Data summary of acidic deposition results requires the conversion of pH values (which are geometric) to their arithmetic form in order to calculate average pH values. It is not appropriate to sum pH values and divide by N to calculate the mean.

Because pH is the negative log of the H+ concentration, we must convert pH values to their equivalent molar concentrations. Let us assume that over 6 weeks we obtained the following weekly precipitation sample pH values: 4.5, 5.0, 3.8, 4.0, 4.2, and 4.8. What was the average pH?

Because pH equals the negative logarithm of each of these values, their corresponding values are $1 \times 10^{-4.5}$, 1×10^{-5}, $1 \times 10^{-3.8}$, 1×10^{-4}, $1 \times 10^{-4.2}$, and $1 \times 10^{-4.8}$ Molar H+. By using the following functions (10, Y^x, exponent, $\pm$, =) on a pocket calculator, we can calculate the molar (H+) concentration for each pH value. The molar concentrations for H+ for each of the six values are: pH 4.5 = 0.000032 M, pH 5 = 0.00001 M, pH 3.8 = 0.000158 M, pH 4 = 0.0001 M, pH 4.2 = 0.000063 M, and pH 4.8 = 0.0000158 M. Summing and dividing by N, the average concentration is 0.00063 M H+. Taking the negative log of this value, the average pH is calculated to be 4.19. Had we simply summed each of the values and divided by N, the average pH would have been 4.38.

Though these values do not appear to be significantly different from each other, they in fact are. We can determine this difference by dividing the molar concentration of pH 4.38 into the molar concentration of pH 4.19 (the higher H+ concentration):

$$\frac{\text{pH } 4.19 = 0.000065 \, M \, \text{H}^+}{\text{pH } 4.38 = 0.000050 \, M \, \text{H}^+} = 1.3 \qquad (13.62)$$

On dividing we find that a pH of 4.19 is 30% higher than a pH of 4.38. Similarly, if we were to compare our average pH value of measured precipitation samples to a theoretical clean air value of 5.7, we would find that a pH of 4.19 is 31.6 times more acidic than a pH of 5.7.

MODELING

As indicated in Chapter 7, air-quality models are widely used to determine whether existing sources or planned new sources are or will be in compliance with air-quality standards.

Using the Gaussian model described in Chapter 7, let us calculate the ground-level concentration (C_x) at some distance downwind (x) for an industrial boiler burning 12 tons of 2.5% sulfur coal/hour with an emission rate of 151 g/sec. Let us assume the following: effective stack height equals 120 m and wind speed equals 2 m/sec. It is 1 hour before sunrise, and the sky is clear. Assuming the centerline distance (Y) is equal to 0, we can use Equation 7.2 to calculate C_x.

From Table 7.1 the atmospheric stability class for the condition described is F. From Figure 7.7 the horizontal dispersion coefficient σ_y for a downwind distance of 10 km for atmospheric stability class F is approximately 270 m; the vertical dispersion coefficient σ_z is approximately 45 m (Figure 7.8).

Therefore:

$$C_x = \frac{151}{3.14\,(270)(45)(2)}\,e^{-1/2}(120/45)^2$$

$$= 1.98 \times 10^{-3}\,e^{-1/2}\,(7.1)$$

$$= \frac{1.98 \times 10^{-3}}{e^{-1/2}\,(7.1)}$$

$$= \frac{1.98 \times 10^{-3}}{34.8}$$

$$= 5.7 \times 10^{-5}\,g/m^3$$

$$= 57\,\mu g/m^3$$

(13.63)

Let us now assume that we have a ground-level source. Since the effective stack height equals 0, Equation 7.2 now becomes

$$C_x = \frac{Q}{\pi \sigma_y \sigma_z \bar{u}}$$

(13.64)

Given a stability class B, a wind speed of 4 m/sec, and an emission rate of 100 g/s, we can calculate downwind concentrations at various receptor distances (e.g., 200 and 1000 m). For a receptor distance of 200 m:

$$C_x = \frac{100}{3.14\,(36)(20.5)(4)}$$

$$= 10.8 \times 10^{-3}\,g/m^3$$

$$= 10.8\,mg/m^3$$

(13.65)

For a receptor distance of 1000 m:

$$C_x = \frac{100}{3.14\,(155)(110)(4)}$$

$$= 4.67 \times 10^{-4}\,g/m^3$$

$$= 467\,\mu g/m^3$$

(13.66)

In previous calculations of downwind concentrations from elevated sources, the effective stack height (H) was given. The effective stack height as indicated in Chapter 7 includes both the physical stack height and plume rise. Plume rise must be calculated from various stack and environmental factors.

Using the plume rise equations of Briggs described in Chapter 7, let us calculate effective stack height (H) for an 80 m high source (h) with a stack diameter of 4 m, stack gas velocity of 14 m/sec, a stack gas temperature of 90°C (363°K), an ambient temperature of 25°C (298°K), and a wind speed at anemometer height of 4 m/sec. Let us also assume class B atmospheric stability conditions.

We first calculate buoyancy flux using the equation:

$$F = 2.45 \, v_s \, D^2 \, (T_s - T)/(T_s) \tag{13.67}$$

$$F = \frac{2.45(14)(16)(65)}{363} \tag{13.68}$$

$$F = 98.27$$

Since F is ≥ 55 m⁴/sec³, we use Equation 7.4:

$$H = h + \frac{38.71(98.27)^{3/5}}{4} \tag{13.69}$$

$$H = 80 + 152$$
$$H = 232 \tag{13.70}$$

Effective stack height closer to the source than the final plume rise can be estimated from:

$$H = h + 160 \, F^{1/3} \, x^{2/3}/\overline{u} \tag{13.71}$$

We now determine effective stack height at a downwind distance of 0.3 and 0.8 km.

$$H = 80 + \frac{160(98.27)^{1/3}(0.3)^{2/3}}{4} \tag{13.72}$$

$$H = 163$$

$$H = 80 + \frac{160(98.27)^{1/3}(0.8)^{2/3}}{4} \tag{13.73}$$

$$= 239$$

Since 239 m exceeds the final plume rise of 232 m, the latter is utilized.

Using Equation 7.6 in Chapter 7, we can calculate the maximum ground-level concentration assuming the ratio σ_z/σ_y is constant with distance.

$$C\max = \frac{2Q\sigma_z}{\pi \bar{u} e H^2 \sigma_y} \tag{13.74}$$

The distance to the maximum concentration is the distance where

$$\sigma_z = (H/2)^{1/2} \tag{13.75}$$

which in the case above:

$$\sigma_z = (232/2)^{1/2}$$
$$\sigma_z = 164\,\text{m} \tag{13.76}$$

From Figure 7.8, the downwind distance for this dispersion coefficient is approximately 1.4 km for B stability class. The corresponding σ_y for stability class B (from Figure 7.7) is approximately 210 m. Therefore,

$$C\max = \frac{2(100)(164)}{3.14(4)(2.718)(232)^2(210)}$$

$$C\max = 85 \times 10^{-6}\,\text{g/m}^3 \tag{13.77}$$

$$= 85\,\mu\text{g/m}^3$$

Computer Calculations

All calculations described above were conducted manually for illustrative purposes. In the real world, determinations of maximum ground-level concentrations, effective stack height, and the distance to maximum ground-level concentrations for a variety of wind speeds and stability classes are calculated using computer software. A variety of models and software packages are available on the Internet as well as from their originators.

One commonly used dispersion model for stationary sources is PTPLU developed by the California Air Resources Board. Modeling information can be obtained from USEPA's Applied Modeling Research Branch Bulletin Board Service. It can be reached at (919) 541-1325 and the system operator at (919) 541-1376. Other model services are available on the Technology Transfer Network (TTN). It can be reached at (919) 541-5742 and the system operator at (919) 541-5384.

Using PTPLU, let us model a source emitting SO_2 with an emission rate of 80 g/sec from a 60 m high, 2.5 m diameter stack. The stack gas temperature is

350°K and velocity 10 m/sec. The ambient temperature is 295°K; anemometer height, 10 m; and mixing height, 1500 m. Options including stack downwash, buoyancy-induced dispersion, and rural dispersion coefficients were used. Concentrations can be expressed in a variety of forms. Either $\mu g/m^3$ or ppbv would be appropriate in this modeling activity.

The model is run for all six stability classes and a variety of wind speeds appropriate for those classes. Footnotes indicate that some output values may not be valid. This is the case for the first three output values under stability class A and B. The maximum effective stack height is seen to be 197 m. Under stability class A, wind speed 1.70 m/sec, the maximum ground-level concentration is predicted to be 316 $\mu g/m^3$ (120 ppbv) and 214 $\mu g/m^3$ (81 ppbv) for class B with 1.70 m/sec wind speed. Note the maximum effective stack height of 197 m is not achieved in any other combinations of stability class and wind speeds used in model calculations. In stability classes A, E, and F, a pattern of decreasing maximum ground-level concentration can be seen to be associated with increasing wind which is not evident in stability classes B, C, and D. Consistent in all stability classes is a decrease in both effective stack height and distance of maximum ground-level concentration with increasing wind speed.

It is interesting to note that the lowest ground-level concentrations are observed under the most stable atmospheric conditions (classes E and F). These are associated both with the lowest effective stack height and the greatest distance to the maximum ground-level concentration.

EMISSIONS

Emissions from sources can be determined by conducting emissions testing of flue gases and calculations based on published emission factors. Let us assume that in obtaining a permit under Title V, a source (a multichamber municipal incinerator that burns 50 ton/day) has to determine its emissions both daily and monthly. The published emission factor for such a source is 17 lb/ton of PM of refuse burned. Assuming the source is uncontrolled, what is its hourly and 24-hour emission rates?

Assuming a production of 17 lb of PM/ton and an average daily incineration of 5 tons of refuse, the emission rate would be:

$$17 \text{ lb/ton} \times 50 \text{ ton/day} = 850 \text{ lb/day} \tag{13.78}$$

The emission rate per hour would be:

$$850 \text{ lb/day} \times 1 \text{ day/24 hours} = 35 \text{ lb/hour} \tag{13.79}$$

Total emissions over the course of a year would be:

$$850 \text{ lb/day} \times 365 \text{ day/year} \times \text{ton/2000 lb} = 155 \text{ ton/year} \tag{13.80}$$

The new source performance standard (NSPS) for sulfuric acid plants is 2 kg/metric ton of 100% acid produced averaged over a 3-hour period. For a 250 ton/day plant, we can calculate the maximum allowable emission rate:

$$2 \text{ kg/mton} \times \frac{1 \text{ mton}}{1.102 \text{ ton}} \times 250 \text{ ton/day} = 454 \text{ kg/} \qquad (13.81)$$

The NSPS standard for high-sulfur coal-fired power plants is 1.2 lb SO_2/million BTU. Let us assume that a new coal-fired power plant burns a 3% sulfur coal. The 1000 MW plant has an efficiency of 35%. We must first calculate the heat input rate for the 35% efficient power plant.

$$H = 1000 \text{ MW} / .35 \times 1000 \text{ kW/MW} \times 3412 \text{ BTU/kWh} \times 24 \text{ hour/day}$$

$$= 2.34 \times 10^{11} \text{ BTU/day}$$

$$(13.82)$$

$$SO_2 \text{ emitted} = \frac{1.2 \text{ lb}}{10^6 \text{ BTU}} \times \frac{2.34 \times 10^{11} \text{ BTU}}{\text{day}} \times \frac{1 \text{ ton}}{2000 \text{ lb}} \qquad (13.83)$$

$$= 140 \text{ ton/day}$$

The 1980 NSPS for new coal-fired power plants requires the use of scrubbers to achieve 90% reduction in SO_2 emission from high-sulfur coal (with an emission rate not to exceed 1.2 lb/10^6 BTU) and a 70% reduction from low-sulfur coal. Using Figure 13.7, we can determine emission rates from power plants burning a variety of percent sulfur coals, BTU/lb coal, and scrubber efficiencies. The minimum scrubber efficiency is 70%, and maximum emissions for the 90% SO_x emission requirement is 1.2 lb SO_2/10^6 BTU. Let us assume that a power plant burns a low-sulfur coal (1% S) with an energy content of 9000 BTU/lb. By drawing a line from the origin in the upper left hand corner to the intersection of the curve for 9000 BTU/lb and 1% sulfur coal, this corresponds to a scrubber efficiency requirement of 70% and an emission rate of 0.6 lb SO_2/10^6 BTU. The SO_2 emission rate for a 1000 MW, 35% efficient plant would be:

$$SO_2 \text{ emission rate} = 0.6 \text{ lb}/10^6 \text{ BTU} \times 2.34 \times 10^{11} \text{ BTU/day} \times 1 \text{ ton}/2000 \text{ lb}$$

$$= 70.2 \text{ ton/day}$$

$$(13.84)$$

The NSPS for nitric acid is 1.5 kg NO_x/mton of 100% acid produced. Assume that uncontrolled emissions are 30 kg NO_x/mton acid. What is the collection efficiency required to meet the standard?

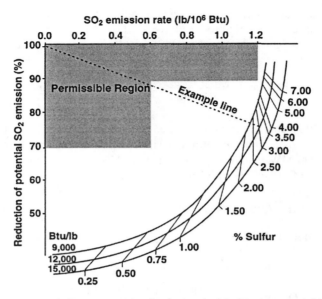

Figure 13.7 Relationship among coal sulfur content, emission rates, and SO_2 reduction requirements under 1979 New Source Performance Standards. (From Molburg, J. 1980. *JAPCA* 30:172. With permission.)

$$\text{collection efficiency} = \frac{\text{uncontrolled emission rate} - \text{emission rate permitted}}{\text{uncontrolled emission rate}}(100)$$

$$(13.85)$$

$$\text{collection efficiency} = \frac{30 \, \text{kg} / \text{mton} - 1.5 \, \text{kg} / \text{mton}}{30}(100) \qquad (13.86)$$

$$= 95\%$$

SOUND/NOISE

The assessment of noise in community and industrial environments is conducted with instruments which measure and record sound pressure. Sound pressure readings are expressed in dimensionless units called decibels (dB) which can be calculated from the following equation:

$$dB = 20 \log_{10} \frac{P_e}{P_r} \qquad (13.87)$$

where P_e = pressure of measured sound (μbar) and P_r = reference pressure (0.0002 μbar) at the threshold of hearing.

If the sound pressure is known, the dB level can be calculated. Let us assume that

$$P_e = 2.0\,\mu\text{bar, then}$$

$$\text{dB} = 20\log_{10}\frac{2.0}{0.0002}$$

$$= 20\log_{10}(10,000) \tag{13.88}$$

$$= 20(4)$$

$$= 80\,\text{dB}$$

Conversely, if we know the sound pressure level in decibels, we can calculate P_e. Let us assume a measured sound pressure level of 73 dB. What is P_e?

$$73\,\text{dB} = 20\log_{10}\frac{P_e}{0.0002\,\mu\text{bar}}$$

$$\text{antilog}\frac{73}{20}(0.0002\,\mu\text{bar}) = P_e$$

$$\text{antilog}3.65(0.0002) = P_e \tag{13.89}$$

$$4467(0.0002) = P_e$$

$$0.89\,\mu\text{bar} = P_e$$

Sound measurements are commonly made multiple times in the same environment. These must be averaged to be meaningful. As is the case of pH, decibel readings are in geometric form. As a result, they cannot be summed and divided by N. Instead, we calculate the Leq or the average equivalent sound level using the following equation.

$$\text{Leq} = 10\log_{10}\frac{\sum_{i=1}^{n}10^{Li/10}}{N} \tag{13.90}$$

Let us assume that we made five sound measurements on the A scale near a street side: 73, 68, 75, 65, and 70 dBA. What was the average of these five readings or Leq? We proceed by calculating $10^{Li/10}$ for each of the five values. These become 19,952,623; 6,309,573; 31,622,777; 3,162,278; and 10,000,000. Summing these (71,047,251) and dividing by $N = 5$, the value becomes 14,094,450. Taking the log of this value and multiplying by 10, Leq = 71.5 dBA.

READINGS

1. Brenchley, D.L., C.D. Turley, and R.F. Yarmic. 1973. *Industrial Source Sampling*. Ann Arbor Science Publishers, Ann Arbor, MI.

2. Cooper, C.D. and F.C. Alley. 1986. *Air Pollution Control: A Design Approach.* PWS Engineering, Boston.

3. Cooper, H.B.H. and A.T. Rossano. 1971. *Source Testing for Air Pollution Control.* Environmental Research & Applications, Inc., Welton, CT.

4. Lodge, J.P. (Ed.). 1988. *Methods of Air Sampling and Analysis.* 3rd ed. Lewis Publishers, Chelsea, MI.

5. Molburg, J. 1980. "A Graphical Representation of the New NSPS for Sulfur Dioxide." *JAPCA* 30:172.

6. Pasquill, F. and F.B. Smith. 1990. *Atmospheric Diffusion.* 3rd ed. Ellis Horwood, New York.

7. Powals, R.J., L.J., Zaner, and K.F. Sporek. 1978. *Handbook of Stack Sampling and Analysis.* Technomic Publishing Co., Inc., Westport, CT.

8. Turner, D.B. 1990. *Workbook of Atmospheric Dispersion Estimates.* 2nd ed. CRC/Lewis Publishers, Boca Raton, FL.

9. USEPA. User's Network for Applied Modeling of Air Pollution.

10. USEPA. 1991. *Air Quality Criteria for Carbon Monoxide.* EPA/600/8-90/045f.

11. Wark, K. and C.F. Warner. 1981. *Air Pollution: Its Origin and Control.* 2nd ed. Harper & Row Publishers, New York.

QUESTIONS

1. The primary 1-hour ambient air-quality standard for CO is 35 ppmv. What is the equivalent concentration in mg/m³ at reference conditions of temperature and pressure (298°K, and 760 mmHg)?

2. The primary ambient air-quality standard for NO_2 (annual average) is 100 μg/m³. What is the equivalent concentration in ppmv at reference conditions and pressure? In ppbv?

3. A rotameter is used to determine the sampling rate when collecting a sample under field conditions. It was calibrated in the laboratory at recorded temperatures of 22°C (295°K) and a pressure of 730 mmHg. The sample was collected at a temperature of 12°C (283°K) and a pressure of 750 mmHg. The sampling rate was 1 L/min; duration equaled 2 hours. What was the volume of air sampled with reference to the calibration temperature and pressure? What was the volume sampled based on reference conditions of 25°C (298°K) and 760 mmHg?

4. SO_2 is emitted from a smokestack at a rate of 160 g/sec. What is the equivalent emission rate in lb/hour?

5. A sample of dry gas was withdrawn from a stack through a sampling train that contained a heated filter, a set of cooled impingers, an air pump and a flow meter. The rate of flow was 30 L/m at calibration conditions of 22°C (295°K) and 750 mmHg.

    a. What was the actual flow rate through the filter at a temperature of 210°C (483°K) and P = 730 mmHg?
    b. What was the total volume of gas sampled at stack conditions in m³? in ft³?
    c. If 1.84 mg of particles were collected on the filter in 30 min, what was the concentration of particles in the stack gas in mg/m³?

6. The density of air at 25°C and 760 mmHg is 0.0739 lb/ft³. What is the equivalent air density in kg/m³?

7. The percent volume of dry gases from a coal-burning facility was 14% CO_2, 79.75% N_2, 6% O_2, and 0.25% SO_2.

    a. What was the dry molecular weight?
    b. What was the wet molecular weight if the moisture content was 8%?

8. Determine the density of air from the Ideal Gas Law for reference conditions. Assume a temperature of 25°C and pressure of 760 mmHg.

9. Determine the median and geometric mean of the following TSP values: 55.2, 74.5, 102.2, 40.3, 45.6, 20.5, 62.4, 80.7, 35.8, 65.8, and 48.9 μg/m³.

10. Calculate the arithmetic mean of the following PM_{10} values: 24.9, 14.9, 38.6, 14.6, 17.8, 32.9, 45.3, 40.3, 7.7, and 22.0 $\mu g/m^3$.

11. Why is the geometric mean calculated for TSP values and the arithmetic mean for PM_{10} values?

12. Obtain a data set of PM_{10} values from your local air pollution control agency. Summarize these data by means of a histogram and a cumulative frequency curve. From the latter determine 50 and 90 percentile values. Use Microsoft Excel to plot your histogram and cumulative frequency curve and to calculate the mean, median, and standard deviation.

13. Weekly samples of wet acidic deposition were collected over a 2-month period. Calculate the average pH value and determine how many times more acidic than the theoretical clean air value of 5.7 it is. Measured values were 3.4, 4.2, 5.0, 3.8, 4.0, 4.4, 4.8, and 3.9.

14. Using Gaussian dispersion equations, and tables and figures in Chapter 7, let us assume that we have a source of SO_2 that is emitting 100 g/sec. Let us further assume that it is a sunny summer afternoon with a wind speed at anemometer height (10 m) of 4 m/sec and that the effective stack height $H = 120$ m.

   a. Determine the stability class and dispersion coefficients for receptor distances of 300 and 1000 m (plume center) and ground-level concentrations at each distance.
   b. Assume that the off-centerline distance is 50 m. Determine ground-level concentrations at 300 and 1000 m.

15. Assume that we have a ground-level source which is emitting 100 g/sec on a clear night with a wind speed of 2 m/sec. What is the ground-level concentration at the center of the plume 200 m downwind?

16. Using the California Air Resources Board's Gaussian model PTPLU described on pages 425–426 and data for input variables, lowest ground-level concentrations were observed under the most stable atmospheric conditions. This appears to be somewhat inconsistent with our understanding that increasing atmospheric stability is responsible for increased ground-level concentrations.

   a. Explain this apparent inconsistency.
   b. In classes E and F, calculations have been made only for wind speeds greater than 3.74 and 5.36 m/sec, respectively. Why?
   c. Assuming wind speeds of relatively similar magnitude in stability classes B to E, what change is evident in the distance of maximum ground-level concentration?
   d. Using the same input parameters as used on pages 425–426, what is the effect of changing the emission rate to 50 g/sec on maximum ground-level concentration, distance of maximum ground-level concentration, and effective stack height?

e. Using the same input parameters as in the computer calculations in the text, change the stack gas temperature to 325°C (598°K) and gas velocity to 8 m/sec.

f. What is the effect on buoyancy flux, maximum ground-level concentration, distance of maximum ground-level concentration, and effective stack height?

17. Using Figure 13.7, determine scrubber efficiencies required for (1) 3.5% S and 11,000 BTU/lb coal and (2) 0.5% S and 6500 BTU/lb coal.

18. A 35% efficient coal-fired power plant burns both low sulfur (0.75% S) and low BTU (7500 BTU/lb) coal. How many tons of coal would be burned daily in a 1000 MW plant with emissions of 0.6 lb/10^6 BTU operating at 50% capacity? What scrubber efficiency is necessary to achieve this emission rate?

19. Exhaust gas from an automobile is determined to have a CO concentration of 1.4% at a temperature of 25°C (298°K) and 760 mmHg. What is the concentration of CO in ppmv? in mg/m³?

20. A solid waste incinerator generates 20 lb of particulate matter (PM) per ton of waste. The maximum 2-hour average flue gas PM concentration permissible is 0.18 g/m³ (corrected to 12% CO_2). If 80 dry standard cubic feet of flue gas are produced per pound of waste incinerated, determine the control equipment collection efficiency required to meet the above emission requirement.

21. Calculate the daily NO_x emissions from a new 120 ton/day nitric acid plant that is emitting NO_x at a rate permitted (1.5 kg/m ton 100% acid) by the NSPS when the plant is operating at full capacity.

22. Given a measured sound pressure (P_e) of 0.785 μbars, what is the equivalent sound pressure level in dB?

23. Given a sound pressure level of 73 dB, what is the equivalent measured sound pressure P_e?

24. In conducting a community sound survey, the following sound pressure levels dBA were measured at 6 sites: 65, 68, 73, 75, 72, and 78 dBA. What was the average sound pressure level in dBA?

25. In conducting the same community sound survey, six readings were also recorded on the C scale and reported as: 69, 72, 77, 79, 76, and 82 dBC. What was the average sound pressure level in dBC?

26. In Problem 25 above, sound readings were higher than those in Problem 24. Why?

27. Assuming that sound obeys the inverse square rule, we anticipate that in doubling the distance from the sound source, sound pressure will decrease by how many decibels from its previous value?

Index

A